Bioanalysis

For further volumes:
http://www.springer.com/series/8091

David D. Nolte

Optical Interferometry for Biology and Medicine

Springer

David D. Nolte
Department of Physics
Purdue University
West Lafayette, IN, USA

ISBN 978-1-4614-0889-5 e-ISBN 978-1-4614-0890-1
DOI 10.1007/978-1-4614-0890-1
Springer New York Dordrecht Heidelberg London

Library of Congress Control Number: 2011940315

Printed on acid-free paper

Springer is part of Springer Science+Business Media (www.springer.com)

Preface

Light is at once the most sensitive and the most gentle probe of matter. It is commonplace to use light to measure a picometer displacement far below the nanometer scale of atoms, or to capture the emission of a single photon from a fluorescent dye molecule. Light is easy to generate using light-emitting diodes or lasers, and to detect using ultrasensitive photodetectors as well as the now ubiquitous digital camera. Light also has the uncanny ability to penetrate living tissue harmlessly and deeply, while capturing valuable information on the health and function of cells. For these reasons, light has become an indispensible tool for biology and medicine. We all bear witness to the central role of light in microscopic imaging, in optical biosensors and in laser therapy and surgery.

Interferometry, applied to biology and medicine, provides unique quantitative metrology capabilities. The wavelength of light is like a meterstick against which small changes in length (or phase) are measured. This meterstick analogy is apt, because one micron is to one meter as one picometer is to one micron – at a dynamic range of a million to one. Indeed, a picometer is detected routinely using interferometry at wavelengths around one micron. This level of interferometric sensitivity has great utility in many biological applications, providing molecular sensitivity for biosensors as well as depth-gating capabilities to optically section living tissue.

Optical Interferometry for Biology and Medicine presents the physical principles of optical interferometry and describes their application to biological and medical problems. It is divided into four sections. The first provides the underlying physics of interferometry with complete mathematical derivations at the level of a junior undergraduate student. The basics of interferometry, light scattering and diffraction are presented first, followed by a chapter on speckle that gives the background for this important phenomenon in biological optics – virtually any light passing through tissue or cells becomes "mottled." Although it presents a challenge to imaging, speckle provides a way to extract statistical information about the conditions of cells and tissues. Surface optics is given a chapter to itself because of the central role played by surfaces in many optical biosensors and their applications.

The next three sections of the book discuss specific applications, beginning with interferometric biosensors, then interferometric microscopy followed by interferometric techniques for bulk tissues. Interferometric biosensors are comprised of many different forms, including thin films, waveguides, optical resonators and diffraction gratings. Microscopy benefits especially from interferometry because layers of two-dimensional cells on plates can be probed with very high sensitivity to measure subtle differences in refractive index of cells and their constituents. Quantitative phase microscopy has become possible recently through application of interferometric principles to microscopy. As cell layers thicken into tissues, imaging becomes more challenging, but coherent techniques like optical coherence tomography (OCT) and digital holography (DH) are able to extract information up to 1 mm deep inside tissue.

While the principles of interferometry are universal, this book seeks always to place them in the context of biological problems and systems. A central role is played by the optical properties of biomolecules, and by the optical properties of the parts of the cell. The structure and dynamics of the cell are also key players in many optical experiments. For these reasons, there are chapters devoted explicity to biological optics, including a chapter on cellular structure and dynamics as well as a chapter on the optical properties of tissues. Throughout the book, biological examples give the reader an opportunity to gain an intuitive feel for interference phenomena and their general magnitudes. It is my hope that this book will be a valuable resource for student and expert alike as they pursue research in optical problems in biology and medicine.

I would like to thank my current students Ran An, Karen Hayrapetyan and Hao Sun for proofreading the final manuscript, and much of this book is based on the excellent work of my former students Manoj Varma, Kwan Jeong, Leilei Peng, Ming Zhao and Xuefeng Wang. My colleagues Ken Ritchie, Brian Todd and Anant Ramdas at Purdue University provided many helpful insights as the book came together into preliminary form. Finally, I give my heartfelt appreciation to my wife Laura and son Nicholas for giving me the time, all those Saturday mornings, to do my "hobby."

West Lafayette, IN, USA David D. Nolte

Contents

Part IV Interferometry of Biological Tissues

Part I
Fundamentals of Biological Optics

Chapter 1
Interferometry

All optical phenomena experience interference effects at some level. Even light from an incandescent light bulb has partial coherence that can lead to constructive or destructive interference when multiple light paths are combined at a detector. The ubiquitous nature of interference underlies many phenomena and techniques in biological optics. This chapter lays the foundation for interferometry applied to biology and medicine. It begins with the simplest form of interference between two waves, explains common forms of interferometers, and ends with a short overview of partial coherence and of higher-order interference effects.

1.1 Two-Wave Interference

The simplest interference involves two independent waves that have the same frequency. This simplest case is also the one most commonly used in applications, in which one wave carries phase information that is compared against a stable reference wave. Phase cannot be detected directly by a detector. Two-wave interference provides a direct way to convert phase to detectable intensity.

1.1.1 Complex-Plane Representation of Plane Waves

Wave phenomena are conveniently represented in the complex plane, shown in Fig. 1.1, as a real-valued amplitude r and a phase angle θ

$$z = r\,\mathrm{e}^{\mathrm{i}\theta} \tag{1.1}$$

D.D. Nolte, *Optical Interferometry for Biology and Medicine*, Bioanalysis 1,
DOI 10.1007/978-1-4614-0890-1_1, © Springer Science+Business Media, LLC 2012

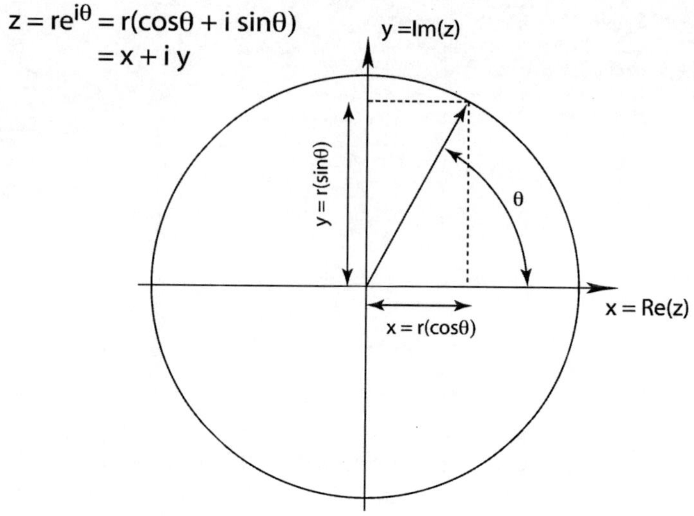

Fig. 1.1 Representation of a complex number $z = x + \mathrm{i}y$ on the complex plane. The x-axis is the real part and the y-axis is the imaginary part

The graphical representation of a complex number plots the real part along the x-axis and the imaginary part along the y-axis using Euler's formula to separate a complex exponential into a real and an imaginary part as

$$r\,\mathrm{e}^{\mathrm{i}\theta} = r(\cos\theta + \mathrm{i}\,\sin\theta) \tag{1.2}$$

The complex representation of a plane wave propagating along the z-axis is

$$E = \tilde{A}\,\exp(\mathrm{i}(kz - \omega t)) \tag{1.3}$$

where

$$\tilde{A} = |A|\mathrm{e}^{\mathrm{i}\phi} \tag{1.4}$$

is the complex amplitude with static phase ϕ, and the dynamical phase of the wave is the dynamic quantity $(kz - \omega t)$. For a plane wave propagating in the z direction, the phase of the wave evolves according to the convention

$$\theta = kz - \omega t + \phi \tag{1.5}$$

where $k = 2\pi/\lambda$ is the "wavenumber" with wavelength λ, ω is the angular frequency of the wave, and ϕ is a fixed (or slowly varying) phase. The wavenumber k describes the spatial frequency, while ω describes the angular frequency of the wave. These two frequencies are linked to the wave velocity c by

$$\omega = ck \tag{1.6}$$

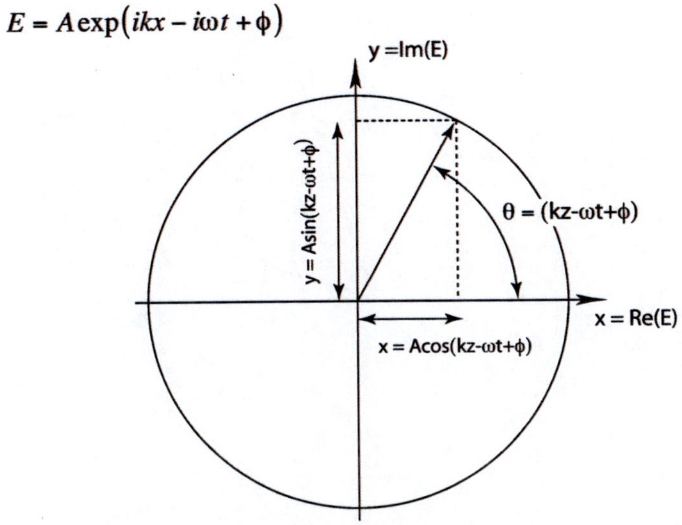

Fig. 1.2 Phasor diagram of the complex amplitude of a plane wave. The phasor rotates counter-clockwise with increasing distance, and clockwise with increasing time

When a wave is represented graphically on the complex plane, it is called a "phasor." A phasor graphical construction is shown in Fig. 1.2. The phasor, which represents the state of a wave, rotates counter-clockwise as the wave travels an increasing distance, and clockwise with increasing time.

When two fields of the same frequency are added, the resultant is a wave with the same spatial and temporal frequencies

$$\begin{aligned} &A_1 \exp(\mathrm{i}(kz - \omega t + \phi_1)) + A_2 \exp(\mathrm{i}(kz - \omega t + \phi_2)) \\ &= [A_1 \exp(\mathrm{i}\,\phi_1) + A_2 \exp(\mathrm{i}\,\phi_2)] \exp(\mathrm{i}(kz - \omega t)) \\ &= |A| \exp(\mathrm{i}\,\phi) \exp(\mathrm{i}(kz - \omega t)) \end{aligned} \tag{1.7}$$

where the net amplitude is

$$|A| = \sqrt{A_1^2 + A_2^2 + 2A_1A_2 \cos(\phi_1 - \phi_2)} \tag{1.8}$$

and the new phase is

$$\phi = \tan^{-1}\left(\frac{A_1 \sin\phi_1 + A_2 \sin\phi_2}{A_1 \cos\phi_1 + A_2 \cos\phi_2}\right) \tag{1.9}$$

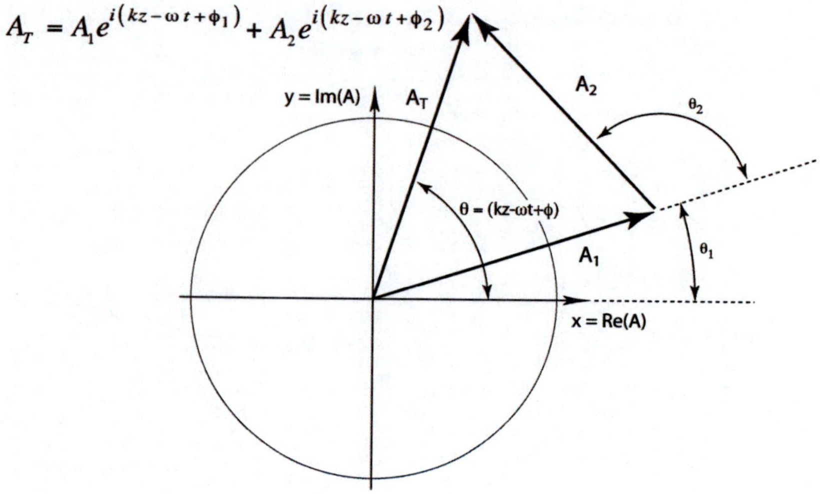

Fig. 1.3 Addition of phasors on the complex plane. If the waves have the same frequency, then the resultant A_T also rotates at that frequency

Phasor addition is shown graphically in Fig. 1.3 for two waves with the same frequencies. The net phasor A_T rotates at the same rate as the individual phasors.

Intensities in a homogeneous dielectric medium are related to electric fields through

$$I = \varepsilon_m \varepsilon_0 v \langle E^2 \rangle = n_m c \varepsilon_0 \langle E^2 \rangle \tag{1.10}$$

where ε_m is the dielectric constant of the medium, $v = c/n_m$ is the wave speed of the medium with a refractive index $n_m = \sqrt{\varepsilon_m}$, and the brackets denote a time average. Often in this book the proportionality between intensity and squared field is set conveniently to unity as $I = |A|^2$. The detected intensity of the resultant is then expressed as

$$\begin{aligned} I &= |A_1 \exp(i\phi_1) + A_2 \exp(i\phi_2)|^2 \\ &= A_1^2 + A_2^2 + A_1 A_2 \exp(i(\phi_1 - \phi_2)) + A_1 A_2 \exp(i(\phi_2 - \phi_1)) \\ &= A_1^2 + A_2^2 + 2A_1 A_2 \cos(\phi_1 - \phi_2) = I_1 + I_2 + 2\sqrt{I_1 I_2} \cos(\phi_1 - \phi_2) \end{aligned} \tag{1.11}$$

The resultant intensity is a function of the two intensities and of the relative phase difference between the fields.

It is common to define an interferometric response function that is normalized by the total input intensity as

$$R = 1 + 2\frac{\sqrt{I_1 I_2}}{(I_1 + I_2)} \cos(\phi_1 - \phi_2) = 1 + 2\frac{\sqrt{\beta}}{1 + \beta} \cos(\phi_1 - \phi_2) \tag{1.12}$$

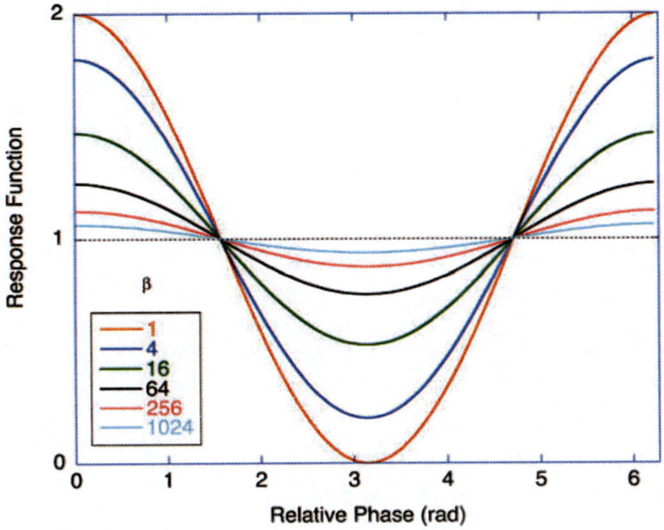

Fig. 1.4 The interferometric response curve as a function of relative phase for a selection of beam intensity ratios. The interference contrast is $C = 1, 0.8, 0.47, 0.25, 0.12$ and 0.062 for $\beta = 1, 4, 16, 64, 256$ and $1{,}024$, respectively

where the ratio of intensities is $\beta = I_1/I_2$. When $I_1 = I_2$, there is perfect constructive interference and perfect destructive interference for the appropriate relative phases. The graph of the response function is shown in Fig. 1.4 for several beam ratios.

Contrast is an important property of interference. The contrast is defined as

Fringe contrast:

$$C(\beta) = 2\frac{\sqrt{I_1 I_2}}{(I_1 + I_2)} = 2\frac{\sqrt{\beta}}{1+\beta} \tag{1.13}$$

Contrast is plotted in Fig. 1.5 as a function of beam ratio β on both a linear and a log scale. Even when the beam intensity ratio is over a thousand to one, there is still a 6% contrast, because the square-root dependence in the numerator partially compensates the denominator. The slow dependence of contrast on beam ratio is the reason why interference effects can be relatively strong even when one of the interfering fields is very weak. This allows interference effects to persist even in high-background conditions that are common in biomedical applications that have strong light scattering backgrounds.

1.1.2 Two-Port Interferometer

The simplest way to combine two waves is with a beam splitter, shown in Fig. 1.6. Two waves are incident on opposite input ports of a beamsplitter, and each are

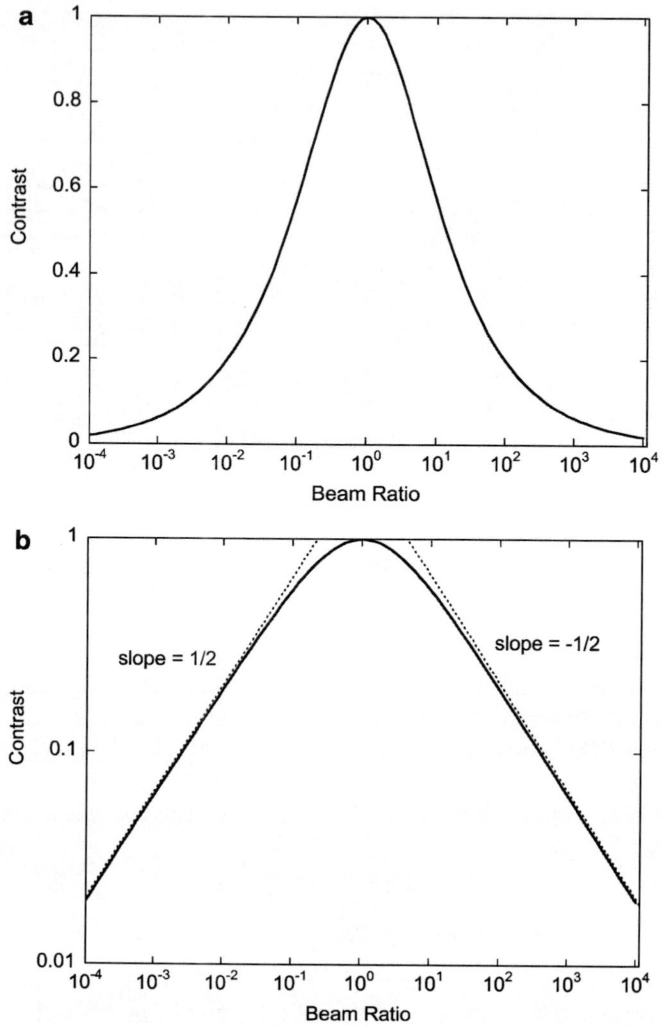

Fig. 1.5 Interference contrast as a function of beam ratio $\beta = I_1/I_2$. The log–log plot shows the slope of ½ which gives a weak dependence on β

combined with the other in the two outputs. The output intensities share the total input energy, but the intensities depend on the relative phase between the two input beams.

The two electromagnetic waves incident on a beamsplitter are

$$
\begin{aligned}
E_s &= E_s^0 \exp(\mathrm{i}\,kx - \mathrm{i}\,\omega t + \phi_s) \\
E_r &= E_r^0 \exp(\mathrm{i}\,kx - \mathrm{i}\,\omega t + \phi_r)
\end{aligned} \tag{1.14}
$$

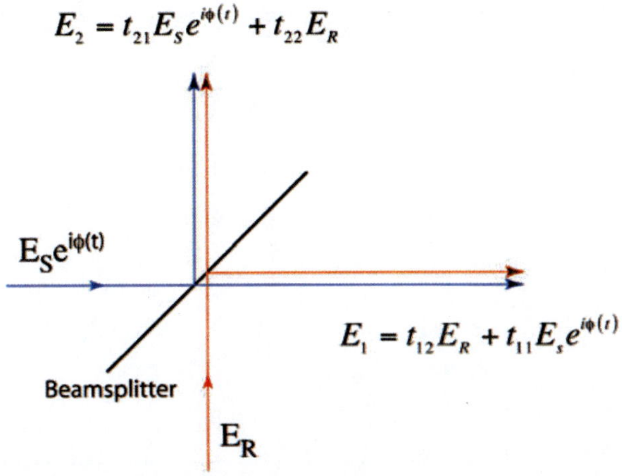

Fig. 1.6 Beam combination by a beamsplitter. A phase-modulated signal wave and a stable reference wave are incident on opposite input ports. Superposed waves exit the two output ports. The total energy exiting the beamsplitter equals the total energy incident on the beamsplitter. However, the relative intensities that exit each port depend on the phase difference between the two input waves

If each wave is split equally by the beamsplitter (50/50 beamsplitter), the two output ports have the fields

$$\begin{aligned} E_1 &= \frac{E_s^0}{\sqrt{2}} \exp(\mathrm{i}\,kx - \mathrm{i}\,\omega t + \phi_s) + \frac{E_r^0}{\sqrt{2}} \exp(\mathrm{i}\,kx - \mathrm{i}\,\omega t + \phi_r) \\ E_2 &= \frac{E_r^0}{\sqrt{2}} \exp(\mathrm{i}\,kx - \mathrm{i}\,\omega t + \phi_r) - \frac{E_r^0}{\sqrt{2}} \exp(\mathrm{i}\,kx - \mathrm{i}\,\omega t + \phi_r) \end{aligned} \tag{1.15}$$

The transformation by the beamsplitter can be expressed in matrix form as

$$\begin{aligned} E_{\text{out}} &= BE_{\text{in}} \\ \begin{pmatrix} E_1 \\ E_2 \end{pmatrix} &= \begin{pmatrix} B_{11} & B_{12} \\ B_{21} & B_{22} \end{pmatrix} \begin{pmatrix} E_s \\ E_r \end{pmatrix} \end{aligned} \tag{1.16}$$

in which the beamsplitter matrix B is a unitary operator that creates the coherent superposition of the input waves at the output ports in a way that conserves total intensity (energy). For a 50/50 beamsplitter

$$B = \begin{pmatrix} B_{11} & B_{12} \\ B_{21} & B_{22} \end{pmatrix} = \frac{1}{\sqrt{2}} \begin{pmatrix} 1 & 1 \\ 1 & -1 \end{pmatrix} \tag{1.17}$$

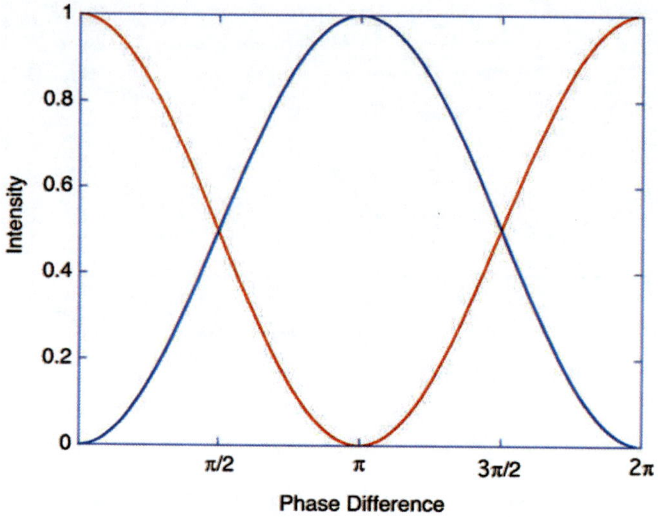

Fig. 1.7 Universal interferometric response curve for a two-mode interferometer. The total energy is shared between the two outputs, with the relative intensities in each channel dependent on the relative phase between the two waves

The measured intensities at the outputs includes the interfering cross terms between the fields, and are

$$\begin{aligned} I_{1,2} &= \left|\frac{E_s^0}{\sqrt{2}}\right|^2 + \left|\frac{E_r^0}{\sqrt{2}}\right|^2 \pm \frac{E_r^0 E_s^{0*}}{2}\exp(\mathrm{i}(\phi_r - \phi_s)) \mp \frac{E_r^0 E_s^0}{2}\exp(\mathrm{i}(\phi_s - \phi_r)) \\ &= \frac{1}{2}\left[I_s + I_r \pm 2\sqrt{I_s I_r}\cos(\phi_r - \phi_s)\right] \end{aligned} \tag{1.18}$$

These conserve total intensity, but the intensity switches back and forth between the two ports as the relative phase $(\phi_r - \phi_s)$ changes. The response curve of a 50/50 beam-combiner is

$$R = 1 \pm \frac{2\sqrt{I_s I_r}}{I_s + I_r}\cos(\phi_r - \phi_s) = 1 \pm C(\beta)\cos(\phi_r - \phi_s) \tag{1.19}$$

which is shown in Fig. 1.7.

There are two common nomenclatures for interferometry known as *homodyne* or *heterodyne* detection. These terms can mean somewhat different things, depending on how the interferometry is being performed. In a two-port interferometer, homodyne relates to a signal and reference waves that both have the same frequency and have a stable relative phase (but with a small phase modulation). On the other hand, heterodyne relates to the case where there is a frequency offset between the waves, or equivalently, if there is a large time-varying phase offset between the waves. These two terms *homodyne* or *heterodyne* are also used for light scattering experiments where the meaning is different. When only scattered light is detected, this is called homodyne detection. But if part of the original wave interferes with the scattered light, even if it is at the same frequency, this is called heterodyne detection. The different usages of these terms may lead to some confusion. However, the context (interferometer vs. light scattering) usually makes it clear.

1.1.3 Homodyne Phase Quadrature

When the signal wave in a two-port interferometer under homodyne detection carries a small time-dependent phase modulation $\phi_s(t) = \phi_0 + \Delta\phi(t)$, interferometry converts this phase modulation (which cannot be observed directly) into an intensity modulation (that can be). The main question is, what relative phase ϕ_0 allows the maximum phase-to-intensity conversion? The most sensitive part of the response curve in Fig. 1.7 to a small phase modulation is at the condition of steepest slope. The slope of the response curve is called the interferometric responsivity. For phase modulation, the responsivity $\mathbb{R}_\phi$ is

Phase responsivity:

$$\mathbb{R}_\phi = \frac{dR}{d\Delta\phi} = \mp C(\beta)\sin(\phi_r - \phi_0) \tag{1.20}$$

For a given small phase modulation amplitude $\Delta\phi$, the intensity response to phase modulation on the signal wave is

$$\Delta I = \mathbb{R}_\phi \Delta\phi_s \tag{1.21}$$

The phase responsivity is maximum when the phase difference between the signal and the reference is at 90°, or $(\phi_r - \phi_0) = \pi/2$. This is the condition known as "phase quadrature." The intensity modulation in phase quadrature is

$$\Delta I = C(\beta)\Delta\phi_s \tag{1.22}$$

and the largest intensity modulations are obtained in phase quadrature at a contrast of unity.

The phasor diagram for homodyne quadrature detection of phase modulation on a signal wave is shown in Fig. 1.8. The reference wave is drawn along the real axis, and the signal wave is 90° out of phase (in phase quadrature) along the imaginary axis. In a phasor diagram, phase modulation on a wave is represented as a small phasor orthogonal to the carrier phasor. In this example, the phase modulation ΔE_s is parallel to the real axis and hence is *in phase* with the reference wave E_r. Therefore, the phase modulation adds in-phase with the reference and is detected as an intensity modulation.

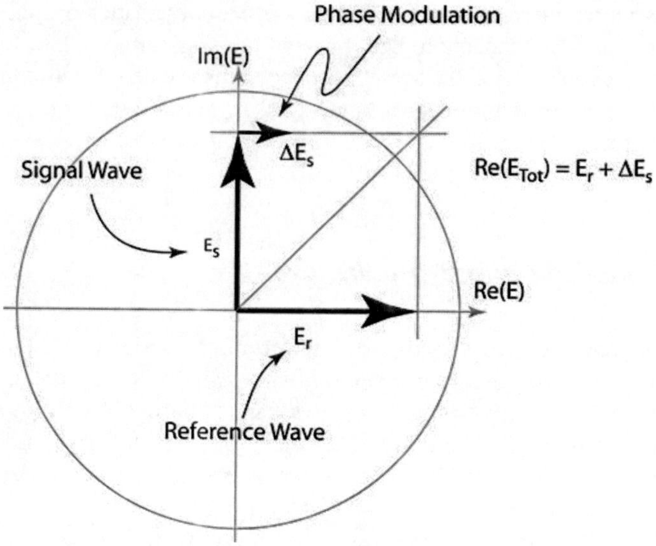

Fig. 1.8 Complex representation of phase quadrature. The phase modulated field ΔE_s on the signal wave is parallel to the reference wave E_r to produce constructive interference

1.1.4 Heterodyne and Beats

When there is a frequency difference between the two interfering waves, the combined wave exhibits beats. This is demonstrated by adding two waves with different frequencies. The relative phase between the waves changes linearly in time, and the system moves continually in and out of phase. The total wave, considering only the temporal part, is

$$E_{\mathrm{Tot}} = E_1 \exp(\mathrm{i}(-\omega_1 t + \alpha_1)) + E_2 \exp(\mathrm{i}(-\omega_2 t + \alpha_2)) \tag{1.23}$$

This is best rewritten in terms of the average frequencies and phases as

$$\begin{aligned} E_{\mathrm{Tot}} &= \exp(\mathrm{i}(-\omega_{\mathrm{m}} t + \alpha_{\mathrm{m}})) \\ &\quad \times [E_1 \exp(\mathrm{i}(-\Delta\omega t + \Delta\alpha)) + E_2 \exp(-\mathrm{i}(-\Delta\omega t + \Delta\alpha))] \end{aligned} \tag{1.24}$$

where the average values and differences are

$$\begin{aligned} \omega_{\mathrm{m}} &= \frac{1}{2}(\omega_1 + \omega_2) \;\; \Delta\omega = \tfrac{1}{2}(\omega_1 - \omega_2) \\ \alpha_{\mathrm{m}} &= \frac{1}{2}(\alpha_1 + \alpha_2) \;\; \Delta\alpha = \tfrac{1}{2}(\alpha_1 - \alpha_2) \end{aligned} \tag{1.25}$$

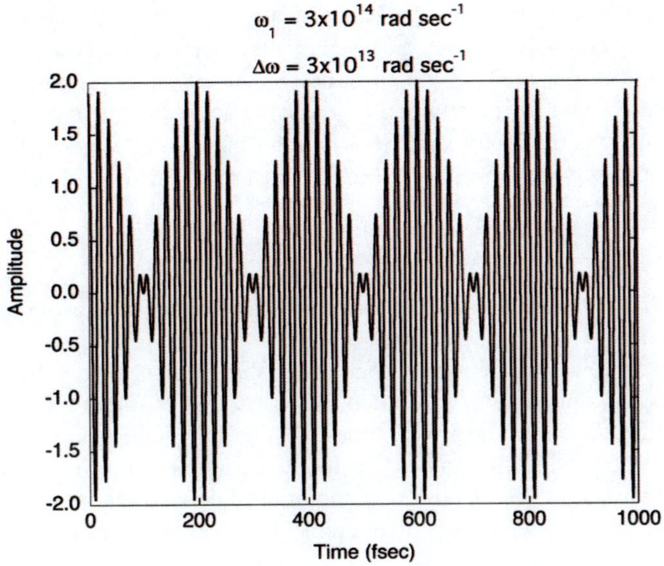

Fig. 1.9 Beat pattern between two waves with a 10% frequency offset

The total wave is now characterized by a high-frequency carrier wave that has the average frequency and phase of the original waves, multiplied by a low-frequency envelope that has the difference frequency, as shown in Fig. 1.9. The total intensity varies with time as

$$I_{\mathrm{Tot}} = 2I[1 + C(\beta)\cos(2\Delta\omega t - 2\Delta\alpha)] \tag{1.26}$$

which, on a spectrum analyzer, has frequency side-lobes at the frequency difference $\pm|\omega_1 - \omega_2|$.

1.1.5 Noise and Detection

All optical systems have noise, and to understand the sensitivity of any optical measurement requires an understanding of that noise. A challenge to optical metrology is the many different contributions to noise, for which one or another may dominate, or several may contribute nearly equally. Pulling apart which noise mechanisms are participating and dominating an optical measurement is sometimes difficult, because different noise contributions may behave in similar ways that makes it difficult to distinguish among them.

There are three generic types of noise that can be distinguished by their intensity dependence. These are: (1) system noise that is intensity independent, often set by electronic noise of the detection and amplification systems; (2) shot noise that is a fundamental quantum noise source caused by the discreteness of the photon; and (3) relative intensity noise (RIN) which is noise that increases proportionally to the light intensity. Signal is conventionally measured as the electrical power dissipated in a detector circuit, which therefore is proportional to the square of the detector current. The total noise power is the quadrature sum of the individual contributions

$$i_{\mathrm{N}}^2 = i_{\mathrm{floor}}^2 + i_{\mathrm{shot}}^2 + i_{\mathrm{RIN}}^2 \tag{1.27}$$

where the first term is the so-called "noise floor" set by the details of the measuring apparatus.

The second noise contribution is shot noise. Shot noise is simply the discreteness of light incident upon a detector, much like the sound of raindrops falling upon a tent or skylight. The discrete arrival times of light, like raindrops, are uncorrelated, meaning that the probability of any photon detection per unit time is independent of any previous detection. Under this condition, known as Poisson statistics, and for moderate to large numbers of photons, the fluctuation in the arrival of the photons is

$$\Delta N = \sqrt{N} \tag{1.28}$$

where N is the mean number of detected photons in some integration time, and ΔN is the standard deviation in the mean value. Expressed in terms of photodetector current, the mean detected current is

$$i_{\mathrm{d}} = \eta_{\mathrm{q}} e \frac{P}{h\nu} = \eta_{\mathrm{q}} e N_{\mathrm{photon}} \mathrm{BW} \tag{1.29}$$

where η_{q} is the quantum efficiency (number of photogenerated electrons in an external circuit relative to the number of incident photons), $h\nu$ is the photon energy, P is the power captured on the detector area, and BW is the detection bandwidth (often identified as the inverse of the detector integration time). The noise current is the variance of the detector current and is

$$i_{\mathrm{shot}}^2 = \left(e\sqrt{N_{\mathrm{photon}}\mathrm{BW}}\right)^2 = e^2 \frac{\eta_{\mathrm{q}} P_{\mathrm{d}}}{h\nu} \mathrm{BW} \tag{1.30}$$

The linear dependence on detection bandwidth assumes the noise is white noise (independent of frequency). In this case, the signal-to-noise ratio (S/N) of shot noise is

$$S/N_{\mathrm{shot}} = \frac{i_{\mathrm{d}}^2}{i_{\mathrm{N}}^2} = \frac{\eta_{\mathrm{q}} P_{\mathrm{d}}}{h\nu \mathrm{BW}} \tag{1.31}$$

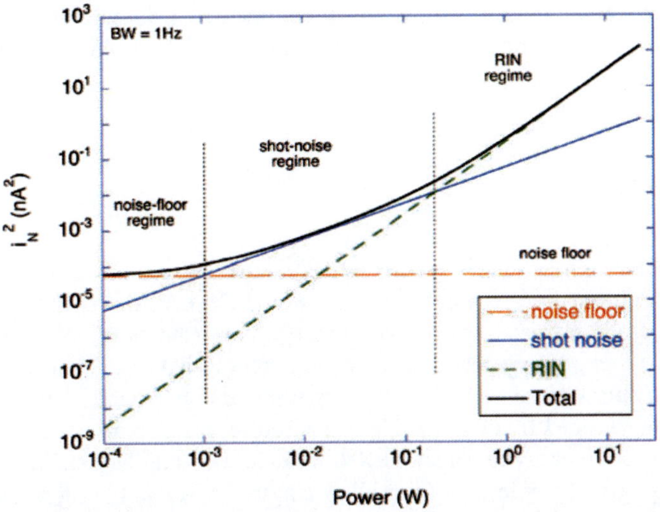

Fig. 1.10 Noise power as a function of detected power for a system that moves from a fixed noise floor (set by the detection electronics), through a shot-noise regime into a relative intensity noise (RIN) regime at high power

which is linear in the detected light power. Higher light powers yield higher signal-to-noise ratios.

Although (1.31) was derived under the assumption of white noise (frequency independence), virtually all experimental systems exhibit 1/*f* noise at sufficiently low frequencies. This 1/*f* noise is ubiquitous, and is the same thing as long-term drift. If the detection bandwidth becomes too small (integration time becomes too large), then long-term drift begins to dominate, and the *S*/*N* ratio can begin to degrade.

The third type of noise is relative intensity noise (RIN). This noise is simply a statement that if there are conventional intensity fluctuations of a light source, then the noise increases proportionally to the detected signal as

$$i_{\mathrm{N}} = C_{\mathrm{RIN}} i_{\mathrm{d}} \tag{1.32}$$

The signal-to-noise ratio is then

$$S/N_{\mathrm{RIN}} = \frac{i_{\mathrm{d}}^2}{i_{\mathrm{N}}^2} = C_{\mathrm{RIN}}{}^2 \tag{1.33}$$

which is independent of intensity.

The noise power of a generic experiment is shown in Fig. 1.10 as a function of detected power. The noise starts at the noise floor (dominated by the electrical system) at low intensity, moves into a shot-noise regime with a linear intensity dependence, and then into the relative-intensity-noise regime that depends on the square of the intensity. Not all systems exhibit a shot-noise regime. If the fixed noise floor is too high, then the shot-noise regime never dominates, and the system moves from the noise floor into the RIN regime with increasing intensity.

1.1.6 Sub-nanometer Noise-Equivalent Displacement

Interferometry is one of the most sensitive metrologies currently known. For instance, the Laser Interferometric Gravitational Observatory (LIGO) is searching for displacements caused by gravitational waves (the weakest of all physical phenomena) using kilometer-long interferometers to detect displacements that are less than an attometer (1×10^{-18} m). To put this into perspective, the radius of the proton is only about 1 fm (1×10^{-15} m). Therefore, the displacements that a gravity wave detector is seeking to detect are a thousand times smaller than the radius of the proton. It is a trillion times (10^{12}) smaller than the wavelength of light, which is the natural yardstick for interferometric measurements. A basic question is: How is such sensitivity achievable?

Interferometric measurements are often expressed as equivalent displacement, either in the displacement of a surface, or a shift of an optical path length. For instance, in biointerferometric measurements of protein on biosensors, the results are often expressed as protein height [1]. The sensitivity of interferometric measurements are naturally expressed in terms of the smallest surface height displacement that can be detected at a selected signal-to-noise ratio. This leads to a figure of merit known as the noise-equivalent displacement (NED) of an interferometric system.

To estimate the smallest surface displacement that can be detected if the system is dominated by shot noise, we can choose a signal-to-noise ratio of 2:1 as the detectability limit, such that

$$S/N_{\text{shot}} = 8\Delta\phi^2 \frac{\eta_q P_s}{h\nu \text{BW}} = 2 \tag{1.34}$$

where the factor of 8 comes from balanced detection with a strong reference intensity. The phase modulation from a surface displacement (in reflection) is

$$\Delta\phi = \frac{4\pi}{\lambda} \Delta d \tag{1.35}$$

The shot-noise limited NED is then

$$\Delta d = \frac{\lambda}{8\pi} \sqrt{\frac{h\nu \text{BW}}{\eta_q P_s}} \tag{1.36}$$

For a detected power of 1 W, a detection bandwidth of 1 Hz, a photon wavelength of 600 m, and a quantum efficiency of 0.7, the NED is an astonishingly small value of $\Delta d = 0.016$ fm. These parameters are not too far from the anticipated performance of the LIGO gravity wave detector. For biosensor applications, on the other hand, detected powers are more typically 10 mW, with detection bandwidths around 100 Hz, which still provides an NED of 1 fm. In practice, many biosensors detect surface layers with a sensitivity of about 1 pm, limited by aspects of the system noise other than shot noise. Detection limits for a biochemical assay are usually not even limited by the system noise, but rather by the performance of the surface chemistry. It is common for chemical variations to limit surface sensitivity to tens of picometers in interferometric assays.

1.2 Interferometer Configuration Classes

An interferometer is any optical configuration that causes light of finite coherence to take more than one path through the system, and where the light paths subsequently cross or combine with a path-length difference. Because of the generality of this definition of an interferometer, there are many ways that interferometers can be designed and implemented. Despite the open-ended definition, interferometers fall into distinct classes. These classes are: (1) wavefront-splitting interferometers and (2) amplitude-splitting interferometers. These two classes are distinguished by whether distinct optical paths are deflected so that they are no longer parallel and hence cross (wavefront-splitting), or whether a single optical path splits to take different paths before crossing or combining (amplitude-splitting). Wavefront-splitting configurations include Young's double-slit interference (Fig. 1.11) and diffraction generally. Amplitude-splitting interferometers include Michelson, Mach–Zehnder and Fabry–Perot configurations, among many others.

1.2.1 Wavefront-Splitting Interferometers: Young's Double Slit

Wavefront-splitting interferometers take parallel or diverging optical paths (hence non-crossing) and cause them to cross. Young's double-slit interferometer is perhaps the most famous of all interferometers because it was the first (1801–1803). It consists of a spatially coherent source (a distant pin hole, or a laser beam) that is split by two apertures into two partial waves that travel different path lengths to a distant observation plane where multiple interference intensity fringes are observed. The Young's double-slit configuration is shown in Fig. 1.12. In the figure there is a thin film in the path of one of the slits, for instance a molecular film accumulating on the functional surface in a biosensor. As the film increases in thickness d, the optical path length increases by $h^{\mathrm{OPL}} = (n - n_{\mathrm{m}})d$, where n is the refractive index of the film, n_{m} is the surrounding medium and d is the

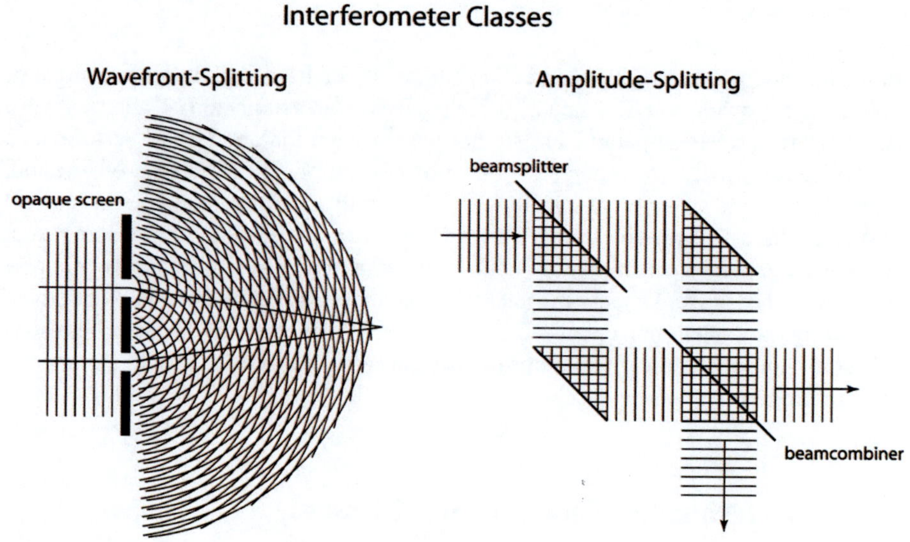

Fig. 1.11 Two interferometer classes. A wavefront-splitting interferometer separates different parts of a common wavefront and causes them to superpose (in this case by diffraction). An amplitude-splitting interferometer divides the field amplitude into two different paths that cross (in this case by a combination of a beamsplitter and mirrors). The wavefront-splitting interferometer is in a Young's double slit configuration. The amplitude-splitting interferometer is in a Mach–Zehnder configuration

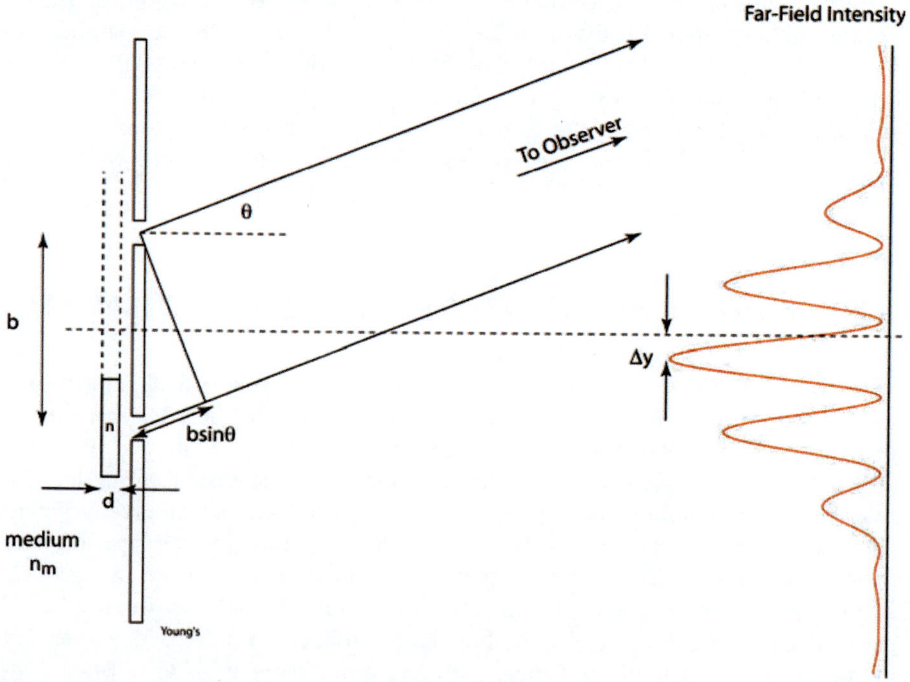

Fig. 1.12 Young's double slit interference configuration. The slit separation *b* is much larger than a slit width. A film of thickness *d* and refractive index *n* in a medium of index n_m is placed in front of one of the slits, shifting the position of the interference fringes

thickness of the film. This optical path length adds to the phase offset between the slits, and hence shifts the far-field fringes.

If the apertures are small (pin holes or narrow slits), and the illumination is even, then the interference pattern on a distant observation screen is

$$I(\theta) = I_0(\theta)[1 + C(\beta)\cos(k_0 b \sin\theta + k_0(n - n_{\mathrm{m}})d)] \tag{1.37}$$

where k_0 is the free-space k-vector, $I_0(\theta)$ is an angle-dependent intensity (arising from the finite width of the slits), and $C(\beta)$ is the fringe contrast (set by the relative intensity from each slit).

The extra phase term

$$\phi = k_0(n - n_{\mathrm{m}})d \tag{1.38}$$

is caused by the dielectric slab in front of the lower slit in Fig. 1.12. If the refractive index changes, or the thickness, or both, then the phase shift is

$$\Delta\phi = k_0 \Delta n d + k_0(n - n_{\mathrm{m}})\Delta d \tag{1.39}$$

and the normalized response function is

$$R = \frac{I(\theta)}{I(0)} \tag{1.40}$$

The sensitivity of the interference pattern to phase modulation in the path of one arm is determined by the interferometric phase responsivity $\mathbb{R}_\phi$, which is defined through

$$\frac{\Delta I}{I} = \mathbb{R}_\phi \Delta\phi = \frac{\mathrm{d}R}{\mathrm{d}\phi}\Delta\phi \tag{1.41}$$

For Young's double slit, the responsivity is

$$\mathbb{R}_\phi = I_0(\theta)C(\beta)\sin(k_0 b \sin\theta + \phi_0) \tag{1.42}$$

which is a maximum when

$$k_0 b \sin\theta + \phi_0 = \pi/2 \tag{1.43}$$

which sets the phase quadrature condition for this configuration at which the responsivity is a linear function of either the refractive index or the thickness for small changes. The diffracted intensity and the responsivity are shown in Fig. 1.13. The responsivity is equal to unity at the half-intensity angles of the diffraction

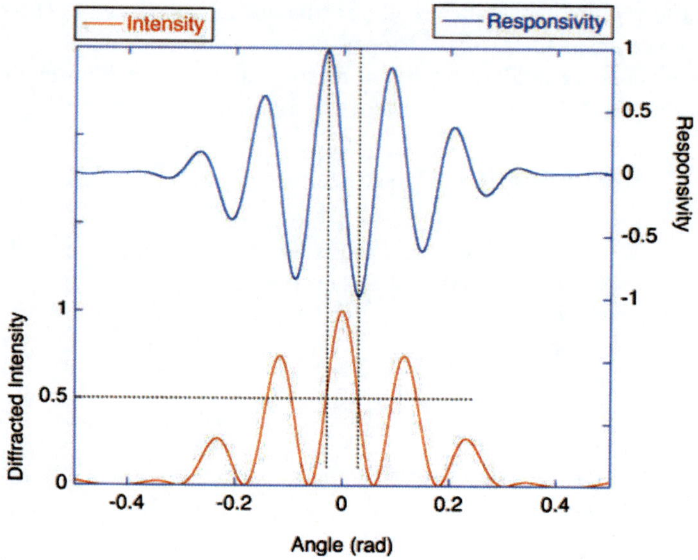

Fig. 1.13 Young's double slit diffracted intensity and the phase responsivity. The slit width is $a = 1.5$ μm. The separation $b = 5$ μm for a wavelength of 0.6 μm

pattern. These half-intensity angles correspond to the conditions of phase quadrature with maximum phase-to-intensity transduction.

Linearity to perturbation is one of the essential signatures of interferometry and is one reason that it is selected for high-sensitivity detection of very small effects. In biological applications, such as the detection of molecular films in biosensors, the chief aim is high sensitivity to small numbers of accumulated molecules. This can only be accomplished by a detection technique that is linearly responsive to the thickness of the film. Optical processes such as polarization nulling or diffraction gratings have a quadratic dependence of signal on the film thickness, which falls to zero much faster than for a linear dependence. Therefore, small-signal molecular sensors are usually operated in the linear regime (quadrature) for maximum responsivity, while large-signal sensors (thick films or particle-based) are usually operated in the quadratic regime to minimize background.

1.2.2 Amplitude-Splitting Interferometers

Amplitude-splitting interferometers split a single amplitude by partial reflection into two or several partial waves that are then directed to cross and interfere. This class includes the Michelson and Mach–Zehnder interferometers. It also includes the Fabry–Perot interferometer that works in an "in-line" configuration.

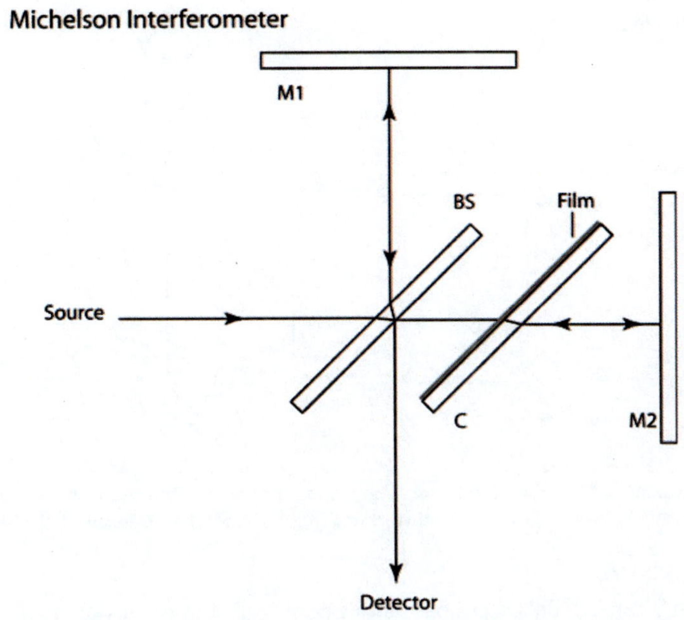

Fig. 1.14 A Michelson interferometer has a beamsplitter that acts both to split the waves and to recombine them after reflection from the mirrors. The compensator, which equalizes the optical path lengths in the two arms, can support an accumulating film that shifts the fringes at the detector. *M1*, *M2* mirrors; *BS* beamsplitter; *C* compensator

1.2.2.1 Michelson Interferometer

The Michelson interferometer was used in the famous Michelson–Morley experiment that found no ether (1887). It is easy to scan in path length or wavelength and is a common configuration used for Fourier-transform spectroscopy [2]. The basic configuration of a Michelson interferometer is shown in Fig. 1.14.

The phase difference caused by an accumulating film on the compensator is

$$\Delta\phi = [k_0 2d \cos\theta']\Delta n + [k_0(n - n_m)2 \cos\theta']\Delta d \qquad (1.44)$$

and the phase responsivity of the Michelson interferometer is

$$\mathbb{R}_\phi = C(\beta)\sin\phi_0 \qquad (1.45)$$

where $\beta \approx 1$ and hence $C \approx 1$ is set by the balance of the beamsplitter, and ϕ_0 is the phase bias. For maximum responsivity, the phase bias should be $\phi_0 = \pi/2$. However, this phase is sensitive to mechanical vibrations and thermal drift, making it difficult to construct stable biosensors from free-space Michelson interferometers. The Michelson interferometer has the advantage of being compact, with a single beamsplitter acting as both a splitter and a combiner. But it has the drawback that

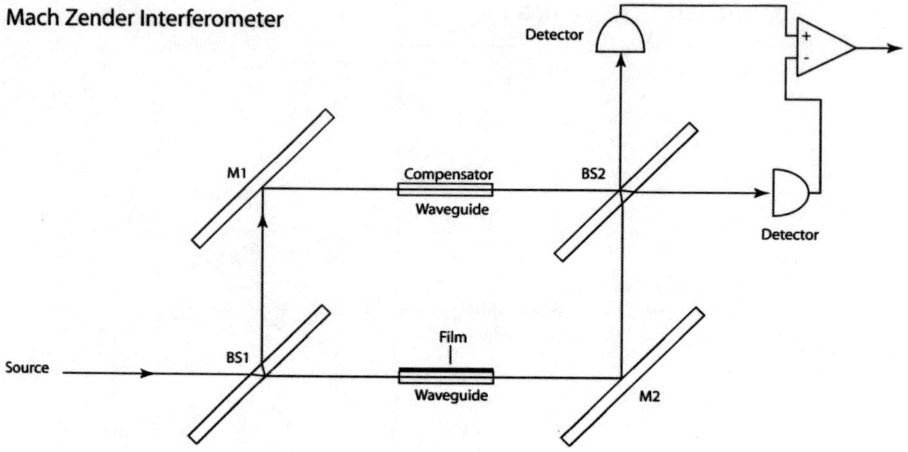

Fig. 1.15 Mach–Zehnder interferometer. The path is split by the first beam splitter, then directed through identical waveguides (one with a thin film) and then recombined at the second beam splitter

there is only one easily accessible output port, with the other quadrature returning counter to the incoming source. This could be picked off using a beamsplitter, but it makes the two output ports asymmetric. A symmetric interferometer that has two symmetric output ports is the Mach–Zehnder interferometer.

1.2.2.2 Mach–Zehnder

A Mach–Zehnder interferometer is shown in Fig. 1.15. A source is split by the first beamsplitter into two waves, one that passes through a sample arm with a solid support for a thin film. The other path has the reference wave that passes through a path-length compensator to match the sample and reference arm path lengths. A second beamsplitter splits each beam, so that each detector receives a contribution from each path. The two output ports are in quadratures, such that the sum of output powers equals the input power. Balanced detectors on both output ports can subtract relative intensity noise of the light source, leaving only the shot noise of the reference wave and the phase difference between the two arms. The phase responsivity of the Mach–Zehnder interferometer is

$$\mathbb{R}_{\phi} = C(\beta)\sin\phi_0 \tag{1.46}$$

where $C(\beta)$ is set by the balance of the beamsplitter, and ϕ_0 is the phase bias, as for the Michelson interferometer.

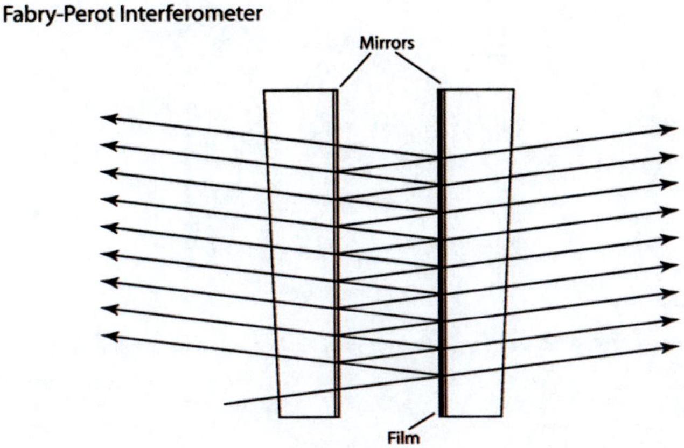

Fig. 1.16 Fabry–Perot interferometer with two high-reflectance mirrors, one of which has an accumulating film that increases the optical path length inside the optical cavity

There are many possible implementations of the Mach–Zehnder interferometer. The drawing in Fig. 1.15 is for a free space setup, but the same principle holds for integrated on-chip waveguide interferometers, and for fiber-based interferometers. Integrated optical interferometers have much more stable path differences and phase bias than free-space optical interferometers.

1.2.2.3 Fabry–Perot

The amplitude-splitting interferometers discussed so far have been two-wave (two-port) interferometers. These lead to the simplest interference effects, with sinusoidal interferograms. However, much higher responsivity to phase perturbations are achieved by having many partial waves interfering. This occurs in the Fabry–Perot interferometer that is composed of two parallel mirrors with high reflectance. An incident wave reflects back and forth many times in the Fabry–Perot cavity, shown in Fig. 1.16, increasing the effective interaction length with a thin dielectric film.

The transmittance and reflectance of a Fabry–Perot cavity are

$$T = \frac{1}{1 + F\sin^2(\phi/2)}$$
$$R = \frac{F\sin^2(\delta/2)}{1 + F\sin^2(\phi/2)} \tag{1.47}$$

where F is the coefficient of finesse

$$F = \left(\frac{2\sqrt{R}}{1-R}\right)^2 \tag{1.48}$$

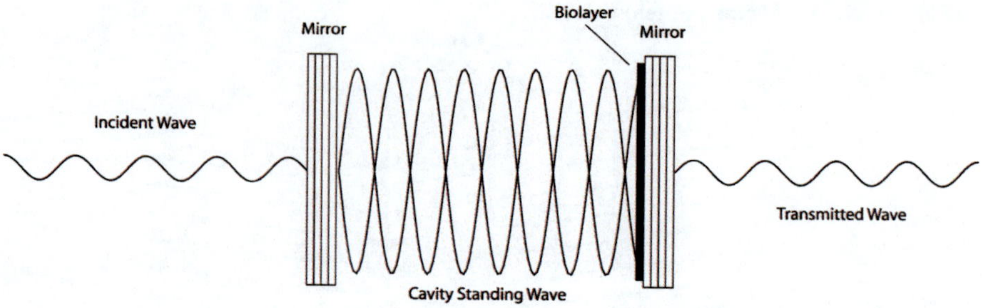

Fig. 1.17 Fabry–Perot structure (with field nodes shown relative to biolayer)

where

$$R = |r_1||r_2| \tag{1.49}$$

is a composite reflectance of both mirrors for asymmetric mirrors with reflection coefficients r_1 and r_2. For $R = 90\%$ the finesse is $F = 30$, while for $R = 99\%$ the finesse is $F = 300$. The phase ϕ is

$$\phi = \frac{4\pi n_c}{\lambda} d \cos\theta' + 2\phi_r + 2k\Delta h^{OPL} \cos\theta'_f \tag{1.50}$$

where n_c is the index inside the cavity, d is the cavity length, θ'_f is the internal angle in the film and θ' is the internal angle in the cavity. For standing-wave interferometers it is important to keep in mind that Δh^{OPL} is not only a function of index and thickness, but also of where the thin layer is located relative to the electric field nodes and anti-nodes. The thin biofilm can be virtually invisible if it is located at a field node. For instance, the biolayer in Fig. 1.17 is placed on a mirror that has an anti-nodal (electric field maximum) at its surface. This important aspect of interferometry will be discussed in detail in Chap. 4 on surface optics. Conversely, the entire cavity can be the biosensor in a Fabry–Perot, as in porous silicon sensors [3, 4], when the average cavity index changes upon volumetric binding of analyte.

A key property of a resonator cavity is the multiple-pass of a photon through the cavity prior to being lost by transmission through one of the mirrors or by photon extinction (scattering or absorption). The time for a single round trip of a photon in the cavity is

$$t_r = 2n_c L/c \tag{1.51}$$

where n_c is the cavity index and L is the cavity length. This round trip produces the characteristic resonance frequency, known as the free spectral range

$$\nu_F = \frac{1}{t_r} = \frac{c}{2L} \tag{1.52}$$

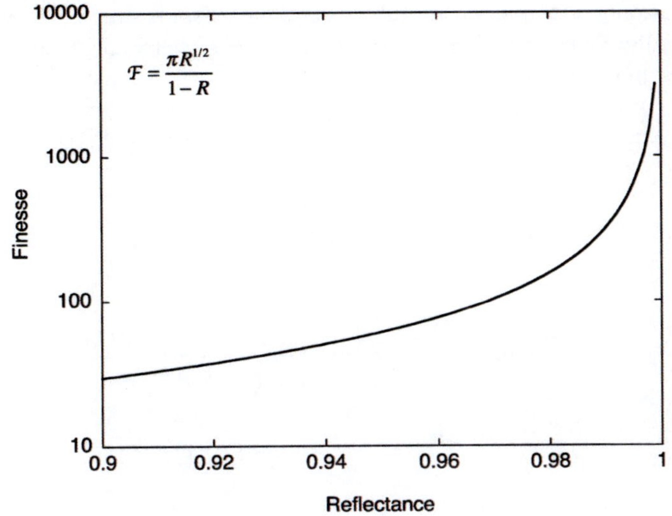

Fig. 1.18 Cavity finesse vs. composite mirror reflectance

The Fabry–Perot spectrum as a function of light frequency consists of evenly spaced transmission peaks (or reflection dips) separated by ν_F. The width $d\nu$ of the resonance is defined by

$$d\nu = \frac{\nu_F}{F} = \frac{1}{2\pi t_p} \tag{1.53}$$

where t_p is the lifetime of the photon in the cavity, and the cavity finesse is related to the coefficient of finesse F (defined in (1.48)) by

$$\mathcal{F} = \frac{\pi}{2}\sqrt{F} \tag{1.54}$$

From this relation, it is clear that the physical interpretation of the cavity finesse is related to the ratio of the photon lifetime to the round trip time

$$\mathcal{F} = 2\pi\frac{t_p}{t_r} = \omega_F t_p \tag{1.55}$$

The finesse is approximately equal to the average number of round-trips of a photon before it is lost from the cavity. A quantity closely related to the finesse is the Q-value of the cavity, defined as the ratio of the resonant frequency to the linewidth

$$Q = \frac{\nu_M}{d\nu} = \frac{\nu_M}{\nu_F}F = MF \tag{1.56}$$

for the Mth resonance of the cavity. The finesse as a function of composite mirror reflectance is shown in Fig. 1.18.

The operating principle for the Fabry–Perot biosensor is simple. Upon binding of an analyte, the round-trip phase of the cavity is changes by $\Delta\phi$ which changes the transmitted intensity to

$$I = \frac{I_{\max}}{1 + F\sin^2(\phi/2 + \Delta\phi/2)} \tag{1.57}$$

The responsivity of the transmitted and reflected intensity to the film (optical path length Δh^{OPL}) is

$$\begin{aligned} \mathbb{R}_{\mathrm{h}}^{\mathrm{T}} &= \mathrm{d}T/\mathrm{d}\Delta h^{\mathrm{OPL}} = \frac{-2kF\,\sin(\phi/2)\cos(\phi/2)}{\left(1 + F\sin^2(\phi/2)\right)^2} \\ \mathbb{R}_{\mathrm{h}}^{\mathrm{R}} &= \mathrm{d}R/\mathrm{d}\Delta h^{\mathrm{OPL}} = \frac{2kF\,\sin(\phi/2)\cos(\phi/2)}{\left(1 + F\sin^2(\phi/2)\right)^2} \end{aligned} \tag{1.58}$$

where the phase ϕ is

$$\phi = \frac{4\pi n_{\mathrm{c}}}{\lambda} d\,\cos\theta' + 2\phi_{\mathrm{r}} \tag{1.59}$$

These responsivities are equal but opposite, with values that scale as the product kF. For a coefficient of finesse of 80 and a wavelength of 840 nm, the responsivity is 6% per nm of optical path length near resonance. The reflectance and transmittance and the corresponding responsivities are shown in Fig. 1.19.

1.2.3 Common-Path Interferometers

The same ultra-high sensitivity to small phase changes that interferometers enjoy makes them notorious for their mechanical sensitivity. Removing mechanical instabilities is often expensive, requiring vibration isolation and expensive optical mounts. Fortunately, there are some interferometer configurations that are called "common-path" interferometers that are intrinsically stable against mechanical vibrations or thermal drift. These common-path interferometers rely on a signal and reference wave that share a common optical path from the sample interaction volume to the detector. While no interferometer can be strictly common path, several configurations approach this ideal.

An example of a common-path interferometer is the in-line amplitude-splitting configuration shown in Fig. 1.20. This uses an eighth-wave dielectric spacer layer to produce a $\pi/2$ phase shift between the top and bottom partial reflections. This establishes the phase quadrature condition that converts the phase of the thin film directly into intensity for far-field detection. The partial waves travel common paths from the surface to the detector, making this a mechanically stable interferometer.

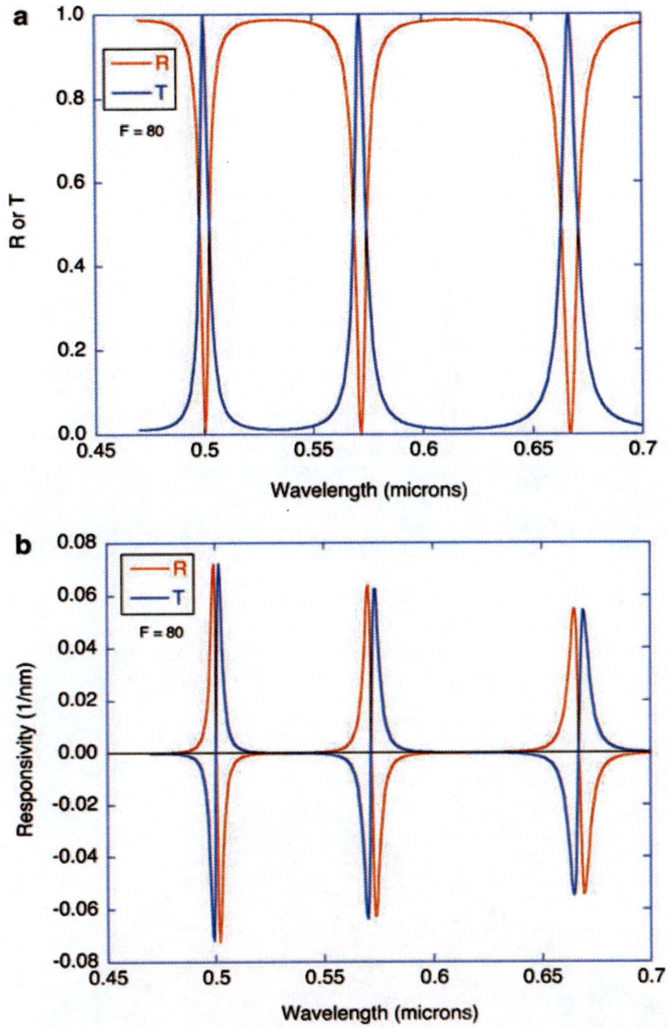

Fig. 1.19 **(a)** Reflectance and transmittance of a Fabry–Perot with a coefficient of finesse equal to 80 ($R = 0.8$) and $L = 2$ μm. **(b)** Height responsivity to a shift in the cavity length

The lower reflector in this case cannot be a perfect reflector, but must have a finite reflection coefficient. An example of a thin film detected with an in-line common-path configuration is shown in Fig. 1.21. The data are for an IgG antibody layer printed on a thermal oxide on silicon. The spot diameter is 100 μm, but the height is only about 1.2 nm [5]. The details of this type of common-path interferometer are described in detail in Chaps. 4 and 5.

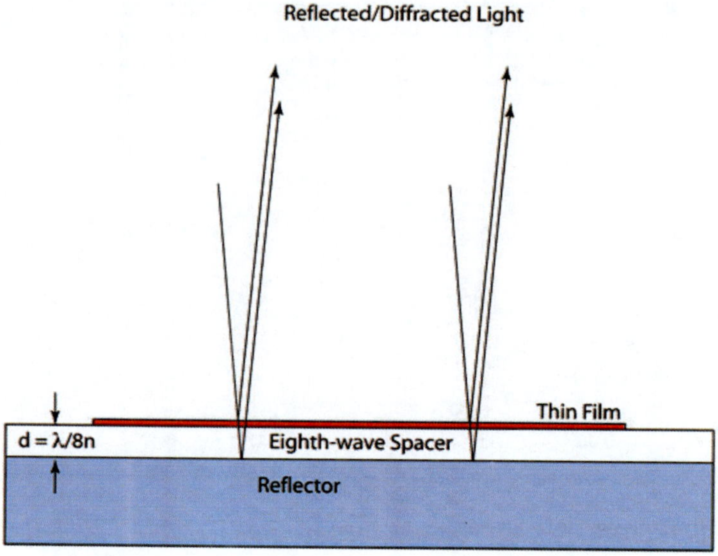

Fig. 1.20 In-line interferometer for thin film detection. The eighth-wave dielectric spacer establishes a $\pi/2$ phase shift between the top reflection and the bottom reflection. This converts the phase of the film directly into intensity in the far field. Both partial waves follow the same path to the far field, making this a common-path configuration with high mechanical stability

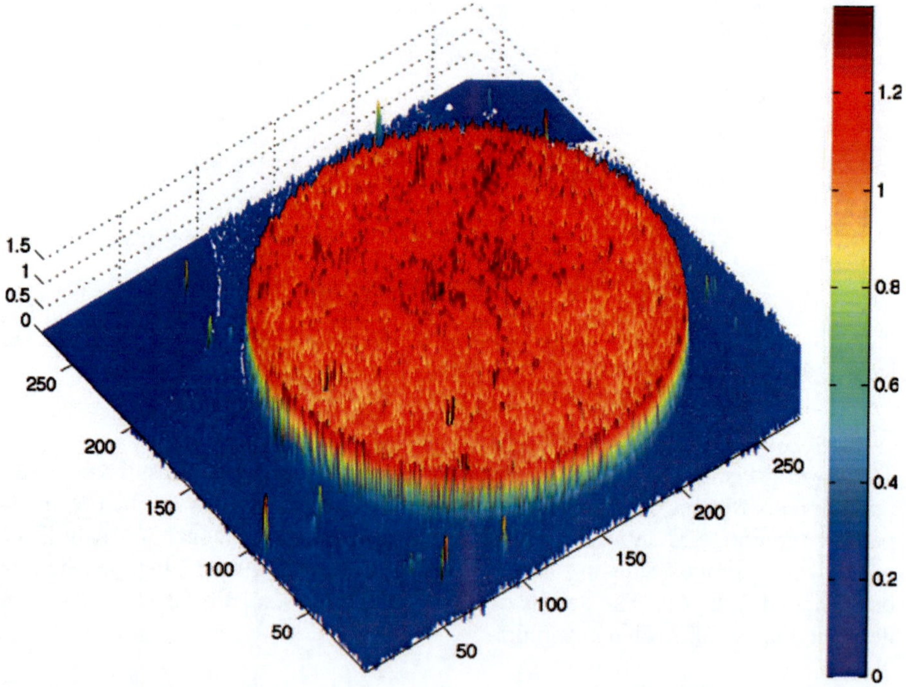

Fig. 1.21 Surface profile of a 100 μm diameter antibody spot with an average height of 1.2 nm, measured using molecular interferometric imaging (MI2). The colorscale is from 0 to 1.4 nm. From [5]

1.3 Holography

Holography is a type of wavefront splitting interferometry that is similar in many ways to Young's-type interferometers. The basic process of holography is the formation of quasi-monochromatic interference fringes between a homogeneous reference wave and a spatially diverse object wave. "Spatially diverse" means that the object wave consists of many different *k*-vector directions. These different *k*-vectors each write interference fringes with the reference wave. The complex pattern of interference fringes is called the hologram.

The goal of holography is to encode (record) and extract (play back) image information, and thus it is primarily an imaging process. However, the holographic process generally occurs away from image planes. Indeed, many holographic configurations have no lenses and hence have no image planes. Nevertheless, image information can be reconstructed using the phase and amplitudes of the recorded interference fringes of the hologram.

Holographic recording can be performed in two distinct ways. A hologram can be recorded as a physical hologram recorded in a holographic film, or it can be recorded as a digital hologram on the surface of a detector array like a charge-coupled device (CCD) as in a digital video recorder or a digital camera. Physical holographic films can be static, like photographic film, or dynamic, like photorefractive films. Static films are recorded once, and then played back by shining a laser through them to reconstruct the original image. Dynamic films change in time and follow the changes in the object beam by continuously diffracting the playback laser beam. To obtain quantitative information from a physical hologram reconstruction, it is necessary to "view" the hologram with an imaging system. On the other hand, the physical hologram can be eliminated and a digital hologram can be recorded directly on the CCD to be reconstructed through numerical inversion algorithms. Digital reconstruction has the advantage that quantitative three-dimensional information can be extracted from a single recorded hologram.

In biomedical applications, rather then being a mere novelty, holography plays an important role by recording phase and 3D information. The positions and motions in three dimensions of cells, or the constituents of cells, can be encoded as holograms and reconstructed physically or digitally. However, coherent light passing through biological tissue rapidly loses its coherence properties due to the strong scattering of the light. Even so, a dwindling fraction of light retains coherence and is image-bearing. If this image-bearing light can be separated from the scattered light, then images from inside the tissue can be extracted from depths far deeper than the mean scattering length. Furthermore, if the original light source has short coherence, then the images can be depth-sectioned from targeted depths, which is the basis of optical coherence imaging (OCI), discussed in Chap. 12.

This section presents the basic physics and mathematics of hologram recording and readout, which lays the foundations for the later biomedical holography applications.

1.3.1 Holographic Gratings

Holography can be viewed as spatial heterodyne detection (in which *k*-vectors differ, rather than temporal frequencies). In the simplest possible case, consider a signal plane wave with a distinct *k*-vector that intersects a coherent reference plane wave with a different *k*-vector. The total field is

$$E_{\text{Tot}} = E_1 \exp\left(\text{i}(\vec{k}_1 \cdot \vec{r} - \omega_1 t + \alpha_1)\right) + E_2 \exp\left(\text{i}(\vec{k}_2 \cdot \vec{r} - \omega_2 t + \alpha_2)\right) \quad (1.60)$$

As in the case of interferometry, this is best rewritten in terms of the average frequencies and phases as

$$\begin{aligned} E_{\text{Tot}} = \exp\left(\text{i}(\vec{k}_{\text{m}} \cdot \vec{r} - \omega_{\text{m}} t + \alpha_{\text{m}})\right)\Big[&E_1 \exp\left(\text{i}(\Delta\vec{k} \cdot \vec{r} - \Delta\omega t + \Delta\alpha)\right) \\ &+ E_2 \exp\left(-\text{i}(\Delta\vec{k} \cdot \vec{r} - \Delta\omega t + \Delta\alpha)\right)\Big] \end{aligned} \quad (1.61)$$

where

$$\begin{aligned} \vec{k}_{\text{m}} &= \tfrac{1}{2}\left(\vec{k}_1 + \vec{k}_2\right) \Delta\vec{k} = \tfrac{1}{2}\left(\vec{k}_1 - \vec{k}_2\right) \\ \omega_{\text{m}} &= \tfrac{1}{2}(\omega_1 + \omega_2) \Delta\omega = \tfrac{1}{2}(\omega_1 + \omega_2) \\ \alpha_{\text{m}} &= \tfrac{1}{2}(\alpha_1 + \alpha_2) \Delta\alpha = \tfrac{1}{2}(\alpha_1 + \alpha_2) \end{aligned} \quad (1.62)$$

The total intensity is

$$I_{\text{Tot}} = |E_1|^2 + |E_2|^2 + 2E_1E_2 \cos\left(2\Delta\vec{k} \cdot \vec{r} - 2\Delta\omega t + 2\Delta\alpha\right) \quad (1.63)$$

This total intensity is composed of a spatial fringe pattern of bright and dark fringes. The fringe pattern has a grating vector

$$\vec{K} = 2\Delta\vec{k} \quad (1.64)$$

The fringe pattern is volumetric in space where the waves overlap. However, it is common to detect the fringes on a detection plane (a CCD chip surface). As a simple case, consider a plane perpendicular to k_1. Then the intensity on this plane is

$$I_{\text{Tot}} = |E_1|^2 + |E_2|^2 + 2E_1E_2 \cos\left(-\vec{k}_2 \cdot \vec{r}_\perp - \Delta\omega t + \Delta\alpha\right) \quad (1.65)$$

where $\vec{r}_\perp$ is the position vector in the detector plane.

In Fig. 1.22, two monochromatic beams are propagating at equal angles off the *x*-axis. This configuration is called off-axis holography. On the *y*-axis, the intensity is

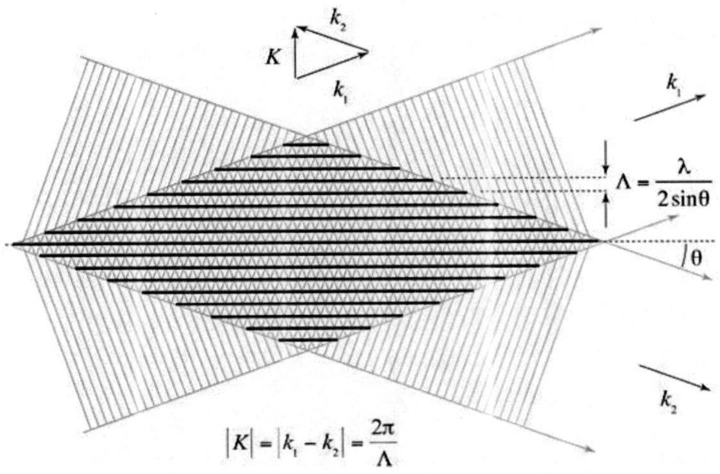

Fig. 1.22 Intersecting coherent plane waves produce an interference fringe pattern with a spatial periodicity that depends on the crossing angle θ

$$I_{\mathrm{Tot}} = |E_1|^2 + |E_2|^2 + 2E_1E_2 \cos\left(\frac{2\pi}{\Lambda}y - \Delta\omega t + \Delta\alpha\right) \tag{1.66}$$

where the fringe spacing and grating vector amplitude are

$$\begin{aligned} \Lambda &= \frac{\lambda}{2}\sin\theta \\ K &= \frac{2\pi}{\Lambda} \end{aligned} \tag{1.67}$$

and θ is half the crossing angle between the beams. Because the interference fringes are real-valued (intensity), there are two counter propagating grating vectors along the vertical direction

$$\begin{aligned} \vec{k}_1 - \vec{K} &= \vec{k}_2 \\ \vec{k}_2 + \vec{K} &= \vec{k}_1 \end{aligned} \tag{1.68}$$

Of course, the simple sinusoidal interference is because both waves are fully coherent monochromatic plane waves. Holography becomes much more interesting when the object wave carries complicated information in the form of multiple waves. In this case, each partial wave writes a hologram with the reference wave, and the full hologram is the combination of all of the partial holograms.

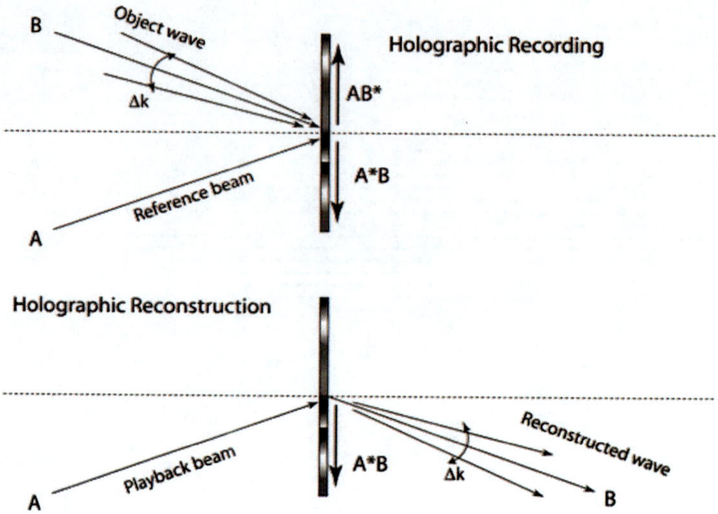

Fig. 1.23 Lens-free holographic recording and reconstruction

1.3.2 Image Reconstruction

One of the chief uses of holography is when phase information and/or volume information is required. Holography is a form of interferometry and records the phase of a signal wave. If the signal wave originates volumetrically, i.e., by scattering from (or inside) three-dimensional targets, then the information of the three-dimensional arrangements of the target is present in the phase of the scattered wave. Holography records this phase, and if the image is reconstructed correctly, the original three-dimensional information can be retrieved (within constraints imposed by the 3D transfer function).

A generic lens-free holographic recording and reconstruction arrangement is shown in Fig. 1.23. The crossing angle between the optic axes of the A (reference) and B (signal) beams defines the spatial carrier frequency (the interference fringes).

The holographic grating is written by the signal field B interfering with the reference wave A, which is assumed here to be a plane wave. The holographic grating is represented as

$$H = \frac{1}{2}\mu(AB^* + BA^*) \tag{1.69}$$

where μ is a grating recording efficiency. This hologram contains all of the complicated phase information of the signal wave B, including the phase associated with the three-dimensional configuration of the target. The spatial carrier frequency of the holographic grating is modulated by the spatial frequency components of the signal wave B. These spatial frequencies of the image are now represented as side-bands in

k-space in the hologram, and these spatial frequency side-bands diffract a play-back wave into multiple angles with different phases that allows the target image to be reconstructed with three-dimensional properties, such as parallax.

When the hologram is illuminated by field A, the diffracted field is

$$\text{Recon} = \eta AH = \frac{\eta\mu}{2}(A^2B^* + BAA^*) \Rightarrow \frac{\eta\mu}{2}B \tag{1.70}$$

where η is a diffraction efficiency. The second term in brackets reproduces the signal field with its phase content and amplitude content (if A represents a plane wave). Hence, by illuminating the hologram with the reference wave, the signal wave is reconstructed. The other term, in A^2, is non-Bragg matched (does not satisfy (1.68)) and does not diffract significantly in a volume hologram. But if the hologram is thin, then the B^* term may also be observed, which is the phase conjugate of the signal wave B.

1.3.3 Image-Domain or Fourier-Domain Holography

While holography can create lens-free imaging systems, it is often preferable to define specific planes in the holographic system. For some applications, it is convenient to place the holographic plane on the image plane. In this configuration, the written hologram is an image-domain hologram, as shown in Fig. 1.24. The read-out process consists of a second imaging system that reconstructs the image-domain hologram onto a new image. In Fig. 1.24, a point on the object plane is imaged to a point on the hologram. The reference beam is incident at an angle and produces interference fringes that are recorded on the hologram. During read-out, the reference beam diffracts from the holographic gratings. If the holographic film is thin, then Raman–Nath diffraction occurs with multiple diffraction orders, most notably the $+1$ and -1 orders, in addition to the zeroth-order diffraction. The $+1$ order diffraction is isolated by a lens (plus apertures (not shown) to block the zero-order) and re-imaged onto an image plane that can be a CCD camera. In Fig. 1.24, the angular bandwidth of the object beam and the crossing angle of the reference beam are exaggerated. For clear separation of the diffraction orders, the reference crossing angle θ_0 should be larger than the angular bandwidth $\Delta\theta$ of the light scattered by the object, giving the condition

$$\theta_0 \gg \Delta\theta \tag{1.71}$$

While the reconstruction in this case looks like it accomplishes only the transfer of an image to itself, it is a useful way to isolate a small coherent wave in the midst of incoherent waves, as in OCI described in Chap. 12. Some applications require the holographic plane to be on the Fourier-transform plane of an optical system, as shown in Fig. 1.25. This has the advantage that small imperfections in the holographic film are not re-imaged onto the detection plane.

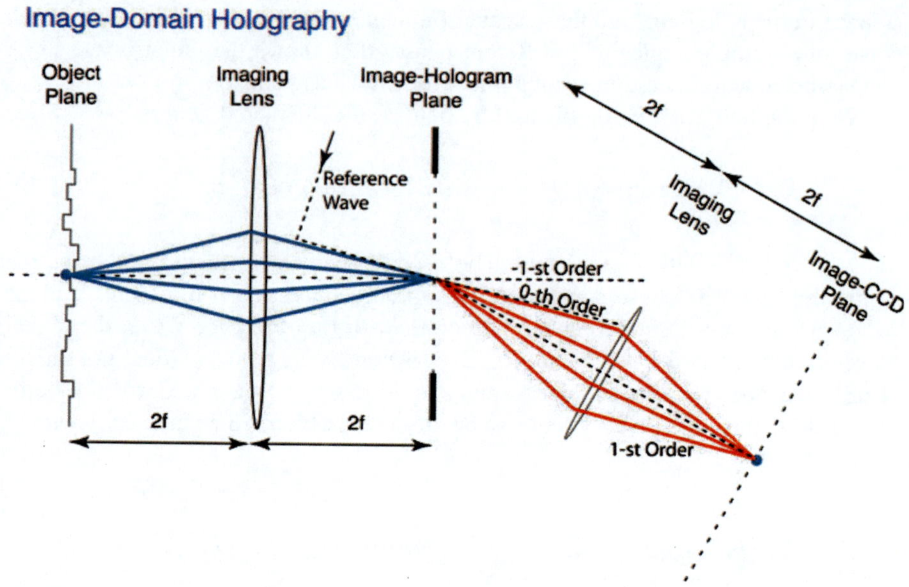

Fig. 1.24 Image-domain hologram configuration. The reference crossing angle θ_0 produces an interference grating superposed on the image field. The first diffraction order of the reference/readout is isolated and re-imaged onto an image plane. In this figure, the angular bandwidth of the object wave and the reference crossing angle are exaggerated relative to common experimental configurations where these angles are usually small

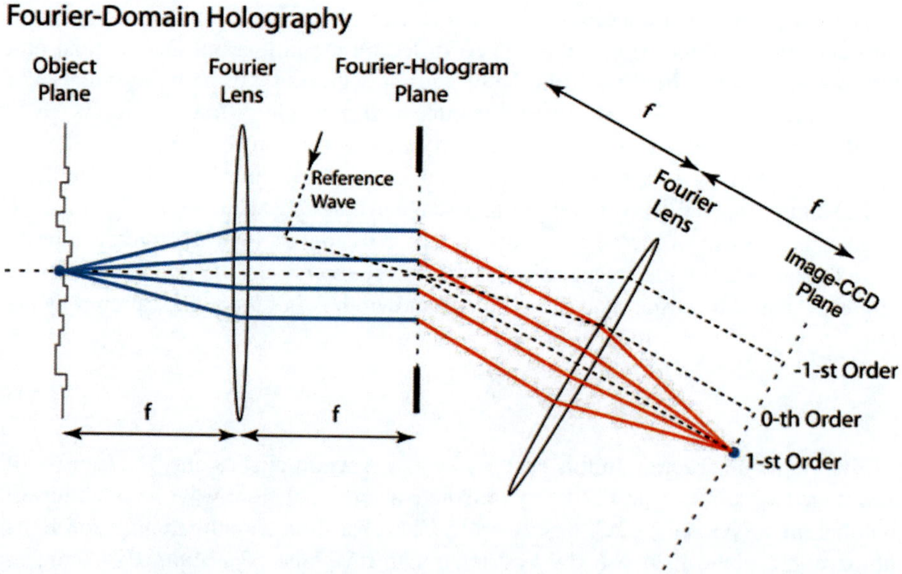

Fig. 1.25 Fourier-domain holography. The hologram plane is the Fourier plane of the lens. The hologram is reconstructed through a second Fourier-transform lens onto the image plane

1.4 Coherence

The study of the physics of the coherent properties of light is a fascinating topic with many surprises and unexpected behavior [6]. In the previous sections of this chapter on interferometry, we have been considering "perfectly coherent" monochromatic light. However, coherence is never absolute. There is no such thing as completely coherent or incoherent light. An incandescent light bulb has residual coherences on short time and space scales. Even black body radiation, if it is passed through a narrow-band filter, has coherence properties. Similarly, laser light, though highly coherent, has space and time scales beyond which its famous coherence properties fade.

All light is partially coherent, which is to say that there are characteristic time and length scales within which interference effects occur and outside of which they do not. These scales are called coherence times and coherence lengths. The more coherent a light source is, the longer the coherence times or lengths (and the easier interferometric experiments are). What determines the degree of coherence of a light source is the range of different waves that add through superposition to make up the light wave. This range of different waves is called "bandwidth." If light is made up of waves of different colors, then this bandwidth is called "spectral bandwidth." If light is made up of different k-vectors of different directions, then this bandwidth is called "angular bandwidth" or "spatial bandwidth." Spectral bandwidth determines the temporal coherence, while angular bandwidth determines spatial coherence. Temporal coherence describes the time scale over which interference effects can be observed, and spatial coherence describes the length scale over which interference effects can be observed.

Consider a light source that has a complex-valued frequency spectrum $s(\omega)$. The field from this source is

$$E(t) = \int_{-\infty}^{\infty} s(\omega) e^{-i\omega t} \, d\omega \tag{1.72}$$

which is just a Fourier transform between $E(t)$ and $s(\omega)$

$$E(t) \underset{FT}{\overset{FT^{-1}}{\rightleftarrows}} s(\omega) \tag{1.73}$$

The temporal coherence time of a pulse is related to the spectral bandwidth of the spectrum, and the spatial coherence length is related to the angular distribution of the field.

Example: Temporal Coherence of a Transform-Limited Gaussian Pulse

Consider a spectrum that has a Gaussian shape and is real-valued

$$s(\omega) = \frac{1}{\sqrt{\sqrt{\pi}\sigma_\omega}} \exp(-(\omega - \omega_0)^2/2\sigma_\omega^2) \tag{1.74}$$

with a power spectrum

$$S(\omega) = |s(\omega)|^2 = \frac{1}{\sqrt{\pi}\sigma_\omega} \exp(-(\omega - \omega_0)^2/\sigma_\omega^2) \tag{1.75}$$

The field is

$$\begin{aligned} E(t) &= \frac{1}{\sqrt{\sqrt{\pi}\sigma_\omega}} \int_0^\infty \exp(-(\omega - \omega_0)^2/2\sigma_\omega^2) \mathrm{e}^{-\mathrm{i}\omega t} \mathrm{d}\omega \\ &= \pi^{1/4} \sqrt{2\sigma_\omega}\, \mathrm{e}^{-\mathrm{i}\omega_0 t}\, \mathrm{e}^{-t^2 \sigma_\omega{}^2/2} = \pi^{1/4} \sqrt{2\sigma_\omega}\, \mathrm{e}^{-\mathrm{i}\omega_0 t}\, \mathrm{e}^{-t^2/2\tau^2} \end{aligned} \tag{1.76}$$

which is a Gaussian pulse of center frequency ω_0 and pulse duration

$$\tau = \frac{1}{\sigma_\omega} \tag{1.77}$$

This shows how a Gaussian spectral field (of width σ_ω and constant phase) produces a temporal pulse of duration $\tau = 1/\sigma_\omega$. In this example with a spectrum of constant phase, the temporal pulse is said to be "transform limited," and the pulse duration is equal to the coherence time. An example of a transform-limited pair is shown in Fig. 1.26. However, we will see that if the spectral phase is not constant, then the pulse duration is longer than the temporal coherence time, and the field is said to be "non-transform limited."

1.5 Spectral Interferometry

Spectral interferometry can be observed at the output of a spectrometer when two identical pulses of duration τ enter the spectrometer with a relative delay of Δt, even when Δt is much larger than the pulse duration τ. At first glance, this would seem to allow interference between pulses that are spatially separated by more than their coherence lengths. However, the individual frequency components of the pulses are interfering, which is observed in the spectral domain (at the output of the spectrometer). This is illustrated in Fig. 1.27 at the top for 25 fs duration pulses at a center

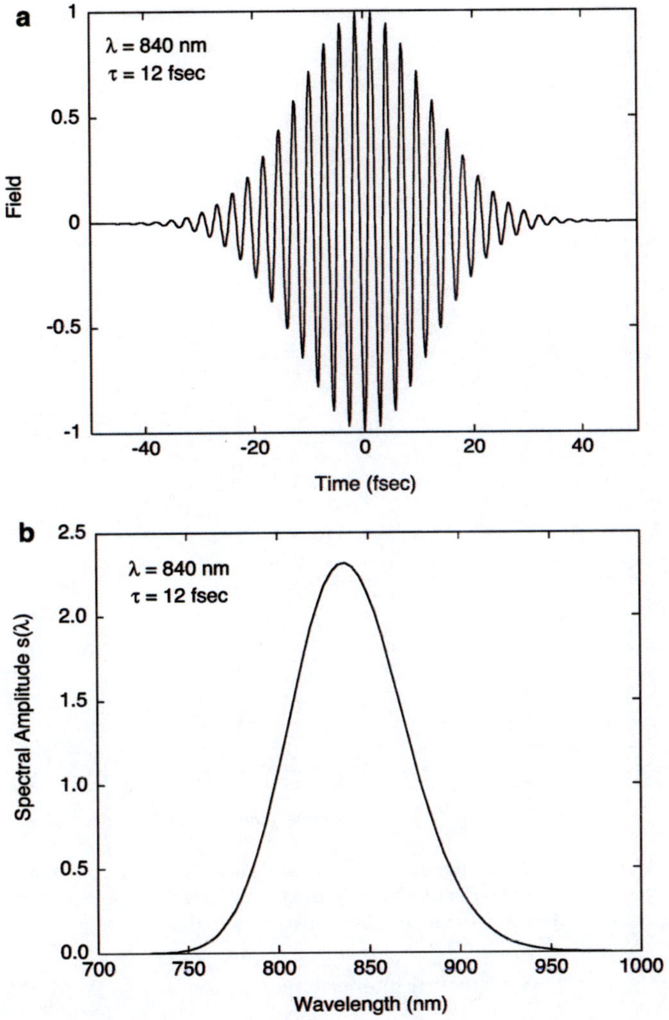

Fig. 1.26 Transform-limited pulse–spectrum pair. The pulse in (**a**) has a duration $\tau_p = 12$ fs. The spectrum in (**b**) has a bandwidth of $\sigma_\lambda = 31$ nm

wavelength of 840 nm, separated by 200 fs. The output of a spectrometer is shown in Fig. 1.27 at the bottom with an interference fringe period of 12 nm.

The mathematical treatment of spectral interferometry begins with two identical bandwidth-limited pulses that are separated by a time larger than the coherence time

$$\frac{1}{2\sqrt{2\pi}\tau}\int_{-\infty}^{\infty}\left[e^{-(t-t_1)^2/2\tau^2}+e^{-(t-t_2)^2/2\tau^2}\right]e^{-i\omega t}\,dt$$

$$=\frac{1}{2}\left[e^{-i\omega t_1}+e^{-i\omega t_2}\right]e^{-\omega^2\tau^2/2}=e^{-\omega^2\tau^2/2}e^{-i\omega\bar{t}}\cos(\omega\Delta t/2) \qquad (1.78)$$

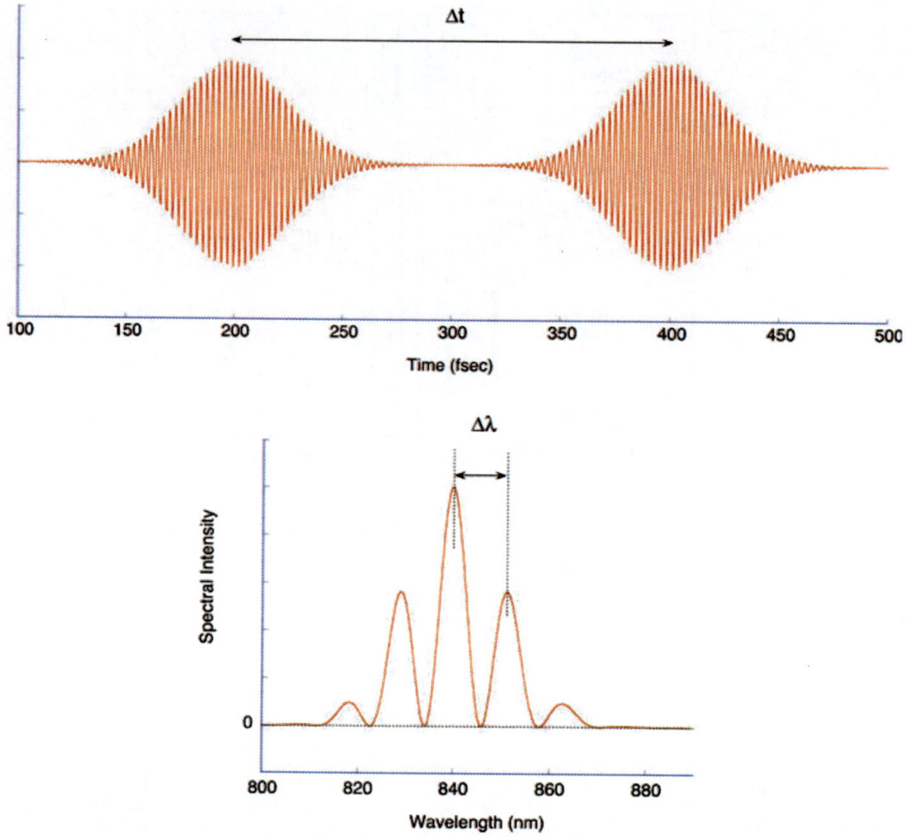

Fig. 1.27 Two pulses that do not overlap in time (shown at *top*) can still interfere in the spectral domain (shown at *bottom*). This is the principle of spectral interference. The periodicity of the fringes in the spectrum relates to the time delay between the pulses

where $\Delta t = (t_2 - t_1)$. The spectral intensity is the modulus squared, that gives

$$S(\omega) = \mathrm{e}^{-\omega^2\tau^2}\frac{1}{2}(1 + \cos(\omega\Delta t)) \tag{1.79}$$

The period of oscillation in the frequency domain is

$$\Delta\omega = 2\pi/\Delta t \tag{1.80}$$

and in terms of wavelength is

$$\Delta\lambda = \frac{\lambda^2\Delta\omega}{2\pi c} = \frac{\lambda^2}{c\Delta t} \tag{1.81}$$

Spectral interferometry is a central feature of spectral-domain optical coherence tomography, which we will see in Chap. 11. By operating in the spectral domain, many different time-delayed pulses can be detected simultaneously, significantly improving the speed of OCT acquisition. The longitudinal range of spectral domain OCT is set by the resolution of the spectrometer. If the pulse delay gets too long, then the interference fringe period in the spectrum becomes smaller than the spectrometer resolution. The maximum range is therefore

$$\text{range} = \frac{c\Delta t_{\max}}{2} = \frac{\lambda^2}{4\Delta\lambda_{\min}} \tag{1.82}$$

where the first factor of 2 is for double-pass in a backscattering configuration, and the second factor of 2 is for Nyquist sampling. As an example, at $\lambda = 840$ nm and with $\Delta\lambda = 0.1$ nm, the maximum range is 2 mm.

1.5.1 Non-transform-Limited Pulses: Broadening

If the phase is not constant across the spectral amplitude, but varies with frequency, then the field is

$$E(t) = \frac{1}{\sqrt{\sqrt{\pi}\sigma_\omega}} \int_{-\infty}^{\infty} \exp(-(\omega - \omega_0)^2/2\sigma_\omega^2) \exp(-i\omega t + \phi(\omega)) \mathrm{d}\omega \tag{1.83}$$

which produces, in general, a broader temporal pulse, as shown in Fig. 1.28. With a varying phase, the pulse is said to be non-transform limited, and there is no set relationship between the pulse duration and the width of the spectrum. Instead, the pulse duration is related to the frequency bandwidth over which the phase remains relatively constant. For instance, if the spectral correlation length is ξ_ω then the temporal pulse duration is

$$\tau = 1/\xi_\omega \tag{1.84}$$

Clearly, one limiting spectral correlation scale is $\xi_\omega = \sigma_\omega$, for which the pulse becomes transform limited. Another limiting value is when $\xi_\omega = 0$, and there is no pulse at all – only a fluctuating field. However – and this is a central concept of coherence – even a fluctuating field has coherence. To extract this coherence, we can turn next to the correlations that exist in any fluctuating field.

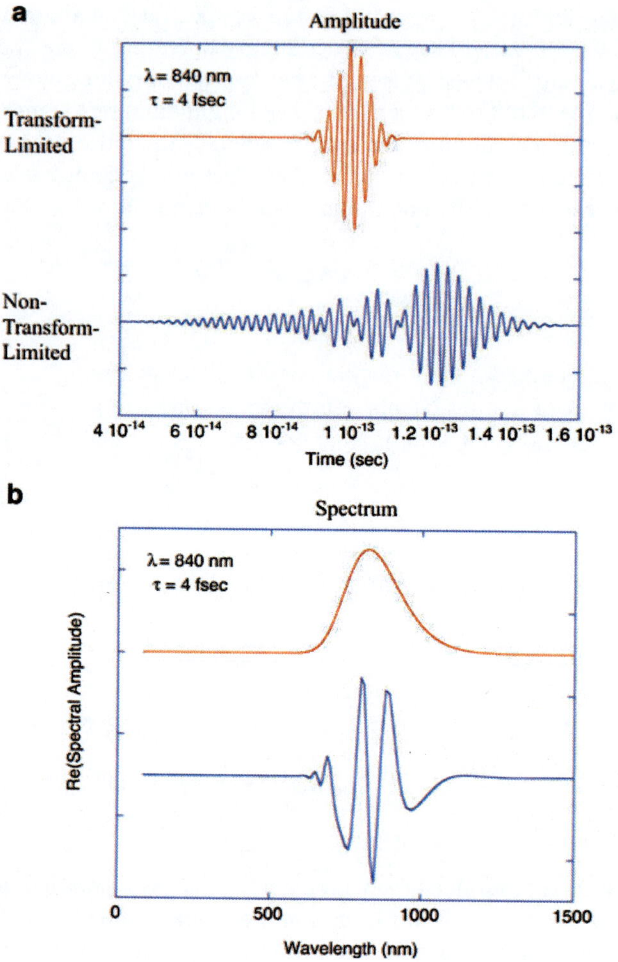

Fig. 1.28 Non-transform-limited field–spectrum pair. The pulse on the top of (**a**) is transform limited, with the associated spectral amplitude at the top in (**b**). The pulse at the bottom of (**a**) is non-transform-limited, with the real-part of the spectral amplitude modulated by the non-constant phase, at the bottom in (**b**)

1.6 Interferometry and Autocorrelation

In the case of finite coherence (either temporal or spatial), interferograms carry the information about the coherence times or lengths of the fields. The information appears as a change in fringe contrast as a relative delay is introduced between two or more paths. Another (equivalent) way of looking at coherence properties of a field uses the concept of autocorrelation. The autocorrelation function of a field $E(t)$ is defined as

$$I_1(\tau) = \int_{-\infty}^{\infty} E(t)E^*(t+\tau)\mathrm{d}t = \langle E(t)E^*(t+\tau)\rangle \tag{1.85}$$

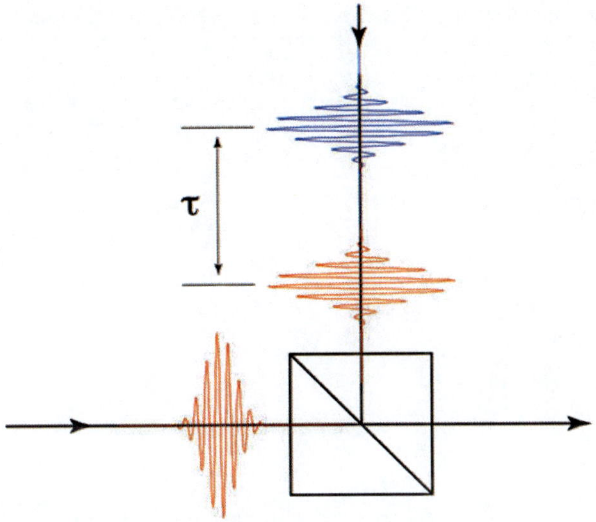

Fig. 1.29 Autocorrelation measurement. A pulse is split and recombined on a beamsplitter with a delay τ, and then measured as a function of delay

where τ is a relative delay. This measures how much of a signal at time t remains at time $t + \tau$. Therefore, the autocorrelation function measures how quickly a signal randomizes (decay of coherence).

Interferometry is a form of correlation. This is seen directly in the self-interference of a pulse that is split, then recombined with a relative delay

$$E_T = \frac{1}{\sqrt{2}}[E(t) + E(t+\tau)] \tag{1.86}$$

shown in Fig. 1.29, which is detected on the detector as the time-integrated intensity

$$\begin{aligned} I_T &= \int_{-\infty}^{\infty} |E(t)|^2 dt + \frac{1}{2}\int_{-\infty}^{\infty} E(t)E^*(t+\tau)dt + \frac{1}{2}\int_{-\infty}^{\infty} E^*(t)E(t+\tau)dt \\ &= I + \frac{1}{2}I_1(\tau) + \frac{1}{2}I_1{}^*(\tau) = I + Re\{I_1(\tau)\} \end{aligned} \tag{1.87}$$

where the first term is the average, and the second term (the interference term) is the real part of the field autocorrelation.

Example: Coherence Time of a Transform-Limited Gaussian Pulse

For the field

$$E(t) = \sqrt{\frac{1}{\pi^{1/2}\tau_p}} e^{-i\omega_0 t} e^{-t^2/2\tau_p{}^2} \tag{1.88}$$

the time-integrated intensity of the delayed and recombined interference is

$$
\begin{aligned}
I_{\mathrm{T}} &= \frac{1}{\pi^{1/2}\tau_{\mathrm{p}}}\int_{-\infty}^{\infty} \mathrm{e}^{-t^2/\tau_{\mathrm{p}}{}^2}\,\mathrm{d}t + \frac{\cos(\omega_0\tau)}{\pi^{1/2}\tau_{\mathrm{p}}}\int_{-\infty}^{\infty} \mathrm{e}^{-t^2/2\tau_{\mathrm{p}}{}^2}\,\mathrm{e}^{-(t+\tau)^{22}/2\tau_{\mathrm{p}}{}^2}\,\mathrm{d}t \\
&= 1 + \frac{\cos(\omega_0\tau)}{\pi^{1/2}\tau_{\mathrm{p}}}\mathrm{e}^{-\tau^2/2\tau_{\mathrm{p}}^2}\int_{-\infty}^{\infty} \exp(-(t^2 + t\tau)/\tau_{\mathrm{p}}^2)\mathrm{d}t \\
&= 1 + \frac{\cos(\omega_0\tau)}{\pi^{1/2}\tau_{\mathrm{p}}}\mathrm{e}^{-\tau^2/2\tau_{\mathrm{p}}^2}\left[\pi^{1/2}\tau_{\mathrm{p}}\,\mathrm{e}^{\tau^2/4\tau_{\mathrm{p}}^2}\right] = 1 + \mathrm{e}^{-\tau^2/4\tau_{\mathrm{p}}^2}\cos(\omega_0\tau) \\
&= 1 + Re\{I_1(\tau)\}
\end{aligned}
\tag{1.89}
$$

where

$$
I_1(\tau) = \mathrm{e}^{-\tau^2/4\tau_{\mathrm{p}}^2}\exp(\mathrm{i}\,\omega_0\tau) \tag{1.90}
$$

and

$$
Re\{I_1(\tau)\} = \mathrm{e}^{-\tau^2/4\tau_{\mathrm{p}}^2}\cos(\omega_0\tau) \tag{1.91}
$$

This has the form of sinusoidal interference fringes periodic in delay τ with a angular frequency ω_0, modulated by a Gaussian function of width $\sqrt{2}\tau_{\mathrm{p}}$, directly related to the correlation time of the pulse.

An important question for interferometric applications is the coherence length over which interference effects are observed for non-transform-limited light sources (when the spectral phase is not constant across the spectrum). The duration of a pulse in this case is not sufficient to define its coherence length. This is clear from the fact that a randomly drifting phase in the spectral amplitude causes the elongated pulse, but the randomly drifting phase limits the range over which interference takes effect. Therefore, the coherence length must reside in a different aspect than the pulse temporal duration.

The autocorrelation function of a pulse is related to the spectral power density $S(\omega)$ through the transform relation

Wiener–Khinchin theorem:

$$
Re\{I_1(t)\} = \int_{-\infty}^{\infty} Re\{E(t)\}Re\{E(t+\tau)\}\mathrm{d}t = FT^{-1}(S(\omega)) \tag{1.92}
$$

This relation between the correlation function and the spectral power density is called the Wiener–Khinchin theorem.

Example: Coherence Length of a Non-transform-Limited Gaussian Spectral Field

The spectral field is

$$s(\omega) = \frac{1}{\sqrt{\sqrt{\pi}\sigma_\omega}} \exp(-(\omega - \omega_0)^2/2\sigma_\omega^2) \exp(\mathrm{i}\,\phi(\omega)) \tag{1.93}$$

where the spectral phase is $\phi(\omega)$. Using the transform relation, the autocorrelation is

$$\begin{aligned} Re\{I_1(\tau)\} &= \frac{1}{\sqrt{\pi}\sigma_\omega} \int_{-\infty}^{\infty} \exp(-(\omega - \omega_0)^2/\sigma_\omega^2) \exp(-\mathrm{i}\,\omega\tau)\mathrm{d}\omega \\ &= \exp(-\tau^2\sigma_\omega^2/4) = \exp(-\tau^2/4\tau_\mathrm{p}^2) \end{aligned} \tag{1.94}$$

where the spectral phase is conjugated and cancels and hence no longer participates in the autocorrelation function. The autocorrelation temporal decay is

$$\xi_\tau = \frac{\sqrt{2}}{\sigma_\omega} \tag{1.95}$$

which is the temporal coherence length. Therefore, the coherence length is related directly to the spectral width of the source spectrum and does not depend on the duration of the pulse in the time domain or on phase variation in the spectral domain. Note that this result applies to general non-transform-limited spectra, because the phase is conjugated in the spectral domain by the autocorrelation definition. Therefore, even if the pulse has severe phase distortion, the coherence length depends only on the spectral width of the source. For example, in Fig. 1.30 the fluctuating field in (a) without an identifiable pulse duration has the autocorrelation in (b) with a well-defined correlation time.

1.7 Intensity–Intensity Interferometry

It is possible to define and measure an intensity–intensity autocorrelation function in analogy to the field–field autocorrelation function. This measures second-order degree of coherence and is a central function in the famous Hanbury Brown–Twiss experiment [7, 8]. Second-order coherence can be useful when a direct reference wave is not available, but interferometric information is still desired. Second-order coherence is often used in the situation of speckle fields and partial spatial coherence. The second-order correlation function is an intensity–intensity correlation function defined by

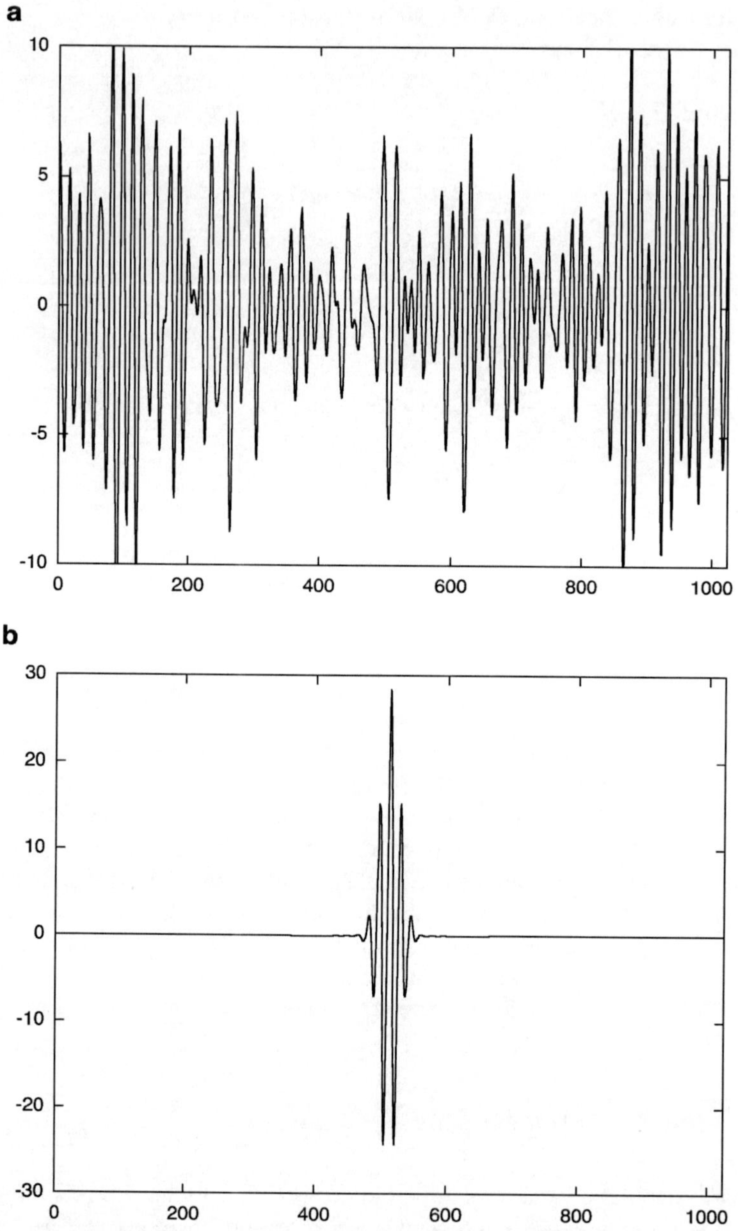

Fig. 1.30 A fluctuating field (**a**) and its autocorrelation (**b**). The field has an indefinite duration, but a finite coherence length

$$I_2(\tau) = \int_{-\infty}^{\infty} I(t)I^*(t+\tau)\mathrm{d}t = \langle I(t)I^x(t+\tau)\rangle \tag{1.96}$$

in which t and τ can stand either for a time delay or a spatial shift for temporal or spatial second-order coherence, respectively.

1.7.1 Degree of Coherence

The degree of coherence of a fluctuating field can be defined easily as a generalization of the first-order coherence function of the field. The first- and second-order degrees of coherence are defined as

$$g^{(1)}(\tau) = \frac{\langle E^*(t)E(t+\tau)\rangle}{\langle E^*(t)E(t)\rangle} \tag{1.97}$$

$$g^{(2)}(\tau) = \frac{\langle I(t)I^*(t+\tau)\rangle}{\langle I\rangle^2} = \frac{\langle E(t)E(t+\tau)E^*(t+\tau)E^*(t)\rangle}{\langle E^*(t)E(t)\rangle^2} \tag{1.98}$$

From the statistical properties of the field and intensity, there is a direct relationship between the first and second-order coherence for fluctuating classical light, given by the Siegert equation

$$g^{(2)}(\tau) = 1 + \beta\left|g^{(1)}(\tau)\right|^2 \tag{1.99}$$

where β is a factor that depends on the experimental geometry and approaches unity for single-mode detection (detection within a single speckle). Examples for fluctuating fields with Lorentzian and Gaussian line shapes are given in Table 1.1.

Table 1.1 Degree of coherence

	Spectrum	$g^{(1)}(\tau)$	$g^{(2)}(\tau)$
Lorentzian	$S(\omega) = \frac{1}{\pi\tau_0}\frac{1}{(\omega_0-\omega)^2+(1/\tau_0)^2}$	$g^{(1)}(\tau) = \exp(-\lvert\tau\rvert/\tau_0)$	$g^{(2)}(\tau) = 1 + \exp(-2\lvert\tau\rvert/\tau_0)$
Gaussian	$S(\omega) = \frac{1}{\sqrt{\pi}\sigma_\omega}\exp\left(-(\omega-\omega_0)^2/2\sigma_\omega^2\right)$	$g^{(1)}(\tau) = \exp(-\sigma_\omega^2\tau^2/2)$	$g^{(2)}(\tau) = 1 + \exp(-\sigma_\omega^2\tau^2)$

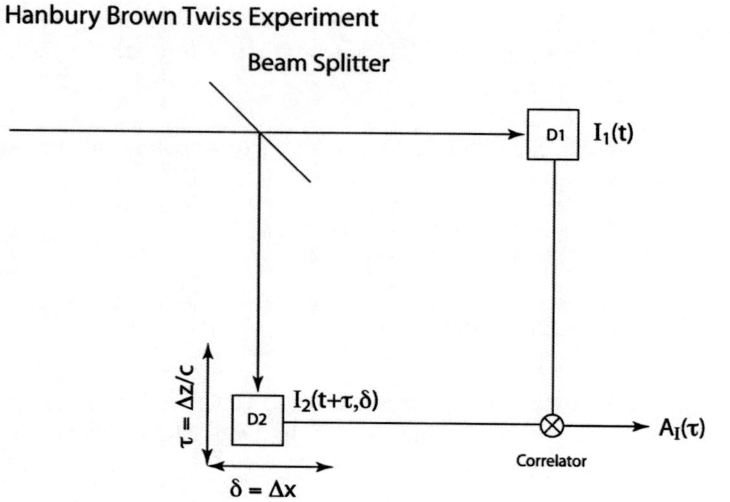

Fig. 1.31 Hanbury Brown–Twiss experiment. A fluctuating field is incident on a beam-splitter and detected as intensity fluctuations by two identical detectors with a variable delay τ and spatial offset δ. The cross-correlation between the two detectors defines the second-order correlation function

1.7.2 *Hanbury Brown–Twiss Interferometry*

One of the first experiments to measure the second-order coherence of light was performed by Hanbury Brown and Twiss (1956). It has a surprisingly simple arrangement, shown in Fig. 1.31, consisting only of a beamsplitter and two detectors with an adjustable position that defines a delay time $\tau = \Delta z/c$. There is no reference wave – only a fluctuating intensity signal delayed relative to itself. However, this simple arrangement is capable of extracting both the first-order and the second-order coherence of the fluctuating light.

The time-varying signals from the two detectors are combined and averaged according to

$$\frac{\langle (I_1(t) - \langle I_1 \rangle)(I_2(t+\tau) - \langle I_2 \rangle) \rangle}{\langle I_1 \rangle \langle I_2 \rangle} = g^{(2)}(\tau) - 1 \tag{1.100}$$

In this way, the second-order coherence is retrieved, and from it the first-order coherence through the relation (1.99). The coherence length of the light is contained in both the first and second-order coherence, reflecting properties of the light source such as Doppler broadening or Lorentzian line shapes.

A similar analysis can be performed with zero delay, but now introducing a lateral shift of the second detector, to give

$$\frac{\langle (I_1(t,0) - \langle I_1 \rangle)(I_2(t,\delta) - \langle I_2 \rangle) \rangle}{\langle I_1 \rangle \langle I_2 \rangle} = g^{(2)}(\delta) - 1 \tag{1.101}$$

This form of the second-order coherence provides a measure of the spatial coherence of the light source. It is in this form that the stellar HBT interferometer is used to detect the radii of nearby stars. The spatial correlation length at the detector is given approximately by

$$\delta_0 = 1.22 L\lambda / D \tag{1.102}$$

where L is the distance to the star (known from parallax measurements) and D is the star diameter. By spatially shifting the second detector and measuring how the intensity correlations between the two detectors decreases, the star diameter is measured. This was first applied to the star Sirius in the constellation Canis Major (1956) that has $L = 8.1 \times 10^{16}$ m and $D = 2.4 \times 10^{9}$ m. The size of the spatial coherence on the Earth caused by Sirius is about 20 m for this case. Therefore, the star Sirius casts a speckle field across the face of the Earth with a speckle size that is well within the range of human experience (even if intensity correlations are not).

Selected Bibliography

Diels, J.-C., Rudolph, W.: Ultrashort Laser Pulse Phenomena: Fundamentals, Techniques and Applications on a Femtosecond Time Scale. Academic, Boston (1996) (A good source of ultrafast pulse measurement and analysis)

Hanbury Brown, R., Twiss, R.Q.: Interferometry of the intensity fluctuations in light. I. Basic theory: the correlation between photons in coherent beams of radiation. Proc. R. Soc. Lond. **A 242**, 300–324 (1957) (The classic paper on second-order coherence)

Hariharan, P.: Optical Interferometry. Elsevier, Amsterdam (2003) (One of the basic texts in interferometry. It defines the many different interferometric configurations)

Hariharan, P.: Optical Holography: Principles, Techniques and Applications. Cambridge University Press, Cambridge (1984) (Another basic text by Hariharan, this time on optical holography)

Mandel, L., Wolf, E.: Optical Coherence and Quantum Optics. Cambridge University Press, Cambridge (1995) (This is the definitive reference on optical coherence)

References

1. Nolte, D.D.: Review of centrifugal microfluidic and bio-optical disks. Rev. Sci. Instrum. 80(10), 101101 (2009)
2. Bell, R.J.: Introductory Fourier Transform Spectroscopy. Acdemic, New York (1971)

3. Chan, S., Fauchet, P.M., Li, Y., Rothberg, L.J., Miller, B.L.: Porous silicon microcavities for biosensing applications. Physica Status Solidi A Appl. Res. **182**(1), 541–546 (2000)
4. Lin, V.S.Y., Motesharei, K., Dancil, K.P.S., Sailor, M.J., Ghadiri, M.R.: A porous silicon-based optical interferometric biosensor. Science **278**(5339), 840–843 (1997)
5. Zhao, M., Wang, X., Nolte, D.D.: Molecular interferometric imaging. Opt. Express **16**(10), 7102–7118 (2008)
6. Mandel, L., Wolf, E.: Optical Coherence and Quantum Optics. Cambridge University Press, Cambridge (1995)
7. Hanbury Brown, R., Twiss, R.Q.: Interferometry of the intensity fluctuations in light. I. Basic theory: the correlation between photons in coherent beams of radiation. Proc. R. Soc. Lond. **A 242**, 300–324 (1957)
8. Hanbury Brown, R., Twiss, R.Q.: Interferometry of the intensity fluctuations in light. II. An experimental test of the theory for partially coherent light. Proc. R. Soc. Lond. **A 243**, 291–319 (1958)

Chapter 2
Diffraction and Light Scattering

Light interacts with matter. This is the foundation for all the rich phenomena and useful applications associated with light and biological media. The interaction takes many forms. Light can be absorbed, or transmitted, or reflected, or scattered. All these processes can participate in interference phenomena in biology and medicine.

The interaction of light with matter (Fig. 2.1) is characterized as a scattering process that converts an incident (initial) optical mode into an outgoing (final) optical mode

$$(\vec{k}_{\mathrm{i}}, \omega_{\mathrm{i}}, \hat{p}_{\mathrm{i}}) \rightarrow (\vec{k}_{\mathrm{f}}, \omega_{\mathrm{f}}, \hat{p}_{\mathrm{f}}) \tag{2.1}$$

A single mode is an infinite plane wave with a frequency ω_{i}, a wave-vector $\vec{k}_{\mathrm{i}}$ and a polarization $\hat{p}_{\mathrm{i}}$. The amplitude of the outgoing mode is related to the amplitude of the incident mode through the scattering function

$$E_{\mathrm{f}}(\omega_{\mathrm{f}}) = S_j^i(\theta, \phi, \omega_{\mathrm{f}} - \omega_{\mathrm{i}}) E_{\mathrm{i}}(\omega_{\mathrm{i}}) \tag{2.2}$$

where the scattering function $S_j^i(\theta, \phi; \omega_{\mathrm{f}} - \omega_{\mathrm{i}})$ is a matrix connecting the input polarizations with the output polarizations of the light field, and θ and ϕ define the scattering direction relative to the incident direction. When the light scattering is elastic, the incident and final frequencies are equal. Inelastic light scattering can also occur, as when scatterers are in motion or when the scattered light is shifted to new frequencies, for instance through Doppler or Raman effects.

Elastic scattering is a general term that can be divided roughly into two separate contributions: geometric ray optics and diffraction. Geometric ray optics involves the reflection or refraction of light. Light rays are reflected or refracted by material density differences when the spatial scale of the inhomogeneity is larger than the wavelength of light. The reflection by a surface, or the bending of a light ray by a change in the index of refraction, is geometric ray optics. The other contribution to scattering is from diffraction. Diffraction is an essential wave phenomenon in which each part of an incident wavefront becomes the source for secondary

D.D. Nolte, *Optical Interferometry for Biology and Medicine*, Bioanalysis 1,
DOI 10.1007/978-1-4614-0890-1_2,

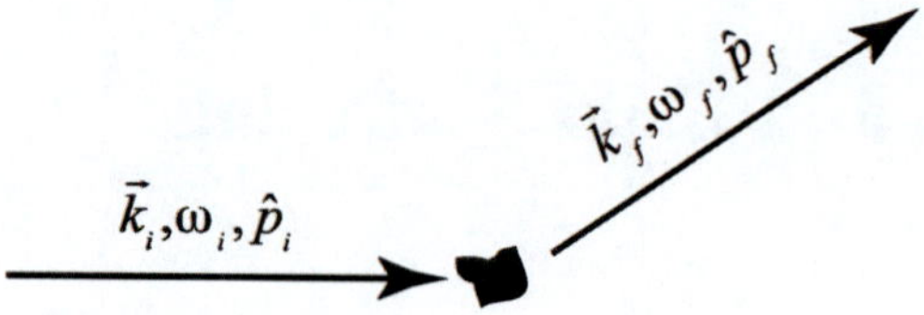

Fig. 2.1 The scattering process converts an incoming optical mode (defined by *k*-vector, frequency ω and polarization *p*) to an outgoing mode

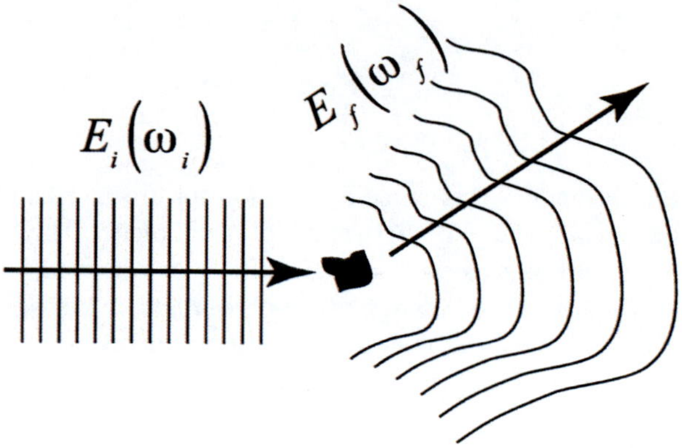

Fig. 2.2 Wave diffraction converts an incident wavefront into a diffracted wavefront

wavelets through Huygen's principle (Fig. 2.2). The secondary wavelets all interfere with each other to produce the observed diffracted light intensities. In this sense, diffraction is the result of wave interference.

2.1 Diffraction

All diffraction is a form of wavefront-splitting interferometry. The interference of partial waves that travel different paths from different parts of a wavefront leads to the complex and beautiful phenomena that fall under the topic of diffraction. Indeed, this summing up of the interference of parts of a wave is the basis for the diffraction integral, which is the main engine of diffraction applications.

2.1.1 Scalar Diffraction Theory

Although electromagnetic waves are vector waves that must satisfy Maxwell's equations at boundaries, a significant simplification is achieved in diffraction theory by treating the field amplitudes as scalar fields. This simplification is often very

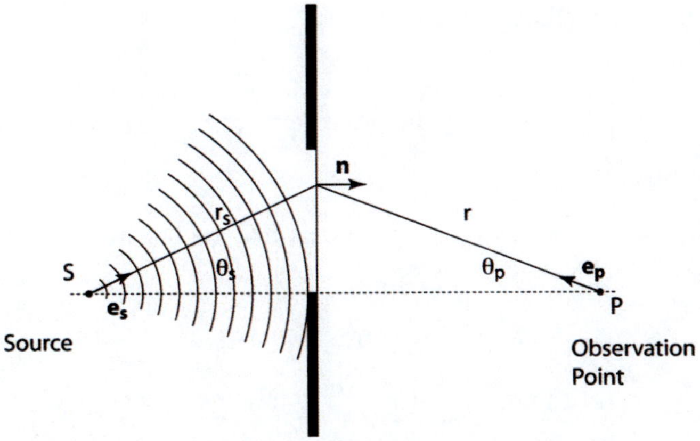

Fig. 2.3 Geometry for the Fresnel–Kirchhoff diffraction approximation

accurate, especially when the diffracting objects are weakly diffracting and produce small perturbations on an incident wave. This is the case if the diffracting objects have small variations in optical path length, and if these variations are on length scales large relative to a wavelength of light. Examples for which scalar diffraction theory fail include Mie scattering by spheres (discussed later in this chapter) because a sphere has a large variation of optical path length, and photonic crystals with photonic bandgaps, because refractive indices vary rapidly on the scale of a wavelength.

The starting point for scalar diffraction is the Helmholtz–Kirchhoff integral for a scalar field

$$\Psi(\mathbf{x}) = \frac{1}{4\pi} \int_{\text{ap}} \left[\Psi \nabla \left(\frac{e^{i\,kr}}{r} \right) - \frac{e^{i\,kr}}{r} \nabla \Psi \right] \cdot \mathbf{n} da \tag{2.3}$$

The simple configuration in Fig. 2.3 shows a source point S creating a wave that is incident on an aperture that diffracts the wave to an observation point P. The incident field is a spherical wave

$$\Psi_{\text{inc}} = A \frac{e^{i\,kr_s}}{r_s} \tag{2.4}$$

and the Helmholtz–Kirchhoff integral becomes

$$\Psi(\mathbf{x}_p) = \frac{1}{4\pi} \int_{\text{ap}} \left[\Psi_{\text{inc}} \nabla \left(\frac{e^{i\,kr}}{r} \right) - \frac{e^{i\,kr}}{r} \nabla \Psi_{\text{inc}} \right] \cdot \mathbf{n} da \tag{2.5}$$

After applying the gradient operator, this is

$$\Psi(\mathbf{x}_p) = \frac{1}{4\pi}\int_{ap}\left[\Psi_{inc}\left(\frac{ik}{r}-\frac{1}{r^2}\right)e^{ikr}\mathbf{e}_p - \frac{e^{ikr}}{r}\left(\frac{ik}{r_s}-\frac{1}{r_s^2}\right)r_s\Psi_{inc}\mathbf{e}_S\right]\cdot\mathbf{n}da \tag{2.6}$$

In the limit of large distances, only the terms that are linearly inverse to distance contribute and the equation becomes

Kirchhoff diffraction integral:

$$\Psi(\mathbf{x}_P) = \frac{iAk}{4\pi}\int_{ap}\frac{1}{rr_s}e^{ik(r+r_s)}(\mathbf{e}_P-\mathbf{e}_S)\cdot\mathbf{n}da \tag{2.7}$$

Often, the incident wave is a plane wave of constant amplitude across the aperture, and this integral reduces to

$$\Psi(\mathbf{x}_P) = -iAk\int_{ap}\frac{1}{r}e^{ikr}\left[\frac{1}{2}(\cos\theta_S+\cos\theta_P)\right]da \tag{2.8}$$

where the term in square brackets is the Fresnel obliquity factor, which is approximately unity for small-angle forward diffraction.

One of the most important features of the Kirchhoff diffraction integral for interferometry is the imaginary number −i in the prefactor. This has the important consequence that the diffracted wave has a $\pi/2$ phase advance relative to the incident wave. This phase shift plays important roles in many aspects of light scattering and diffraction. The mathematical source of this phase shift is the gradient in (2.5) on the dynamic phase of the wave. All extended sources of radiation experience this phase shift upon diffraction. However, point sources do not have this phase shift. This phase difference between extended and point sources is the key element in the origin of refractive index.

A final simplifying step in the development of scalar diffraction integrals takes the observation point far from a planar aperture, known as the Fraunhofer approximation. The geometry in this approximation is shown in Fig. 2.4. This leads to the Fraunhofer diffraction integral

Fraunhofer diffraction integral:

$$\Psi(\mathbf{x}_P) = -i\frac{Ae^{ikR}}{\lambda R}\int_{ap}e^{-ik(x\sin\theta+y\sin\phi)}dxdy \tag{2.9}$$

where R is the (large) distance from the aperture to the observation point, and $x/R=\sin\theta$, $y/R=\sin\phi$.

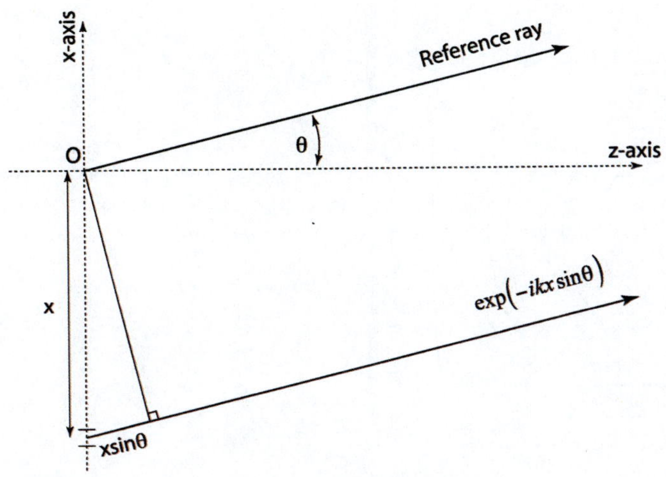

Fig. 2.4 Fraunhofer diffraction geometry. All rays are parallel and inclined at an angle θ. The reference ray from the origin defines zero phase. The path length difference is $x\sin\theta$, with x increasing vertically and θ positive in the counter-clockwise direction

There are an endless number of configurations of sources and apertures that appear in diffraction problems. Several of the most common will be presented as examples that can be extended to more complicated cases that may be encountered in biological applications.

2.1.2 *Fraunhofer Diffraction from Apertures and Gratings*

Some types of molecular biosensors are based on Fraunhofer diffraction from apertures and gratings. This section presents several basic examples of diffraction that are used in later chapters in this book. Some of the examples include the diffraction effects of thin biolayers, and the molecular responsivities of these diffraction structures are defined.

Example: Diffraction from a Single Square Slit

The single square slit is the simplest example, and is always the starting point to build up to more complicated diffraction patterns. The slit has a width equal to a, shown in Fig. 2.5. The Fraunhofer integral is

$$E_P = -i\frac{E_0 e^{ikR}}{\lambda R}\int_{-a/2}^{a/2} e^{-ikx\sin\theta}\,dx \tag{2.10}$$

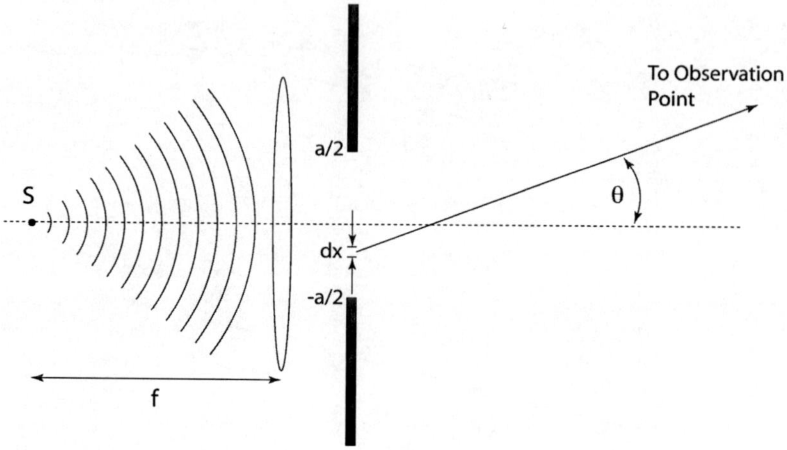

Fig. 2.5 Fraunhofer diffraction geometry for a single slit. The lens converts a point source to a plane wave

where E_0 is the field amplitude in the aperture. The Fraunhofer integral is evaluated to be

$$\begin{aligned} E_{\mathrm{P}} &= -\mathrm{i}\frac{E_0\,\mathrm{e}^{\mathrm{i}kR}}{\lambda R}\int_{-a/2}^{a/2}\mathrm{e}^{-\mathrm{i}kx\sin\theta}\,\mathrm{d}x \\ &= \mathrm{i}\frac{E_0\,\mathrm{e}^{\mathrm{i}kR}}{\lambda R}\,\frac{2}{2ik\,\sin\theta}\left[\mathrm{e}^{-(\mathrm{i}ka\,\sin\theta)/2} - \mathrm{e}^{\mathrm{i}(ka\,\sin\theta)/2}\right] \\ &= -2\mathrm{i}\frac{E_0\,\mathrm{e}^{\mathrm{i}kR}}{\lambda R}\,\frac{1}{k\,\sin\theta}\sin\left(\frac{ka}{2}\sin\theta\right) = -\mathrm{i}a\frac{E_0\,\mathrm{e}^{\mathrm{i}kR}}{\lambda R}\mathrm{sinc}\left(\frac{ka}{2}\sin\theta\right) \end{aligned} \tag{2.11}$$

giving the field

$$E_{\mathrm{P}} = -\mathrm{i}E_0\,\mathrm{e}^{\mathrm{i}kR}\left(\frac{a}{\lambda R}\right)\mathrm{sinc}(ka\,\sin\theta/2) \tag{2.12}$$

and intensity

$$I_{\mathrm{P}} = P_0\left(\frac{a}{\lambda R}\right)^2\mathrm{sinc}^2(ka\,\sin\theta/2) \tag{2.13}$$

Example: Diffraction from a Slit with a Partial Biolayer

Diffraction is one way to detect thin molecular films, such as biomolecules captured by high-affinity capture molecules (like antibodies on surfaces). As an example, consider a rectangular aperture that is half covered by a thin film of refractive index n and thickness d. We want to consider how the presence of the film changes the diffracted intensity in the far-field. The field (2.12) is the starting point for this solution. The answer can be written down by inspection as

$$E_{\mathrm{P}} = -\mathrm{i}E_0\,\mathrm{e}^{\mathrm{i}kR}\frac{a}{2\lambda R}\operatorname{sinc}\left(\frac{ka}{4}\sin\theta\right)\left[\mathrm{e}^{\mathrm{i}(ka\sin\theta)/4} + \mathrm{e}^{\mathrm{i}k(n-1)d}\,\mathrm{e}^{-\mathrm{i}(ka\sin\theta)/4}\right] \quad (2.14)$$

which is the sum of fields from two apertures of width $a/2$, one of which has the extra phase $\delta = k(n-1)d$. The diffracted intensity is

$$I_{\mathrm{P}} = 2I_0\left(\frac{a}{2\lambda R}\right)^2\operatorname{sinc}^2\left(\frac{ka}{4}\sin\theta\right)\left[1+\cos\left(\frac{ka}{2}\sin\theta-\delta\right)\right] \quad (2.15)$$

which still has the basic sinc^2 envelope, with an extra term (the second term in the square brackets) that causes a small shift in the far-field diffraction. This small shift provides a measure of the thickness of the film and is one way to use optical interferometry (in this case diffraction) to detect biolayers in an optical biosensor.

The performance of diffraction-based optical biosensors is characterized by the responsivity that is defined as the intensity change per optical path length difference of the biolayer $h = (n-1)d$. The responsivity of this diffraction-based biosensor is

$$\mathbb{R}_\delta = \frac{\mathrm{d}I_{\mathrm{P}}}{\mathrm{d}\delta} = -\frac{I_0}{2}\left(\frac{a}{\lambda R}\right)^2\operatorname{sinc}^2(ka\sin\theta/4)\sin\left(\frac{ka}{2}\sin\theta\right) \quad (2.16)$$

which continues to have the sinc^2 envelope, but now with the extra sine term at the end. The angular responsivity on the detection x-axis is shown in Fig. 2.6. The shift in the far-field diffraction caused by the biolayer leads to an asymmetric intensity change. If only the total intensity is measured, then the biolayer effect would not be detected. However, by placing a split detector on the detection plane, and differencing the left and right detected intensities, then the difference signal is linearly proportional to the thickness of the biolayer. This diffraction-based biosensor configuration is closely related to phase-contrast detection on bio-optical compact disks (BioCDs) [1].

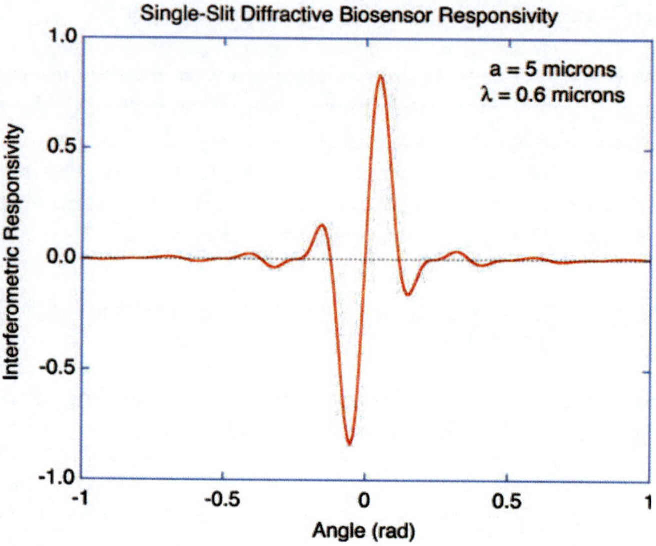

Fig. 2.6 Rectangular-aperture diffractive biosensor responsivity. A rectangular aperture is half-covered by a molecular film of thickness *d* and refractive index *n*. The interferometric responsivity is the change in the diffracted intensity per phase shift caused by the film. The film causes an asymmetric far-field diffraction pattern with a responsivity approaching unity

Example: Diffraction by a Circular Aperture

One of the most common apertures encountered in experimental optics is the circular aperture. The Fraunhofer diffraction integral is expressed in polar coordinates (r, Φ) on the detection plane

$$E(r,\Phi) = -\mathrm{i}\sqrt{S_0}\frac{\mathrm{e}^{\mathrm{i}kR}}{\lambda R}\int_0^a\int_0^{2\pi}\mathrm{e}^{-\mathrm{i}(kr'r/R)\cos(\varphi-\Phi)}\,\mathrm{d}\varphi r'\,\mathrm{d}r' \tag{2.17}$$

integrated over (r', φ) on the aperture plane, where

$$R = \sqrt{L^2 + r^2} \tag{2.18}$$

and L is the distance to the screen (or is the focal length of a Fourier-transform lens). This integral is re-expressed using Bessel functions

$$J_{\mathrm{m}}(u) = \frac{\mathrm{i}^{-m}}{2\pi}\int_0^{2\pi}\mathrm{e}^{\mathrm{i}(mv+u\cos v)}\,\mathrm{d}v \tag{2.19}$$

The diffraction integral becomes

$$E(r,\Phi) = -\mathrm{i}\sqrt{S_0}\frac{\mathrm{e}^{\mathrm{i}kR}}{\lambda L}2\pi\int_0^a J_0(kr'r/R)r'\,\mathrm{d}r' \tag{2.20}$$

that is evaluated using

$$\int_0^u u'J_0(u')\mathrm{d}u' = uJ_1(u) \tag{2.21}$$

to yield

$$E(r) = -\mathrm{i}\sqrt{S_0}\frac{\mathrm{e}^{\mathrm{i}kR}}{\lambda L}2\pi a^2\left(\frac{R}{kar}\right)J_1\left(\frac{kar}{R}\right) = -\mathrm{i}\sqrt{S_0}\frac{\mathrm{e}^{\mathrm{i}kR}}{\lambda L}\pi a^2\left[\frac{2J_1\left(\frac{kar}{R}\right)}{\left(\frac{kar}{R}\right)}\right]$$
$$\approx -\mathrm{i}\sqrt{S_0}\frac{\mathrm{e}^{\mathrm{i}kR}}{\lambda L}\pi a^2\left[\frac{2J_1(ka\sin\theta)}{ka\sin\theta}\right] \tag{2.22}$$

where $r/R = \sin\theta$, and the value of $J_1(x)/x = 1/2$ as x goes to zero. The angular intensity is approximately (for small angles θ)

$$I(\theta) = I_0\left(\frac{2J_1(ka\sin\theta)}{ka\sin\theta}\right)^2 \tag{2.23}$$

that has an oscillatory behavior qualitatively similar to the sinc squared function of a rectangular aperture.

Example: Diffraction by Multiple Square Slits

Multiple slits form a diffraction grating. The diffraction from each slit is given by (2.12). This is modulated by a periodic part determined by the periodic spacing Λ of the multiple slits. For the periodic part, the total field is

$$E = E_P\,\mathrm{e}^{-\mathrm{i}\omega t}\,\mathrm{e}^{\mathrm{i}kr_1}\left[1 + \mathrm{e}^{\mathrm{i}\phi} + \left(\mathrm{e}^{\mathrm{i}\phi}\right)^2 + \left(\mathrm{e}^{\mathrm{i}\phi}\right)^3 + \cdots + \left(\mathrm{e}^{\mathrm{i}\phi}\right)^{N-1}\right] \tag{2.24}$$

where $\phi = k\Lambda\sin\theta$, and where $E_p = -\mathrm{i}E_0 e^{\mathrm{i}KR}(a/\lambda R)\mathrm{sinc}(Ka\sin\theta/2)$ is the field from a single slit located at r_1. The term in brackets is a geometric series with the value

$$\frac{(\mathrm{e}^{\mathrm{i}N\phi} - 1)}{(\mathrm{e}^{\mathrm{i}\phi} - 1)} = \mathrm{e}^{\mathrm{i}(N-1)\phi/2}\left(\frac{\sin N\phi/2}{\sin\phi/2}\right) \tag{2.25}$$

The total field is then

$$E = E_P\mathrm{e}^{-\mathrm{i}\omega t}\,\mathrm{e}^{\mathrm{i}[kr_1+(N-1)\phi/2]}\left(\frac{\sin N\phi/2}{\sin\phi/2}\right) \tag{2.26}$$

If the array is referenced to its center, then the periodic factor is

$$E = E_P\, e^{ikR - i\omega t}\left(\frac{\sin N\phi/2}{\sin \phi/2}\right) \tag{2.27}$$

with a total field

$$E = E_P\, e^{ikR - i\omega t}\left(\frac{\sin N\phi/2}{\sin \phi/2}\right)\operatorname{sinc}(ka \sin\theta/2) \tag{2.28}$$

and an intensity

$$I = I_0\left(\frac{\sin^2 N\phi/2}{\sin^2 \phi/2}\right)\operatorname{sinc}^2(ka \sin\theta/2) \tag{2.29}$$

Note that this is the product of the rectangular grating diffraction pattern and the diffraction pattern of N point sources placed in a regular array with a spacing Λ. Because diffraction is equivalent to a Fourier transform, the convolution of two aperture functions on the object plane becomes the product of the diffraction patterns in the far-field.

Example: Diffraction of a Gaussian Beam by Multiple Slits

A Gaussian beam with radius w_0 illuminating a multiple slit grating diffracts as a Gaussian beam. The field is

$$E_G(\theta) = -iE_0 \operatorname{sinc}\left(\frac{ka \sin\theta}{2}\right)\sum_{m=-M}^{M}\exp\left(-2\left(\frac{\pi w_0}{\Lambda}\right)^2 \sin^2(\theta - \theta_m)\right) \tag{2.30}$$

for a periodicity Λ and a slit width $a < \Lambda$. The conditions on the diffraction orders are

$$\begin{aligned} \sin\theta_m &= m\lambda/\Lambda \\ M &= \operatorname{trunc}(\Lambda/\lambda) \end{aligned} \tag{2.31}$$

The periodic part can be given its own function definition as

$$P_G(\theta, w_0/\Lambda) = \sum_{m=-M}^{M}\exp\left(-2\left(\frac{\pi w_0}{\Lambda}\right)^2 \sin^2(\theta - \theta_m)\right) \tag{2.32}$$

which is encountered any time a Gaussian beam is diffracted from a periodic grating. It replaces the periodic function in (2.29)

$$P_G(\theta, w_0/\Lambda) \approx \left(\frac{\sin^2 Nk\Lambda \sin\theta/2}{\sin^2 k\Lambda \sin\theta/2} \right) \tag{2.33}$$

where the effective number of slits is

$$N = \pi w_0/\Lambda \tag{2.34}$$

When there is no overlap between the diffraction orders, the intensity is

$$\begin{aligned} I_G(\theta) &= I_0 \operatorname{sinc}^2\left(\frac{ka \sin\theta}{2}\right) \sum_{m=-M}^{M} \exp\left(-\left(\frac{2\pi w_0}{\Lambda}\right)^2 \sin^2(\theta - \theta_m)\right) \\ &= I_0 \operatorname{sinc}^2\left(\frac{ka \sin\theta}{2}\right) [P_G(\theta, w_o/\Lambda)]^2 \end{aligned} \tag{2.35}$$

The term in the summation is a series of Gaussian beams with angular widths given by

$$\Delta\theta = \frac{\Lambda}{2\pi w_0} \tag{2.36}$$

Example: Diffraction by a Periodic Biolayer Grating

One class of diffraction-based biosensors uses a periodic grating of capture molecules on a surface. When exposed to a sample containing the target biomolecules, these bind in a stripe pattern, shown in Fig. 2.7. The thin nature of the biomolecular stripes imparts a periodic phase modulation on a reflected optical wave. If we assume a sinusoidal phase modulation, the transmitted field is

$$E_r = E_0 \exp[ik_0(n-1)d(1 - \cos(Kx + \phi))] \tag{2.37}$$

Using the Bessel function identity

$$\exp[i\delta \cos(Kx + \phi)] = \sum_{m=-\infty}^{\infty} J_m(\delta) \exp[im(Kx + \phi + \pi/2)] \tag{2.38}$$

the reflected field just after reflection (in the near field) can be written as

$$E_r = E_0 \, e^{i\delta_0} \sum_{m=-\infty}^{\infty} J_m(\delta) \exp[im(Kx + \phi + \pi/2)] \tag{2.39}$$

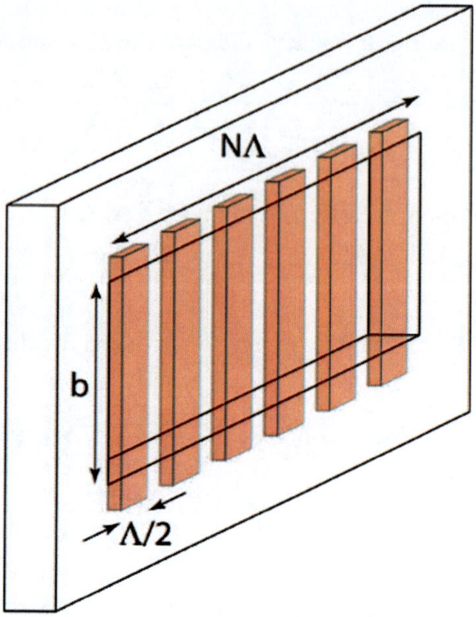

Fig. 2.7 A periodic square grating of N stripes on a rectangular aperture

where $\delta = k_0(n-1)d$. The Fraunhofer diffraction integral for the mth diffraction order selects out each of the Bessel functions as

$$E_m = E_0\,\mathrm{e}^{\mathrm{i}\delta_0}\,\mathrm{e}^{\mathrm{i}(mKx+m\phi+m\pi/2)}J_m(k_0(n-1)d) \tag{2.40}$$

The Bessel function is expanded as

$$J_m(x) = \frac{x^m}{2^m m!} \tag{2.41}$$

and the first-order diffraction is then

$$E_{\pm 1} = E_0\,\mathrm{e}^{\mathrm{i}\delta_0}\,\mathrm{e}^{\mathrm{i}(\pm Kx\pm\phi\pm\pi/2)}k_0(n-1)d/2 \tag{2.42}$$

with a diffraction efficiency given by

$$\eta = \frac{k_0^2(n-1)^2d^2}{4} \tag{2.43}$$

for the sinusoidal grating.

If the grating is a square grating (commonly encountered with protein patterning using lithography) as shown in Fig. 2.7, then the diffracted intensity is given by (2.15) modulated with the periodic envelope function of (2.27)

$$I_P = 2I_0\left(\frac{\Lambda}{2N\lambda R}\right)^2\left(\frac{\sin(Nk_0\Lambda\sin\theta/2)}{\sin(k_0\Lambda\sin\theta/2)}\right)^2 \mathrm{sinc}^2(k\Lambda\sin\theta/4)$$
$$[1+\cos(k\Lambda\sin\theta/2-\delta)] \tag{2.44}$$

The diffraction efficiency of the square grating is

$$\eta = \left(\frac{2}{\pi}\right)^2\frac{1-\cos\delta}{1+\cos\delta} \approx \left(\frac{2}{\pi}\right)^2\frac{k_0{}^2(n-1)^2d^2}{4} \tag{2.45}$$

which can be compared with (2.43).

2.1.3 *Linear vs. Quadratic Response and Detectability*

An important issue for the detectability of small signals is whether linear sensitivity (when signal intensity is proportional to the quantity to be detected), or quadratic sensitivity (when signal intensity is proportional to the squared value of the quantity) gives the best ability to detect small quantities. The answer to this question is generally not fundamental, but depends on details of the noise sources, which in turn usually depend on intensity. For example, diffracted intensity depends on the square of the phase modulation induced by the biolayer, while interferometric intensity (in quadrature) is linear in the phase modulation. Because phase modulation caused by a biolayer is typically less than 1%, linear detection has a much larger absolute modulation caused by a biolayer than diffraction does. On the other hand, interferometric linear detection has a much higher background (lower contrast) that can cause more noise. This is a basic trade-off between the two types of biosensors: linear detection with high-background and low-contrast vs. quadratic detection with low-background and high-contrast. As a general rule, linear detection is more sensitive for detection of very sub-monolayer films in the weak-signal limit, while quadratic detection can have better signal-to-noise in the strong-signal limit.

To make these arguments more quantitative, consider three contributions to the noise

$$I_N^2 = C_{RIN}^2 I_B^2 \mathrm{BW} + h\nu I_B \mathrm{BW} + C_0 \mathrm{BW} \tag{2.46}$$

where BW refers to the detection bandwidth, and I_B is the background intensity incident on the detector. The first term is relative intensity noise (RIN), the second term is shot noise and the third term is a system noise floor (usually electronic noise). It is important to remember that these noise contributions are frequency dependent, usually with a $1/f$ behavior at low detection frequencies, evolving into white noise at higher detection frequencies. The linear dependence on signal bandwidth is strictly true only for white noise. The signal is characterized by

$$\Delta I_S = I_0 C^{(L)} \Delta\phi + I_0 C^{(Q)} \Delta\phi^2 \tag{2.47}$$

where the first term is the linear dependence and the second term is quadratic, and the intensity I_0 is related to the average intensity on the detector. The signal-to-noise ratio is then

$$S/N = \frac{(\Delta I_S)^2}{I_N^2} = \frac{I_0^2 \left(C^{(L)}\Delta\phi + C^{(Q)}\Delta\phi^2\right)^2}{C_R^2 I_B^2 \text{BW} + h\nu I_B \text{BW} + C_0 \text{BW}} \tag{2.48}$$

which can be interpreted in different limits. If RIN is dominant, then the *S/N* ratio is independent of intensity. If shot noise is dominant, then the *S/N* ratio increases linearly with intensity. If the system noise floor dominates, then the *S/N* ratio increases quadratically with intensity. This general behavior of the *S/N* ratio was discussed and shown in Fig. 1.10.

The relative importance of linear sensitivity vs. quadratic sensitivity depends on the intensity dependence of the signal, and the relationship between I_0 and I_B. To illustrate this, consider the classic performance of the two-port interferometer. The background intensity is given by

$$I_B = \frac{I_0}{2}(1 + m\cos\phi) \tag{2.49}$$

and the signal is obtained as

$$\begin{aligned} I_B + \Delta I_S &= \frac{I_0}{2}(1 + m\cos(\phi + \Delta\phi)) \\ &= \frac{I_0}{2} + I_0\frac{m}{2}[\cos\phi\cos\Delta\phi - \sin\phi\sin\Delta\phi] \\ &= \frac{I_0}{2} + I_0\frac{m}{2}\cos\phi\left(1 - \frac{1}{2}\Delta\phi^2\right) - I_0\frac{m}{2}\sin\phi\Delta\phi \end{aligned} \tag{2.50}$$

The linear and quadratic coefficients are

$$\begin{aligned} C^{(L)} &= -\frac{m}{2}\sin\phi \\ C^{(Q)} &= -\frac{m}{4}\cos\phi \end{aligned} \tag{2.51}$$

The *S/N* ratio for the simple interferometer is

$$S/N = \frac{m^2\left(\sin\phi\Delta\phi + \frac{1}{2}\cos\phi\Delta\phi^2\right)^2}{C_R^2(1 + m\cos\phi)^2\text{BW} + 2h\nu(1 + m\cos\phi)\text{BW} + C_0\text{BW}} \tag{2.52}$$

When the detection is limited by RIN, then for optimized quadratic sensitivity to exceed optimized linear sensitivity the condition

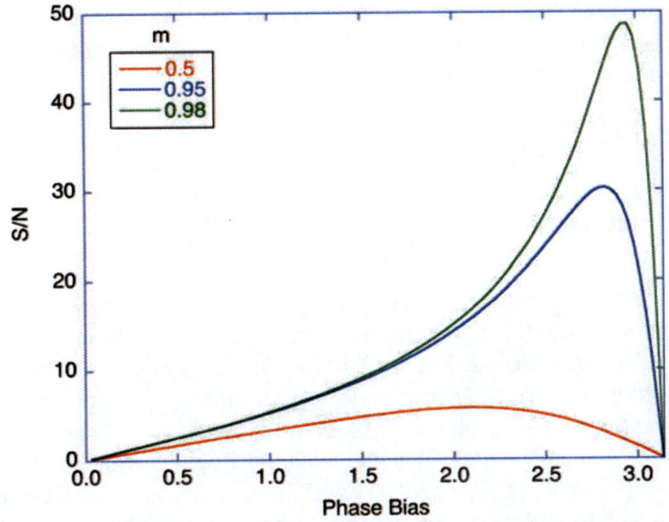

Fig. 2.8 *S/N* ratio for $\Delta\phi = 0.002$ and a 0.01% RIN for a two-wave interferometer as a function of phase bias for three contrasts of $m = 0.5$, 0.95 and 0.98. A phase bias near π reduces the background noise while still allowing a (small) linear response

$$\frac{\frac{m}{2}\Delta\phi^2}{C_R(1-m)\sqrt{BW}} > \frac{m\Delta\phi}{C_R\sqrt{BW}} \tag{2.53}$$

should hold, which gives

$$\Delta\phi_{quad} > 2(1-m) \tag{2.54}$$

and a quadratic detection would be favored over linear only if the contrast m of the interferometer can be constructed to be larger than

$$m > 1 - \frac{k(n-n_m)d}{2} \tag{2.55}$$

or $m > 0.998$ for 1 nm of protein in a water ambient. Otherwise linear detection in quadrature is favored.

A different approach to this analysis sets a maximum practical contrast m, and considers what phase bias ϕ of the interferometer gives the largest *S/N* ratio. This yields phase biases that are between $\pi/2$ and π, but approaching π as m approaches unity, because of the suppression of the noise by suppressing the background intensity. Indeed, most diffraction-based biosensors have a slight structural bias that introduces a small linear dependence in addition to the quadratic dependence of diffraction on the magnitude of the grating. This small linear dependence can dominate the signal for small grating amplitudes, while giving a strong *S/N* ratio because of the low background. This is shown in Fig. 2.8 for a two-mode interferometer for contrasts of $m = 0.50$, 0.95 and 0.98. A phase bias near π reduces the background noise while still allowing a (small) linear response. Diffraction-based biosensors are discussed in Chap. 7.

2.2 Fourier Optics

Fourier optics provides a particularly useful point of view of diffraction. The foundation of Fourier optics is the recognition that the Fraunhofer integral is equivalent to a mathematical Fourier transform. Furthermore, a lens can perform the equivalent of a Fourier transform under appropriate object and screen distances relative to the focal length of the lens. This allows the far-field Fraunhofer condition to be realized on focal planes of lenses in compact optical systems. Fourier optics combines the power of Fourier analysis with the ease of use of lenses, leading to a fruitful approach to understand image formation as well as diffraction phenomena.

To begin, an object is considered to be composed of a superposition of multiple spatial frequencies with periodicities Λ and associated K-vectors. For quasi-planar objects, the K-vectors lie in the object plane. An example of a single spatial frequency is shown in Fig. 2.9, in which the object amplitude is

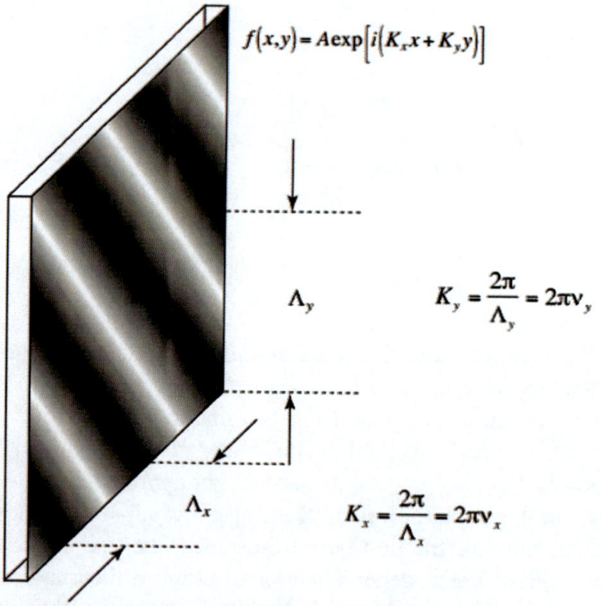

Fig. 2.9 A planar object consisting of a single spatial periodicity. The periodicities in the x and y directions are Λ_x and Λ_y, with the associated spatial frequencies ν_x and ν_y

$$f(x,y) = A\ \exp\left[\mathrm{i}(K_x x + K_y y)\right] \tag{2.56}$$

where the components of the K-vector are

$$\begin{aligned} K_x &= \mathbf{K}\cdot\hat{\mathbf{x}} = K\ \cos\,\phi = \frac{2\pi}{\Lambda}\cos\,\phi = \frac{2\pi}{\Lambda_x} = 2\pi\nu_x \\ K_y &= \mathbf{K}\cdot\hat{\mathbf{y}} = K\ \sin\,\phi = \frac{2\pi}{\Lambda}\sin\,\phi = \frac{2\pi}{\Lambda_y} = 2\pi\nu_y \end{aligned} \tag{2.57}$$

The spatial frequencies that appear in the Fraunhofer integral and the Fourier integral are

$$\nu_x = \frac{1}{\Lambda_x} \qquad \nu_y = \frac{1}{\Lambda_y} \tag{2.58}$$

To see the connection between spatial periodicities and the Fourier transform, consider a single harmonic grating

$$f(x,y) = A\ \exp[\mathrm{i}(K_x x + K_y y)] = A\ \exp[\mathrm{i}2\pi(\nu_x x + \nu_y y)] \tag{2.59}$$

The diffraction integral is

$$\begin{aligned} E_{\mathrm{d}}(\theta_x,\theta_y) &= \iint f(x,y)\exp\left[-\mathrm{i}\frac{2\pi}{\lambda}(\sin\,\theta_x x + \sin\,\theta_y y)\right]\mathrm{d}x\mathrm{d}y \\ &= \iint A\exp\left[\mathrm{i}2\pi(\nu_x x+\nu_y y)\right]\exp\left[-\mathrm{i}\frac{2\pi}{\lambda}(\sin\,\theta_x x + \sin\,\theta_y y)\right]\mathrm{d}x\mathrm{d}y \\ &= A\delta\left(\left(\nu_x - \frac{\sin\,\theta_x}{\lambda}\right),\left(\nu_y - \frac{\sin\,\theta_y}{\lambda}\right)\right) \end{aligned} \tag{2.60}$$

which defines a delta function at the diffraction angles given by

$$\begin{aligned} \sin\,\theta_x &= \lambda\nu_x \\ \sin\,\theta_y &= \lambda\nu_y \end{aligned} \tag{2.61}$$

This can be interpreted physically as a mapping of a unique scattering angle to a unique point on the observation (Fourier) plane. In other words, all rays leaving the object with a given angle converge to a single point on the Fourier plane. This viewpoint becomes especially easy to visualize when a lens performs the Fourier transform (see Fig. 2.12 below).

2.2.1 Fresnel Diffraction

If $f(x, y)$ is a combination of harmonics

$$f(x,y) = \iint F(\nu_x, \nu_y)\exp[\mathrm{i}2\pi(\nu_x x + \nu_y y)]\mathrm{d}\nu_x\,\mathrm{d}\nu_y \tag{2.62}$$

then the transmitted field just after the screen is

$$E_s(x,y,z) = \iint F(\nu_x, \nu_y)\exp[\mathrm{i}2\pi(\nu_x x + \nu_y y)]\exp[\mathrm{i}k_z z]\mathrm{d}\nu_x\,\mathrm{d}\nu_y \tag{2.63}$$

where

$$k_z = \sqrt{k^2 - k_x^2 - k_y^2} = 2\pi\sqrt{\frac{1}{\lambda^2} - \nu_x^2 - \nu_y^2} \tag{2.64}$$

For small angles this is approximately

$$k_z z = \frac{2\pi z}{\lambda}\sqrt{1 - \lambda^2\nu_x^2 - \lambda^2\nu_y^2} \approx \frac{2\pi z}{\lambda} - \pi\lambda z(\nu_x^2 + \nu_y^2) \tag{2.65}$$

The transmitted field can then be written as

$$\begin{aligned} E_s(x,y,z) &= \iint F(\nu_x, \nu_y)\exp[\mathrm{i}2\pi(\nu_x x + \nu_y y)]\exp(\mathrm{i}kz)\exp[-\mathrm{i}\pi\lambda z(\nu_x^2 + \nu_x^2)]\mathrm{d}\nu_x\,\mathrm{d}\nu_y \\ &= \iint F(\nu_x, \nu_y)\exp[\mathrm{i}2\pi(\nu_x x + \nu_y y)]H(\nu_x, \nu_y)\mathrm{d}\nu_x\,\mathrm{d}\nu_y \end{aligned} \tag{2.66}$$

where

$$H(\nu_x, \nu_y) = \exp(\mathrm{i}kz)\exp[-\mathrm{i}\pi\lambda z(\nu_x^2 + \nu_x^2)] \tag{2.67}$$

is known as the free-space propagator in the Fresnel approximation.

The Green's function solution (response to a delta function) for an impulse function on the object plane is the inverse Fourier transform of the free-space propagator

$$G(x,y,z) = \frac{1}{\mathrm{i}\lambda z}\exp(\mathrm{i}kz)\exp\left[\mathrm{i}k\frac{x^2 + y^2}{2z}\right] \tag{2.68}$$

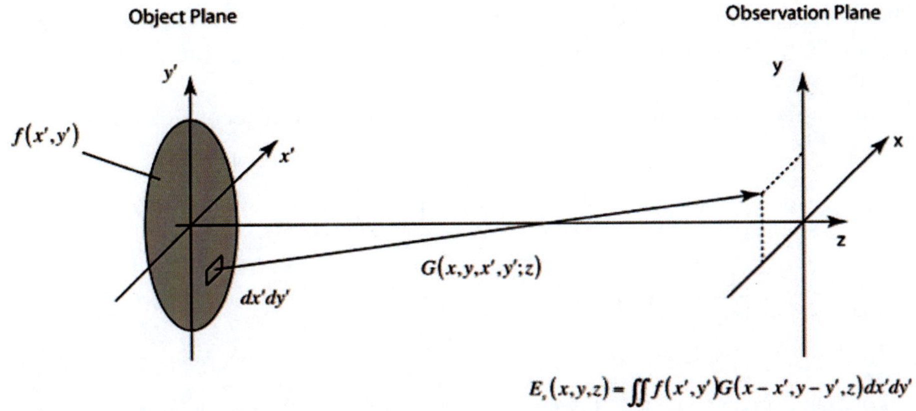

Fig. 2.10 Fresnel diffraction geometry. A delta function on the object plane has an associated Green's function. Convolution of the object function $f(x', y')$ with the Green's function gives the field at the observation plane

so that the general scattered field at z is a convolution of the object function over the Green's function

$$
\begin{aligned}
E_s(x,y,z) &= \iint f(x',y')G(x-x',y-y',z)\mathrm{d}x'\,\mathrm{d}y' \\
&= \frac{1}{\mathrm{i}\lambda z}\exp(\mathrm{i}kz)\iint f(x',y')\exp\left[\mathrm{i}k\frac{(x-x')^2+(y-y')^2}{2z}\right]\mathrm{d}x'\,\mathrm{d}y'
\end{aligned}
\tag{2.69}
$$

This is the Fresnel integral that can be used for general scattering problems, including holography and lens-free imaging. Holographic reconstruction in digital holography uses the Fresnel integral explicitly to reconstruct three-dimensional aspects of an object. The scattering geometry with the object and observation plane are shown in Fig. 2.10.

2.2.2 *Optical Fourier Transforms*

The Fresnel regime converts to the Fraunhofer (far-field) regime when the distance to the observation plane is much larger than the size of the scattering object. The cross-over condition is captured quantitatively in terms of the Fresnel number

$$N_F = \frac{a^2}{\lambda L} \tag{2.70}$$

where a is the size of the scattering object and L is the distance to the detection plane. The scattering is in the Fresnel regime when $N_F > 1$ (but still $L \gg \lambda$), and in the Fraunhofer regime when $N_F < 1$. The cross-over is not a sharp threshold, so it is best to satisfy $N_F \ll 1$ to ensure that the scattering is the Fraunhofer regime. As an example, a 10 μm object illuminated with visible light scatters to the Fraunhofer regime for an observation distance greater than a millimeter.

In the far-field regime at a large distance L from the object screen the Fresnel integral becomes

$$\begin{aligned} EF(x,y) &= \frac{1}{\mathrm{i}\lambda L}\exp(\mathrm{i}kL)\exp\left[\mathrm{i}k\frac{x^2+y^2}{2L}\right]\iint f(x',y')\exp\left[-\mathrm{i}k\frac{xx'+yy'}{L}\right]\mathrm{d}x'\,\mathrm{d}y' \\ &= \frac{1}{\mathrm{i}\lambda L}\exp(\mathrm{i}kL)\exp\left[\mathrm{i}k\frac{x^2+y^2}{2L}\right]F(\nu_x,\nu_y) \end{aligned} \tag{2.71}$$

in which $F(\nu_x, \nu_y)$ is the Fourier transform of $f(x, y)$, and where the spatial frequencies of the object are identified as

$$\nu_x = \frac{x}{\lambda L} \qquad \nu_y = \frac{y}{\lambda L} \tag{2.72}$$

This result has the important consequence that the far-field diffraction pattern is proportional to the Fourier transform of the object modulation function. It is an easy way to do a Fourier transform – just diffract to the far-field. However, it is usually convenient to work with lenses, and these too perform an optical Fourier transform.

A single simple (thin) lens of focal length f can be viewed mathematically as imposing a quadratic phase profile on an incoming wave by the transmission function

$$t(x,y) = \exp\left[-\mathrm{i}\pi\frac{(x^2+y^2)}{\lambda f}\right] \tag{2.73}$$

The geometry of the Fourier lens is shown in Fig. 2.11 with the distance L from the object to the lens, and with the observation plane a distance f from the lens. The angle-to-point transformation that is achieved by a Fourier lens is illustrated in Fig. 2.12. Using (2.73) in (2.71) gives

$$E_F(x,y) = \frac{1}{\mathrm{i}\lambda f}\exp[\mathrm{i}k(L+f)]\exp\left[-\mathrm{i}\pi\frac{(x^2+y^2)(L-f)}{L\lambda f}\right]F\left(\frac{x}{\lambda f},\frac{y}{\lambda f}\right) \tag{2.74}$$

This is proportional to the Fourier transform of $f(x, y)$, but with a quadratic phase factor. When the object distance $L = f$, then

$$E_F(x,y) = \frac{1}{\mathrm{i}\lambda f}\exp(\mathrm{i}k2f)F\left(\frac{x}{\lambda f},\frac{y}{\lambda f}\right) \tag{2.75}$$

which is the desired Fourier transform with a simple phase factor.

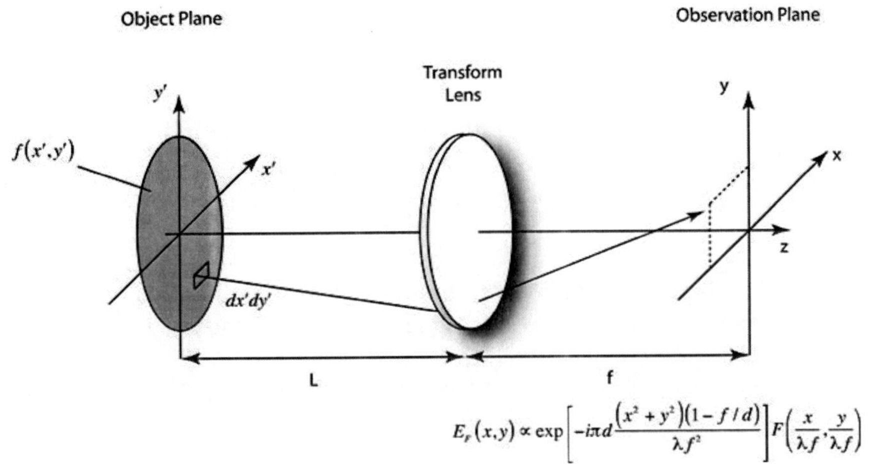

Fig. 2.11 Configuration of a Fourier lens. When the object, lens, and observation plane are in a 1*f*–1*f* condition (both the object plane and the image plane are a distance *f* from the lens), then the field on the observation plane (now called the Fourier plane) is the Fourier transform of the object amplitude

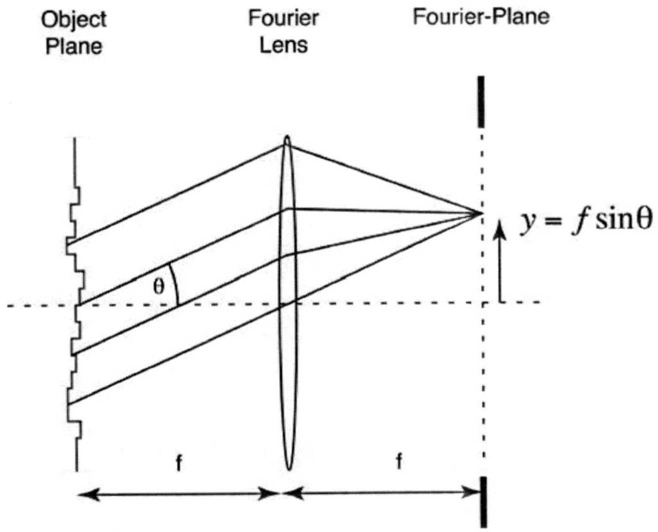

Fig. 2.12 A Fourier lens in a 1*f*–1*f* configuration. All rays emitted from the object at a given angle are focused by the lens to a point (actually a point spread function) on the Fourier plane

2.2.3 *Gaussian Beam Optics*

In laser optics, one of the most common intensity distributions is the Gaussian beam. This is a beam that has the transverse intensity profile given by

$$I(x, y) = S_0 \exp(-(x^2 + y^2)/w_0^2) \tag{2.76}$$

with an integrated power

$$P = \iint I(x, y)\mathrm{d}x\,\mathrm{d}y = S_0 \pi w_0^2 \tag{2.77}$$

Because a Gaussian beam has a limited lateral extent, it self-diffracts into the far-field, even without any aperture to diffract off of. This is an important part of the principle of wave propagation. No matter how a field is spatially modulated, whether by a physical aperture, or simply because fields are inhomogeneous, these fields propagate as partial waves with relative phases and relative path lengths that cause interference and hence diffraction.

For a Gaussian beam at its focal plane, the dimensionless field amplitude is

$$E = \sqrt{I} = \sqrt{S_0} \exp(-(x^2 + y^2)/2w_0^2) \tag{2.78}$$

and the diffraction integral is

$$E_\mathrm{d} = -\mathrm{i}\sqrt{S_0} \int_{-\infty}^{\infty} \exp(-(x^2 + y^2)/2w_0^2) \mathrm{e}^{\mathrm{i}kx \sin\theta} \mathrm{e}^{\mathrm{i}ky \sin\psi}\, \mathrm{d}x\,\mathrm{d}y \tag{2.79}$$

This has the diffraction field

$$E(\theta, \Psi) = -\mathrm{i}\sqrt{S_0} \frac{kw_0^2}{R} \exp\left(-\frac{k^2 w_0^2}{2}(\sin^2\theta + \sin^2\psi)\right) \tag{2.80}$$

which is again a Gaussian, but now expressed as an angular distribution in angles θ and Ψ. When Fourier transformed using a lens of focal distance f, the field at the Fourier plane is

$$E(x', y') = -\mathrm{i}\sqrt{S_0} \frac{2\pi w_0^2}{f\lambda} \exp\left(-\frac{2\pi^2 w_0^2}{(f\lambda)^2}(x'^2 + y'^2)\right) \tag{2.81}$$

illustrated in Fig. 2.13 as the focal waist of a Gaussian beam. The intensity at the Fourier plane is

$$I(x', y') = S_0 \left(\frac{2\pi}{f\lambda}\right)^2 w_0^4 \exp\left(-\left(\frac{2\pi}{f\lambda}\right)^2 w_0^2 (x'^2 + y'^2)\right) \tag{2.82}$$

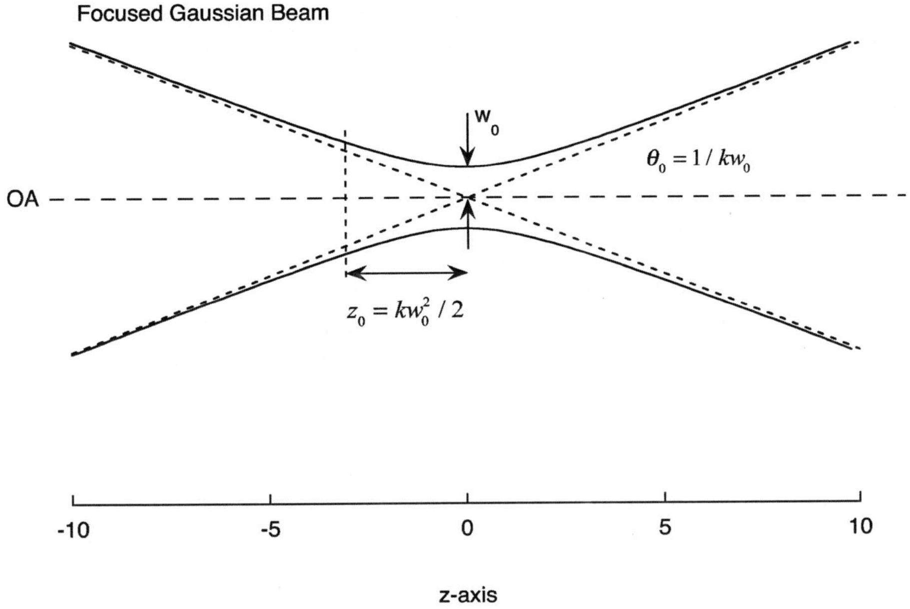

Fig. 2.13 Profile of a focused Gaussian beam. The beam waist w_0 occurs at the focal plane at $z = 0$. At distances large relative to the Rayleigh range z_0 the beam converges or diverges with angle θ_0. The beam evolution is self diffraction

with a beam radius given by

$$w_0' = \frac{f\lambda}{2\pi w_0} \tag{2.83}$$

This result is of special importance in Fourier optics. It expresses the inverse relationship between the widths of Fourier-transform pairs of Gaussians. Tightly focused beams are transformed into broad beams, and vice versa. This inverse relation also holds generally for smooth non-Gaussian beams, but with different numerical factors.

2.3 Dipoles and Rayleigh Scattering

The origin of all light-scattering phenomena is the interaction of the electric and magnetic fields of a light wave with electric and magnetic multipole moments of electrons in atoms and molecules. For optical interactions, the strongest interactions are usually through the electric dipole interaction, which tends to be stronger than the magnetic dipole interaction by the ratio $e^2/\hbar c = 1/137$. However, in cases in which the electric dipole interaction is not allowed for symmetry reasons, magnetic dipole or other multipole moments can dominate. In this book, we consider electric dipole interaction exclusively.

All molecules have a molecular polarizability, that is, a tensor relation between the applied field and the structure and symmetry of the molecule

$$p^i = \varepsilon_{\mathrm{m}} \alpha^i_j E^j \tag{2.84}$$

where α^i_j is the molecular polarizability tensor and ε_{m} is the isotropic dielectric constant of the surrounding medium. Because protein molecules lack general symmetry, the tensor polarizability is simplified through configurational averaging to a scalar property relating the induced dipole moment to the applied electric field.

$$\mathbf{p} = \varepsilon_{\mathrm{m}} \alpha \mathbf{E} \tag{2.85}$$

Many protein molecules are globular in structure, such as the immunoglobulins, and to lowest approximation may be viewed as dielectric spheres. The polarizability of a dielectric sphere of radius a is

Polarizability of a dielectric sphere:

$$\alpha = \frac{(\varepsilon - \varepsilon_{\mathrm{m}})}{(\varepsilon + 2\varepsilon_{\mathrm{m}})} 4\pi\varepsilon_0 a^3 \tag{2.86}$$

where ε is the dielectric function of the sphere. For comparison, the polarizability of a metallic sphere is

Polarizability of a metallic sphere:

$$\alpha = 4\pi\varepsilon_0 a^3 \tag{2.87}$$

Given an induced dipole moment p, the resulting dipole field is

Dipole field:

$$\mathbf{E}(\mathbf{r}) = \frac{3(\mathbf{p}\cdot\mathbf{r})\mathbf{r} - r^2\mathbf{p}}{4\pi\varepsilon_{\mathrm{m}}\varepsilon_0 r^5} \tag{2.88}$$

which falls off as the inverse cube of the distance from the dipole. The dipole field of the dipole induced by an incident field is the origin of molecular scattering and ultimately of the refractive index of a layer of biological molecules.

The first approximation for understanding molecular scattering is scattering in the Rayleigh limit. The incident electric and magnetic fields in a dielectric medium with $k_{\mathrm{m}} = n_{\mathrm{m}} k_0$ are

$$\begin{aligned} \mathbf{E}_{\mathrm{inc}} &= \boldsymbol{\varepsilon}_0 E_0 \, \mathrm{e}^{ik_{\mathrm{m}}\hat{n}_0\cdot\mathbf{x}} \\ \mathbf{B}_{\mathrm{inc}} &= \frac{\hat{n}_0}{v_{\mathrm{m}}} \times \mathbf{E}_{\mathrm{inc}} \end{aligned} \tag{2.89}$$

where $\mathbf{n}_0$ is the unit vector in the direction of the incident field. The scattered fields from the induced dipole are

$$\mathbf{E}_{sc} = k_m^2 \frac{e^{ikr}}{4\pi\varepsilon_m\varepsilon_0 r}[(\mathbf{n} \times \mathbf{p}) \times \mathbf{n}]$$
$$\mathbf{B}_{sc} = \frac{1}{v_m}\mathbf{n} \times \mathbf{E}_{sc} \tag{2.90}$$

where $\mathbf{n}$ is the unit vector in the direction of the scattered wave. The Poynting vector is

$$\mathbf{S} = \frac{1}{\mu_0}\mathbf{E} \times \mathbf{B} \tag{2.91}$$

and the differential scattering cross-section is constructed from the Poynting vector along the scattering direction

$$\frac{d\sigma}{d\Omega}(\theta, \phi) = \frac{r^2|\mathbf{S}(\theta, \phi)|^2}{\frac{1}{2}v\varepsilon\varepsilon_0|\mathbf{E}_0|^2} = \frac{k_m^4}{(4\pi\varepsilon\varepsilon_0)^2 E_0^2}|\mathbf{p}|^2 \frac{1}{2}(1 + \cos\theta) \tag{2.92}$$

where θ is measured relative to the incident k-vector direction.

Example: Small Dielectric Sphere in Vacuum

In the case of scattering by a small dielectric sphere, the induced dipole is

$$\mathbf{p} = 4\pi\varepsilon_0\left(\frac{\varepsilon - 1}{\varepsilon + 2}\right)a^3\mathbf{E}_{inc} \tag{2.93}$$

with a scattering cross-section

$$\frac{d\sigma}{d\Omega} = k^4 a^6 \left|\frac{\varepsilon - 1}{\varepsilon + 2}\right|^2 |\varepsilon^* \cdot \varepsilon_0|^2 \tag{2.94}$$

The scattering cross-section is decomposed into expressions for each incident polarization relative to the scattering plane

$$\frac{d\sigma_\parallel}{d\Omega} = \frac{k^4 a^6}{2}\left|\frac{\varepsilon - 1}{\varepsilon + 2}\right|^2 \cos^2\theta$$
$$\frac{d\sigma_\perp}{d\Omega} = \frac{k^4 a^6}{2}\left|\frac{\varepsilon - 1}{\varepsilon + 2}\right|^2 \tag{2.95}$$

When these are averaged over polarization, they become

Rayleigh differential cross-section:

$$\frac{d\sigma}{d\Omega} = k^4 a^6 \left|\frac{\varepsilon - 1}{\varepsilon + 2}\right|^2 \frac{1}{2}(1 + \cos^2\theta) \tag{2.96}$$

with the familiar angular distribution of scattered light with symmetry between forward-scattering and back-scattering. The differential cross-section is integrated over all scattering angles to obtain the total cross-section

$$\sigma = \int \frac{d\sigma}{d\Omega} d\Omega = \frac{8\pi}{3} k^4 a^6 \left|\frac{\varepsilon - 1}{\varepsilon + 2}\right|^2 \tag{2.97}$$

The scattering cross-section depends on the sixth power of the radius and on the fourth power of the frequency. The cross-section has units of area, as seen when it is expressed as

Rayleigh total cross-section:

$$\sigma = \left(\frac{8}{3}(ka)^4 \left|\frac{\varepsilon - 1}{\varepsilon + 2}\right|^2\right) \pi a^2 \tag{2.98}$$

The effective scattering area of a dielectric sphere is proportional to the cross-sectional area of the sphere, but reduced by the factor in parentheses in front. For biological molecules with a radius of 1 nm and a dielectric constant of 2 the reduction is approximately 2×10^{-10}. The effective cross-section for such a molecule is $\sigma = 10^{-23}$ cm^2.

Example: Small Dielectric Sphere in Water

There are many applications in which the dielectric sphere is in a water environment. It is important in this case to be clear about the vacuum k-vector k_0, and the medium k-vector, which is $k_m = n_m k_0$. In addition, the dielectric water medium modifies the polarization as

$$\alpha = \frac{(\varepsilon - \varepsilon_m)}{(\varepsilon + 2\varepsilon_m)} 4\pi\varepsilon_0 a^3 \tag{2.99}$$

Therefore, the differential cross-section in water is

$$\frac{d\sigma}{d\Omega} = \varepsilon_m^2 k_0^4 a^6 \left|\frac{\varepsilon - \varepsilon_m}{\varepsilon + 2\varepsilon_m}\right|^2 \frac{1}{2}(1 + \cos^2\theta) \tag{2.100}$$

Table 2.1 Rayleigh scattering

In air	In medium
$\alpha = \frac{(\varepsilon-1)}{(\varepsilon+2)} 4\pi\varepsilon_0 a^3$	$\alpha = \frac{(\varepsilon-\varepsilon_m)}{(\varepsilon+2\varepsilon_m)} 4\pi\varepsilon_0 a^3$
$\mathbf{p} = \left(\frac{\varepsilon-1}{\varepsilon+2}\right) 4\pi\varepsilon_0 a^3 \mathbf{E}_{inc}$	$\mathbf{p} = \left(\frac{\varepsilon-\varepsilon_m}{\varepsilon+2\varepsilon_m}\right) 4\pi\varepsilon_m\varepsilon_0 a^3 \mathbf{E}_{inc}$
$\frac{d\sigma}{d\Omega} = k^4 a^6 \left\lvert\frac{\varepsilon-1}{\varepsilon+2}\right\rvert^2 \frac{1}{2}(1+\cos^2\theta)$	$\frac{d\sigma}{d\Omega} = \varepsilon_m^2 k_0^4 a^6 \left\lvert\frac{\varepsilon-\varepsilon_m}{\varepsilon+2\varepsilon_m}\right\rvert^2 \frac{1}{2}(1+\cos^2\theta)$
$\sigma = \left(\frac{8}{3}(ka)^4 \left\lvert\frac{\varepsilon-1}{\varepsilon+2}\right\rvert^2\right)\pi a^2$	$\sigma = \left(\frac{8}{3}(k_m a)^4 \left\lvert\frac{\varepsilon-\varepsilon_m}{\varepsilon+2\varepsilon_m}\right\rvert^2\right)\pi a^2$

with the corresponding total cross-section (real dielectric function: no absorption)

$$\sigma = \left(\frac{8}{3}(k_m a)^4 \left|\frac{\varepsilon - \varepsilon_m}{\varepsilon + 2\varepsilon_m}\right|^2\right)\pi a^2 \tag{2.101}$$

Selected formulas for Rayleigh scattering are given in Table 2.1.

2.4 Refractive Index of a Dilute Molecular Film

Refractive index is a macroscopic property of a material that relates the phase velocity of a light wave to frequency and wavelength through the expression

$$v_p = \frac{\omega}{k} = \frac{c}{n(\omega)} \tag{2.102}$$

In this section, we approach the derivation of the refractive index using concepts from diffraction theory.

An apparent paradox arises in relation to the phase shifts experienced upon scattering from a dipole and the phase shifts that are detected in the far-field that are attributed to, or interpreted as, refractive index. The central paradox is that the scattered wave from a dipole is in phase with the incident field. If the scattered field and incident field have the same phase, then why is a phase shift detected in the far-field? The answer to this paradox comes from diffraction theory (scalar Kirchhoff diffraction suffices) due to the difference between a continuous field distribution compared with the discrete scatterer. An outline of the physics is shown in Fig. 2.14. Two phasor diagrams are shown, one in the near-field and one in the far-field. In the near-field, the local field and the scattered field are in phase. However, as the local field propagates to the far-field, it acquires a $\pi/2$ phase shift, while the scattered field does not. Therefore, in the far-field, the scattered field is in phase quadrature with the continuous field, which is manifested as a phase shift. Mathematically, integrating the diffraction integral over a finite-size continuous distribution yields a factor of -i, while a discrete delta-function scatterer simply produces a spherical wave that propagates to the far-field without any phase shift.

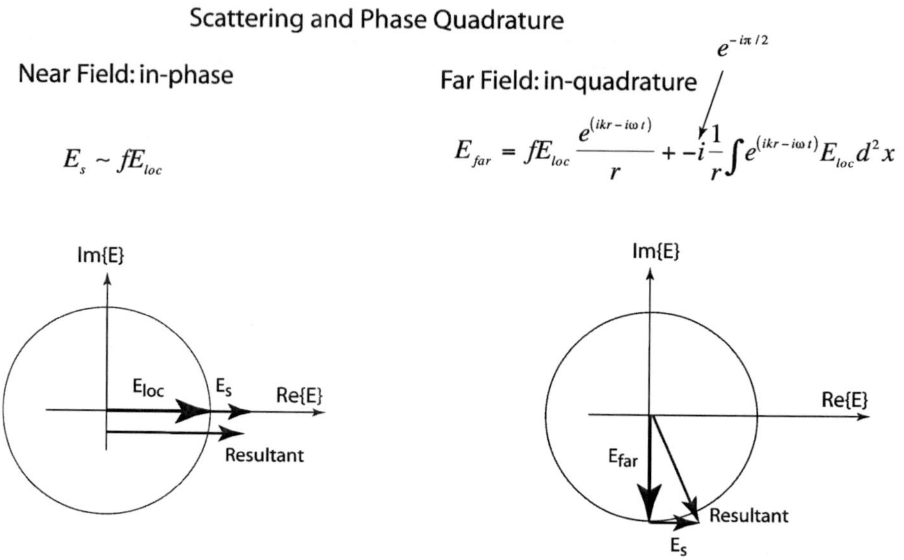

Fig. 2.14 Light scattering from a molecule and the $\pi/2$ phase shift. In the near field, the scattered wave is in-phase with the incident field. In the far-field, a $\pi/2$ phase shift has occurred that places the scattered wave in phase-quadrature with the original wave, representing a phase modulation related to the refractive index

2.4.1 *Phase Shift of a Single Molecule in a Focused Gaussian Beam*

A single molecule in a tightly focused Gaussian beam will scatter light and impart a small phase shift on the direct wave in the far-field. This phase shift is a direct consequence of diffraction and interference. An interesting question is whether this single-molecule phase shift might be detectable experimentally. The induced dipole from (2.86) on the optic axis and on the focal plane of the Gaussian beam is

$$\mathbf{p} = \left(\frac{\varepsilon - 1}{\varepsilon + 2}\right) a^3 \mathbf{E}_0 \tag{2.103}$$

where the field on the optic axis is $E_0 = \sqrt{S_0}$. The local scattered electric field is then

$$\mathbf{E}_{\mathrm{sc}} = k^2 \frac{\mathrm{e}^{ikr}}{r} \left(\frac{\varepsilon - 1}{\varepsilon + 2}\right) a^3 \sqrt{S_0} \tag{2.104}$$

This field is transformed to the Fourier plane using a lens of focal length f and diameter $D > \frac{1}{2\pi}\frac{f\lambda}{w_0}$. The power collected by this lens is

$$P_{\text{scat}} = \frac{\pi}{4} D^2 \frac{k^4}{f^2} \left(\frac{\varepsilon - 1}{\varepsilon + 2} \right)^2 a^6 S_0 \tag{2.105}$$

and the field at the Fourier plane is

$$E_{\text{scat}} = \frac{k^2}{f} \left(\frac{\varepsilon - 1}{\varepsilon + 2} \right) a^3 \sqrt{S_0} \tag{2.106}$$

The interfering fields (the original field plus the scattered field) at the Fourier plane are now

$$\begin{aligned} E &= E_{scat} + E_0 \\ &= \frac{k\sqrt{S_0}}{f} \left[ka^3 \left(\frac{\varepsilon - 1}{\varepsilon + 2} \right) - iw_0^2 \exp\left(-\frac{2\pi^2 w_0^2}{(f\lambda)^2} (x'^2 + y'^2) \right) \right] \end{aligned} \tag{2.107}$$

The phase shift on the optic axis of the detected Gaussian field is given by

$$\tan \phi \approx \frac{ka^3}{w_0^2} \left(\frac{\varepsilon - 1}{\varepsilon + 2} \right) \tag{2.108}$$

For a molecule of radius $a = 3$ nm with $\varepsilon = 1.43^2 = 2$ at a wavelength of 500 nm and a beam radius of 0.5 μm, this phase shift is about 1×10^{-7} radians.

To estimate how detectable this phase shift is, consider a situation when the $\pi/2$ phase shift between the Gaussian field and the scattered field is shifted to zero to place it in the condition of constructive interference. Then the intensity at the Fourier plane is

$$\begin{aligned} I = \frac{k^2 S_0}{f^2} &\left[w_0^4 \exp\left(-\frac{4\pi^2 w_0^2}{(f\lambda)^2} (x'^2 + y'^2) \right) \right. \\ &\left. + 2w_0^2 ka^3 \left(\frac{\varepsilon - 1}{\varepsilon + 2} \right) \exp\left(-\frac{2\pi^2 w_0^2}{(f\lambda)^2} (x'^2 + y'^2) \right) \right] \end{aligned} \tag{2.109}$$

When this is integrated over the Fourier plane, it yields

$$\int I \, \mathrm{d}a = S_0 \left[\pi w_0^2 + 4\pi ka^3 \left(\frac{\varepsilon - 1}{\varepsilon + 2} \right) \right] \tag{2.110}$$

and the relative modulation is

$$\frac{\Delta I}{I} = \frac{4ka^3}{w_0^2} \left(\frac{\varepsilon - 1}{\varepsilon + 2} \right) \tag{2.111}$$

which is four times the value of the phase shift along the optic axis. One factor of two is from the interference cross terms, and the other is from the integration over the Gaussian profile. Therefore, for typical values of the parameters, the relative intensity modulation from a single macromolecules is about one part per million.

The approximate shot noise that this corresponds to is a photon number of 1×10^{12}, which is about 400 nJ. If this energy is detected in a millisecond detection time, the power at the detector would be about 400 μW, which is an easy power level to achieve with conventional photodiodes. Therefore, a single macromolecule could be detected interferometrically under shot-noise-limited detection conditions.

2.4.2 *Phase Shift from a Dilute Collection of Molecules*

The effective phase shift of a collection of molecules in a Gaussian beam can be obtained by extending the single-molecule result. The scattered field for a molecule off the optic axis is

$$E'_{\text{scat}} = \frac{k^2}{f}\left(\frac{\varepsilon - 1}{\varepsilon + 2}\right) a^3 \sqrt{S_0} \exp(-r^2/2w_0^2) \exp(ikxx'/f) \exp(ikyy'/f) \qquad (2.112)$$

where the primed coordinates are in the Fourier (detection) plane, and the unprimed coordinates are in the focal plane. The contribution of a distribution of molecules is obtained using the integration

$$\int_0^\infty 2\pi r\, \mathrm{e}^{-r^2/2w_0^2}\, \mathrm{d}r = 2\pi w_0^2 \qquad (2.113)$$

to give, for a surface density $N_A = N/\pi w_0^2$, the effective number of scatterers

$$\frac{N}{\pi w_0^2} 2\pi w_0^2 = 2N \qquad (2.114)$$

The phase is then

$$\tan \phi \approx 2N \frac{ka^3}{w_0^2}\left(\frac{\varepsilon - 1}{\varepsilon + 2}\right) \qquad (2.115)$$

This is turned into an effective index of refraction for a layer with a thickness equal to $2a$ as

$$\phi \approx 2N \frac{ka^3}{w_0^2}\left(\frac{\varepsilon - 1}{\varepsilon + 2}\right) = k(n_{\text{av}} - 1)2a \qquad (2.116)$$

or

$$n_{\mathrm{av}} = 1 + N\frac{a^2}{w_0^2}\left(\frac{\varepsilon - 1}{\varepsilon + 2}\right) = 1 + \pi \mathrm{N}_{\mathrm{A}} a^2 \left(\frac{\varepsilon - 1}{\varepsilon + 2}\right) \tag{2.117}$$

where N_{A} is the area number density (particles per area). This is restated in terms of the volume fraction f_{v} using

$$f_{\mathrm{v}} = \frac{(4/3)\pi a^3}{2a} N_{\mathrm{A}} = \frac{2}{3}\pi a^2 N_{\mathrm{A}} \tag{2.118}$$

to give

$$n_{\mathrm{av}} = 1 + \frac{3}{2} f_{\mathrm{v}} \left(\frac{\varepsilon - 1}{\varepsilon + 2}\right) \tag{2.119}$$

which is consistent with the Maxwell Garnett effective medium in the Rayleigh limit, which is described next.

2.5 Local Fields and Effective Medium Approaches

Effective medium theories seek to convert a distribution of discrete and stochastic elements into an effective homogeneous medium that behaves with appropriately averaged properties. For instance, the optical effects of a dilute dispersion of particles on a surface (such as biomolecules in optical biosensors) can be treated as if the particles constituted a thin layer having an average refractive index. These approaches can never capture the "full" physics, especially when fluctuations dominate the optical behavior, but they give a heuristic approach, usually with analytic formulas that can be evaluated and compared against experiment.

2.5.1 Local Fields and Depolarization

The scattering from a dilute collection of scatterers as a function of increasing scatterer density is straightforward until multiple scattering becomes significant. In the case of discrete scatterers, the scattered field from each is added to the total field self consistently as a new incident field that is scattered by the collection. To treat this problem explicitly using self-consistent approaches is numerically exhaustive, such as through the discrete dipole approximation [2]. A simpler approach, that works well when the medium is dense, is a mean field approach that uses the mean local fields of neighboring induced dipoles as a depolarization field experienced by an individual scatterer.

In the case of a spherical dielectric particle, the polarizability is the well-known result

$$\alpha = 4\pi\varepsilon_0 a^3 \frac{(\varepsilon - \varepsilon_m)}{(\varepsilon + 2\varepsilon_m)} \tag{2.120}$$

The uniform field inside the particle is reduced from the field nearby as

$$\mathbf{E}_{in} = \mathbf{E}_{near} + \frac{\mathbf{P}}{3\varepsilon_0} \tag{2.121}$$

If the material has high symmetry or is isotropic, then the average near-field is zero, and the induced dipole moment of a molecule at the center is

$$\mathbf{p} = \alpha\varepsilon_m(\mathbf{E} + \mathbf{E}_{in}) \tag{2.122}$$

For a collection of dipoles with number density N, the polarizability is

$$\mathbf{P} = N\alpha\varepsilon_m\left(\mathbf{E} + \frac{\mathbf{P}}{3\varepsilon_0}\right) = \chi\varepsilon_0\mathbf{E} \tag{2.123}$$

Solving for χ yields

$$\chi = \frac{1}{\varepsilon_0} \frac{N\alpha\varepsilon_m}{1 - \dfrac{N\alpha\varepsilon_m}{3\varepsilon_0}} \tag{2.124}$$

Using the relationship $\varepsilon = 1 + \chi$ yields the Clausius–Mossotti relation

$$\frac{\varepsilon - 1}{\varepsilon + 2} = \frac{\varepsilon_m}{3\varepsilon_0} \sum_j N_j\alpha_j \tag{2.125}$$

This expression includes the local fields generated by nearby molecules. The Clausius–Mossotti relationship is most accurate for dilute systems such as gases. However, it still holds as a heuristic relationship even for denser liquids and solids.

2.5.2 *Effective Medium Models*

There are many different approaches to effective medium approximations. Each is equivalent in the dilute Rayleigh limit, but each is slightly different in the case of mixtures of different materials. Furthermore, different approximations relate to different situations. For instance, small spherical inclusions distributed inside a homogeneous medium are best approximated by Maxwell Garnett formula, while two uniformly mixed phases are best approximated by the Bruggeman formula.

2.5.2.1 Maxwell Garnett

When the collection of scatterers is no longer dilute, but is not sufficiently dense or symmetric to follow the Clausius–Mossotti relation, then a working alternative is the Maxwell Garnett approximation. The Maxwell Garnett model assumes a collection of small particles (inclusions) of material with bulk index n with *radius much smaller than a wavelength* embedded in a medium with *a thickness much larger than the diameters of the spheres.*

The average field is

$$\langle \mathbf{E} \rangle = (1-f)\langle \mathbf{E}_\mathrm{m} \rangle + \sum_k f_k \langle \mathbf{E}_k \rangle \tag{2.126}$$

where $\langle \mathbf{E}_\mathrm{m} \rangle$ is the average field in the medium. The volume fraction of each particle sums to the total volume fraction

$$\sum_k f_k = f \tag{2.127}$$

The average polarization is

$$\langle \mathbf{P} \rangle = (1-f)\langle \mathbf{P}_\mathrm{m} \rangle + \sum_k f_k \langle \mathbf{P}_k \rangle \tag{2.128}$$

which is related to the average susceptibility tensor by

$$\langle \mathbf{P} \rangle = \varepsilon_0 \overleftrightarrow{\chi}_\mathrm{av} \langle \mathbf{E} \rangle \tag{2.129}$$

The total self-consistent field now satisfies

$$(1-f)(\overleftrightarrow{\varepsilon}_\mathrm{av} - \overleftrightarrow{\varepsilon}_\mathrm{m})\langle \mathbf{E}_\mathrm{m} \rangle + \sum_k f_k (\overleftrightarrow{\varepsilon}_\mathrm{av} - \overleftrightarrow{\varepsilon})\langle \mathbf{E}_k \rangle = 0 \tag{2.130}$$

The next assumption is the central assumption of the Maxwell Garnett model. It states that the uniform field inside the kth particle is related to the external uniform field in the medium through the tensor relation

$$\mathbf{E}_k = \overleftrightarrow{\lambda}_k \mathbf{E}_\mathrm{m} \tag{2.131}$$

Table 2.2 Geometric factors

Prolate spheroids			Oblate spheroids		
b/a	L_1	$L_2 = L_3$	a/b	L_1	$L_2 = L_3$
0 (needle)	0	0.5	0 (disk)	1.0	0.0
0.2	0.056	0.472	0.2	0.750	0.125
0.4	0.134	0.433	0.4	0.588	0.206
0.6	0.210	0.395	0.6	0.478	0.261
0.8	0.276	0.362	0.8	0.396	0.302
1.0	0.333	0.333	1.0	0.333	0.333

for the tensor $\overleftrightarrow{\lambda}$. The principal components of the $\overleftrightarrow{\lambda}$ tensor along the principal axes of the ellipsoid are

$$\lambda_j = \frac{\varepsilon_m}{\varepsilon_m + L_j(\varepsilon - \varepsilon_m)} \tag{2.132}$$

and the anisotropy factors L_j for spheroids are given in Table 2.2.

The self-consistent field equation is now

$$(1-f)(\overleftrightarrow{\varepsilon}_{av} - \overleftrightarrow{\varepsilon}_m) + \sum_k f_k(\overleftrightarrow{\varepsilon}_{av} - \overleftrightarrow{\varepsilon})\overleftrightarrow{\lambda}_k = 0 \tag{2.133}$$

which is still a tensor expression with a sum over the individual particles in the medium.

For a collection of randomly distributed spheroids, the tensor equation is converted to a scalar equation for ε_{av} by integrating over all the particle orientations. This gives

$$\varepsilon_{av} = \frac{(1-f)\varepsilon_m + f\beta\varepsilon}{1 - f + f\beta} \tag{2.134}$$

which is the final Maxwell Garnett equation.

The most important special cases for Maxwell Garnett are for spheres

$$\beta = \frac{3\varepsilon_m}{\varepsilon + 2\varepsilon_m} \tag{2.135}$$

$$\varepsilon_{av} = \varepsilon_m \left[1 + \frac{3f\left(\frac{\varepsilon - \varepsilon_m}{\varepsilon + 2\varepsilon_m}\right)}{1 - f\left(\frac{\varepsilon - \varepsilon_m}{\varepsilon + 2\varepsilon_m}\right)} \right] = \varepsilon_m \left[\frac{\varepsilon(1+2f) + 2(1-f)\varepsilon_m}{\varepsilon(1-f) + (2+f)\varepsilon_m} \right] \tag{2.136}$$

An alternative expression for the Maxwell Garnett result is

$$\frac{\varepsilon_{av} - \varepsilon_m}{\varepsilon_{av} + 2\varepsilon_m} = f\frac{\varepsilon - \varepsilon_m}{\varepsilon + 2\varepsilon_m} \tag{2.137}$$

which captures the relationship of this model to the Clausius–Mossotti relation.

2.5.2.2 Bruggemann

The Maxwell Garnett formula is not symmetric with respect to the inclusions and the matrix. It specifically applies to spheroids in a homogeneous medium. However, in many situations, it is difficult to identify which material is the inclusion and which is the matrix. In such a case, a slightly different effective medium model can be used that is symmetric between the two phases. This is the Bruggeman model that applies for a two-phase mixture

$$f\frac{\varepsilon - \varepsilon_{\rm av}}{\varepsilon + 2\varepsilon_{\rm av}} = (f-1)\frac{\varepsilon_{\rm m} - \varepsilon_{\rm av}}{\varepsilon_{\rm m} + 2\varepsilon_{\rm av}} \tag{2.138}$$

For dilute systems (of either phase) this gives the same result as Maxwell Garnett to lowest order. This expression is easily extended to multi-component systems by

$$\sum_j f_j \frac{\varepsilon_j - \varepsilon_{\rm av}}{\varepsilon_j + 2\varepsilon_{\rm av}} = 0 \tag{2.139}$$

where the f_j are the individual fractions of the different components.

2.6 Mie Scattering

Many approaches to optical biosensors use nanoparticles and gold and glass beads, which can be large and outside of the Rayleigh limit. When their size approaches an appreciable fraction of a wavelength of light, they enter into the Mie, or resonant scattering, regime. Of special interest are spherical particles and their light-scattering properties.

2.6.1 Spherical Particles

The relationship between the incident fields and the scattered fields is

$$\begin{pmatrix} E_{\|{\rm s}} \\ E_{\perp{\rm s}} \end{pmatrix} = \frac{{\rm e}^{ik(r-z)}}{-ikr}\begin{pmatrix} S_2 & 0 \\ 0 & S_1 \end{pmatrix}\begin{pmatrix} E_{\|{\rm i}} \\ E_{\perp{\rm i}} \end{pmatrix} \tag{2.140}$$

where parallel $\|$ and perpendicular $\perp$ relate to the scattering plane. The scattering matrix elements are

$$S_1 = \sum_n \frac{2n+1}{n(n+1)}(a_n\pi_n + b_n\tau_n) \tag{2.141}$$

$$S_2 = \sum_n \frac{2n+1}{n(n+1)} (a_n \tau_n + b_n \pi_n) \tag{2.142}$$

and the series are terminated after a sufficient number of terms. In the forward direction $S_1(0°) = S_2(0°) = S(0°)$.

The angle-dependent functions are

$$\pi_n = \frac{P_n^1}{\sin\theta} \qquad \tau_n = \frac{\mathrm{d}P_n^1}{\mathrm{d}\theta} \tag{2.143}$$

where

$$P_n^1 = -\frac{\mathrm{d}P_n}{\mathrm{d}\theta} \tag{2.144}$$

is the derivative of the Legendre polynomials P_n.

The Mie coefficients are

$$a_n = \frac{m\psi_n(mx)\psi'_n(x) - \psi_n(x)\psi'_n(mx)}{m\psi_n(mx)\xi'_n(x) - \xi_n(x)\psi'_n(mx)} \tag{2.145}$$

$$b_n = \frac{\psi_n(mx)\psi'_n(x) - m\psi_n(x)\psi'_n(mx)}{\psi_n(mx)\xi'_n(x) - m\xi_n(x)\psi'_n(mx)} \tag{2.146}$$

where $m = n/n_\mathrm{m}$ is the relative refractive index of the particle relative to the surrounding medium, and the functions are defined as

$$\psi_n(\rho) = \rho j_n(\rho) \qquad \xi_n(\rho) = \rho h^{(1)}{}_n(\rho) \tag{2.147}$$

The spherical Bessel functions are

$$j_n(\rho) = \sqrt{\frac{\pi}{2\rho}} J_{n+1/2}(\rho) \qquad y_n(\rho) = \sqrt{\frac{\pi}{2\rho}} Y_{n+1/2}(\rho) \tag{2.148}$$

These are combined into the spherical Hankle functions

$$h_n^{(1)}(\rho) = j_n(\rho) + \mathrm{i}y_n(\rho) \tag{2.149}$$

The scattering cross-section is

$$C_\mathrm{sca} = \frac{2\pi}{k^2} \sum_n (2n+1)(|a_n|^2 + |b_n|^2) \tag{2.150}$$

and the extinction cross-section is

$$C_{\text{ext}} = \frac{2\pi}{k^2} \sum_n (2n+1)\text{Re}(a_n + b_n) \tag{2.151}$$

For a particle that is small relative to the wavelength, the scattering matrix elements are

Rayleigh particle scattering coefficients:

$$\begin{aligned} S_1 &= -\text{i}x^3 \tfrac{m^2-1}{m^2+2} \\ S_2 &= -\text{i}x^3 \tfrac{m^2-1}{m^2+2} \cos\theta \end{aligned} \tag{2.152}$$

where $x = n_{\text{m}}k_0a = k_{\text{m}}a$. These may also be expressed in terms of polarizability α as

$$S_1 = \frac{-ik_{\text{m}}^3\alpha}{4\pi\varepsilon_0} \tag{2.153}$$

using the polarizability

$$\alpha = 4\pi\varepsilon_0 a^3 \frac{\varepsilon_{\text{sph}} - \varepsilon_{\text{m}}}{\varepsilon_{\text{sph}} + 2\varepsilon_{\text{m}}} \tag{2.154}$$

Note that the scattered far-field of the Rayleigh particle is purely real and positive if the particle dielectric constant is purely real. The exciting local field (assume a focused Gaussian beam) is also real, but acquires a $\pi/2$ phase shift upon free-space propagation to the far-field. Therefore, a nonabsorbing Rayleigh particle in a nonabsorbing medium induces a phase shift on the combined scattered and original wave in the far-field.

2.6.2 Effective Refractive Index of a Dilute Plane of Particles

The effective refractive index of a dilute collection of Mie scatterers detected in transmission is obtained by generalizing (2.107) to a collection of particles. This is a mean scattered field approach, and differs in its predictions from effective medium models like Maxwell Garnett. The mean scattered field approach is better suited to describe effective scattering properties of particles on surfaces. The total field in the far-field of a single particle on the optic axis is

$$E(\theta) = \frac{k\sqrt{S_0}}{f}\left[\frac{S(\theta)}{-ik^2} - \text{i}w_0^2 \exp(-k^2w_0^2\theta^2/2)\right] \tag{2.155}$$

When there is an area density of spherical scatterers, the integral over the particle density weighted by the Gaussian incident field is

$$N_A \int_0^\infty e^{-r^2/2w_0^2} 2\pi r dr = 2\pi N_A w_0^2 \tag{2.156}$$

then the total far-field becomes

$$E(\theta) = \frac{k\sqrt{S_0}}{f}\left[2\pi N_A w_0^2 \frac{S(\theta)}{-ik^2} - iw_0^2 \exp\left(-\frac{2\pi^2 w_0^2}{\lambda^2}\theta^2\right)\right] \tag{2.157}$$

The field on the optic axis ($\theta = 0°$) is

$$E(0) = \frac{k\sqrt{S_0}}{f}(-iw_0^2)\left[1 - \frac{2\pi N_A}{k^2} S(0)\right] \tag{2.158}$$

If this is compared with the extinction of a wave in a homogeneous medium with refractive index n_{av}

$$E(0) = E_0 e^{ikd(n_{av}-1)} \approx E_0(1 + ikd(n_{av} - 1)) \tag{2.159}$$

the average refractive index is

$$n_{av} = 1 + i\frac{2\pi \mathrm{N_V}}{k^3} S(0) \tag{2.160}$$

where $\mathrm{N_V} = N_A/d$ is the equivalent volume density. The real and imaginary parts of the effective refractive index $n = n' + i\kappa$ are

$$\begin{aligned} n'_{av} &= 1 - \frac{2\pi \mathrm{N_V}}{k^3} Im\{S(0)\} \\ \kappa_{av} &= \frac{2\pi \mathrm{N_V}}{k^3} Re\{S(0)\} \end{aligned} \tag{2.161}$$

In terms of the volume fraction

$$f_V = \frac{4\pi}{3}\mathrm{N_V} a^3 \tag{2.162}$$

these are

$$\begin{aligned} n'_{av} &= 1 - \frac{3f_V}{2k^3 a^3} \mathrm{Im}\{S(0)\} \\ \kappa_{av} &= \frac{3f_V}{2k^3 a^3} \mathrm{Re}\{S(0)\} \end{aligned} \tag{2.163}$$

In the Rayleigh limit, with

$$S(0) = -ik^3a^3\frac{m^2-1}{m^2+2} \tag{2.164}$$

in a medium with index n_m and relative index $m = n/n_\mathrm{m}$, these are

$$\begin{aligned} n'_\mathrm{av} &= n_\mathrm{m} + \frac{3}{2}f_\mathrm{V}\mathrm{Re}\left\{\frac{m^2-1}{m^2+2}\right\} \\ \kappa_\mathrm{av} &= \frac{3}{2}f_\mathrm{V}\mathrm{Im}\left\{\frac{m^2-1}{m^2+2}\right\} \end{aligned} \tag{2.165}$$

The effective index is linear in the volume fraction f_V in this limit. For denser dispersions, local screening modifies these values and more general approaches must be used, such as generalized Mie theory [3]. Equation (2.165) is consistent with Maxwell Garnett in the dilute limit.

2.7 Nanoparticle Light-Scattering

Nanoparticles have become essential tools for biomedical research. They are used in two main applications: (1) light scattering or emitting labels to image molecular and cellular processes and (2) drug delivery vehicles. In the labeling application, the distinct signature of the light scattering provides specific identification of the nanoparticle density and distribution within a biological sample. In the drug application, light scattering can track the transport and clearing of the drug delivery vehicles.

There are many varieties of biomedical nanoparticles. The simplest are nanobeads that are dielectric spheres that scatter light through Rayleigh or Mie scattering. These beads can be transparent, or dyed with dye molecules (chromophores) having identifiable absorption spectra. Beads also can be fluorescent emitters that are detected using fluorescence interference microscopy. Among the brightest nanoparticle emitters are semiconductor quantum dots. The emission wavelengths of quantum dots are tuned by controlling the size of the semiconductor particle using quantum confinement effects to shift the electron quantum states. Gold and silver nanoparticles scatter light resonantly (and hence strongly) through surface plasmon oscillations that are excited by the incident light fields. The optical properties of these nanoparticles are tuned by selecting a wide variety of sizes and shapes, from nanorods to nanostars.

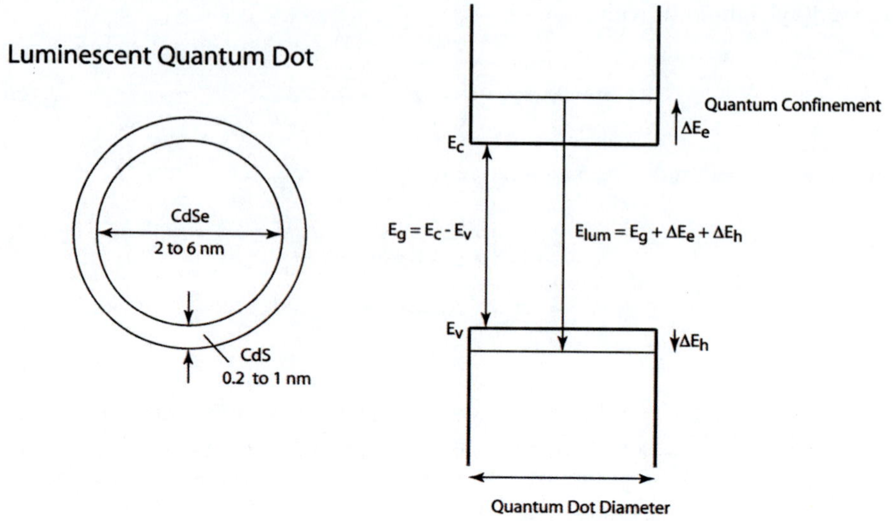

Fig. 2.15 Luminescent quantum dot structure. A CdSe nanocrystal is encased in a CdS shell to prevent nonradiative surface recombination. The photon energy of the luminescence is determined by the bandgap and by quantum confinement of the electrons in the conduction band and the holes in the valence band

2.7.1 *Quantum Dots*

Quantum dots are semiconductor nanocrystals typically 2–10 nm in diameter. Semiconductors have distinct and strong optical properties because of the bandgap between valence states occupied by holes and conduction states occupied by electrons. When an electron in the conduction band falls across the bandgap to fill a hole in the valence band, light is emitted with a photon energy equal to the energy that the electron loses in the process. For large nanocrystals, the energy is approximately equal to the bandgap between the top of the valence band and the bottom of the conduction band

$$\hbar\omega_g = \frac{hc}{\lambda_g} = E_g = E_c - E_v \tag{2.166}$$

One of the important properties of quantum dots is the size dependence of the emission wavelength. This is caused by quantum confinement effects on the electrons and holes. In quantum mechanics, when a particle is confined to a finite volume, there is a quantum energy associated with the confinement. The stronger the confinement, the larger the confinement energy. Therefore, as the quantum dots get smaller, the emission wavelength shifts to the blue (higher energy).

The structure of a luminescent quantum dot is illustrated in Fig. 2.15 for a CdSe nanocrystal encased in a CdS shell. The shell reduces surface recombination that

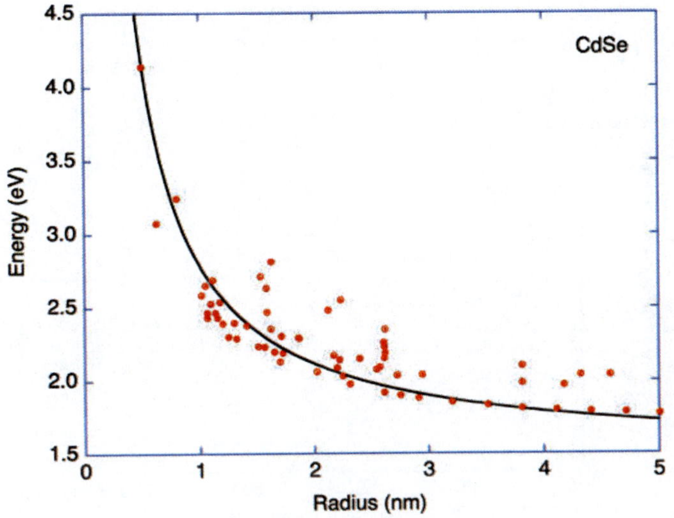

Fig. 2.16 CdSe luminescence energy from assorted experiments as a function of the quantum dot radius. Reproduced with permission from [5]

lowers the luminescence efficiency of the quantum dot. The emitted photon energy is determined by the bandgap of the nanocrystal plus the quantum confinement.

For a spherical quantum dot of radius R, the emission energy of the photon is

$$E \approx E_g + \frac{\hbar^2 \pi^2}{2R^2}\left(\frac{1}{m_e^*} + \frac{1}{m_h^*}\right) - \frac{1.786e^2}{4\pi\varepsilon_0 \varepsilon R} - 0.248E_{Ry} \tag{2.167}$$

where the first term is the crystalline bandgap, the second term is the quantum confinement energy for the electrons and holes, and the third and fourth terms are a binding energy between the electron and hole caused by the Coulomb interaction between the particles [4]. The bound state of the electron and the hole is called an exciton, and the binding energy is the solid-state analog of the hydrogen atom. In (2.167) m_e^* is the electron effective mass, m_h^* is the hole effective mass, ε is the dielectric constant, and E_{Ry} is the Rydberg energy of the exciton. The emission energy of CdSe quantum dots as a function of radius is shown in Fig. 2.16 compared with the numerical result of (2.167). Examples of absorption and emission for CdSe quantum dots are shown in Fig. 2.17.

2.7.2 *Gold and Silver Nanoparticles*

Gold and silver nanoparticles are among the most commonly used particles in labeling applications of light scattering and interferometry. These particles show

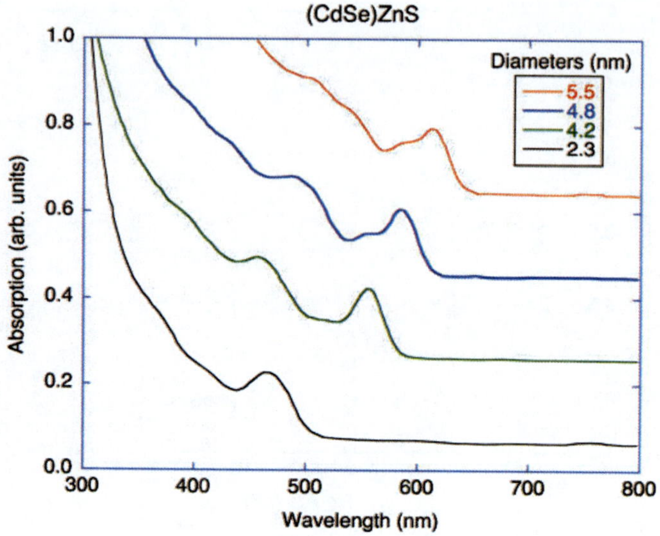

Fig. 2.17 Absorbance of CdSe quantum dots clad with ZnS as a function of wavelength for increasing diameters of (**a**) 2.30, (**b**) 4.20, (**c**) 4.80, and (**d**) 5.50 nm. (Reproduced with permission from [6])

strong plasmonic resonances with plasmonically enhanced absorption and light scattering. In the Rayleigh limit, the particles contribute to enhanced effective-medium properties, while in the large Mie-scattering limit, they provide single-particle detectability and finite-particle statistics in applications such as imaging and laser scanning.

The plasmonic enhancements of gold and silver arise from the nearly free-electron behavior of these noble metals. The dielectric functions for a free-electron gas, including the bulk dielectric function $\varepsilon_b = 6 + i1.6$, is

$$\varepsilon_1 = \mathrm{Re}(\varepsilon_b) - \frac{\omega_p^2}{\omega^2 + \gamma^2}$$
$$\varepsilon_2 = \mathrm{Im}(\varepsilon_b) + i\omega_p^2 \frac{\gamma/\omega}{\omega^2 + \gamma^2} \tag{2.168}$$

For gold, the plasmon energy is approximately $\hbar\omega_p = 7.5\,\mathrm{eV}$, and the damping rate is approximately $\hbar\gamma = 0.25\,\mathrm{eV}$. The free-electron functions for gold are shown in Fig. 2.18. The real part of the dielectric function has a zero near 400 nm. This is not near $\omega_p = 7.5$ eV at a wavelength of 165 nm because of the background dielectric function arising from interband transitions among the gold orbitals. There is also a significant imaginary component caused by damping (scattering) of the free electrons and from the interband transitions.

Free-electron models capture the rough behavior of real noble metals, but fail to capture the contributions from interband absorption and the associated effects on

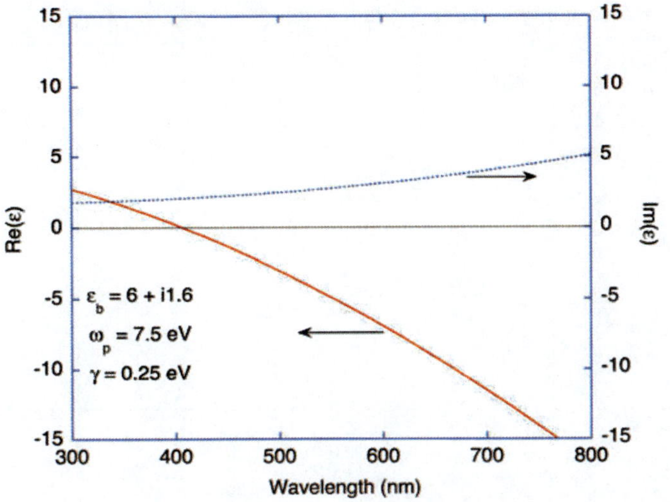

Fig. 2.18 Real and imaginary components of the dielectric functions of gold assuming a free-electron model

the refractive index. A more accurate parameterization for gold has been provided in [7], including the role of the finite size of gold particles. As the particle size decreases, the electron scattering by the surface increases, and the overall damping of the plasmon resonance increases. This is parameterized as

$$\gamma_p(d) = \frac{1}{1/\gamma_p(\infty) + 1/Ad} \tag{2.169}$$

where A is a scattering parameter and d is the particle diameter in nanometers. The total dielectric function can be expressed as

$$\begin{aligned} \varepsilon(\lambda) = \varepsilon_\infty &- \frac{1}{\lambda_p^2(1/\lambda^2 + i/\gamma_p\lambda)} \\ &+ \sum_{i=1,2} \frac{A_i}{\lambda_i}\left[\frac{e^{i\phi_1}}{1/\lambda_i - 1/\lambda - i/\gamma_i} + \frac{e^{-i\phi_1}}{1/\lambda_i + 1/\lambda + i/\gamma_i}\right] \end{aligned} \tag{2.170}$$

The first term is the plasmon contribution, while the second term is the contribution from interband absorption. The parameters that best fit the data in [8] are given in Table 2.3.

The real and the imaginary parts of the dielectric functions are plotted in Fig. 2.19. Strong finite-size effects on the imaginary part occur for particle radii smaller than 40 nm. However, the real part is only significantly affected for radii smaller than 10 nm.

Table 2.3 Parameters in the parameterization of the gold dielectric function by Etchegoin [7]

Data parameters	Value
ε_∞	1.53
λ_p	145 nm
$\gamma_p(\infty)$	17,000 nm
A	170
A_1	0.94
ϕ_1	$-\pi/4$
λ_1	468 nm
γ_1	2,300 nm
A_2	1.36
ϕ_2	$-\pi/4$
λ_2	331 nm
γ_2	940 nm

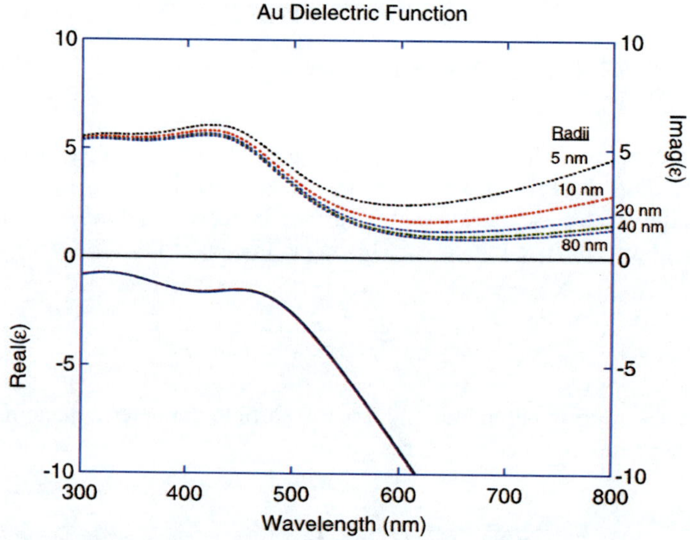

Fig. 2.19 Dielectric function of gold parameterized [7] and fit to the data [8]. The surface plasmon resonance of a spherical nanoparticle in vacuum occurs when $\mathrm{Re}(\varepsilon) = -2$

The dipole moment of a sphere was given in (2.93). When the sphere is embedded in a medium with dielectric constant ε_m, it is

$$\mathbf{p} = 4\pi\varepsilon_m\varepsilon_0\left(\frac{\varepsilon - \varepsilon_m}{\varepsilon + 2\varepsilon_m}\right)a^3\mathbf{E}_{inc} \tag{2.171}$$

Therefore, when $\varepsilon_{gold} = -2\varepsilon_m$ a spherical nanoparticle has a plasma resonance that occurs in both the absorption and the scattering cross-sections. The absorption and scattering efficiencies of gold nanoparticles are calculated numerically using

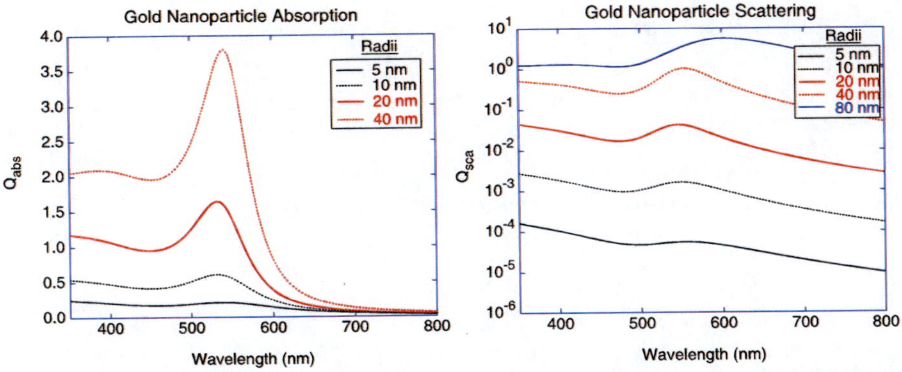

Fig. 2.20 Numerical simulations of the absorption and scattering efficiencies of gold nanoparticles calculated using Mie theory and the gold dielectric function of (2.170)

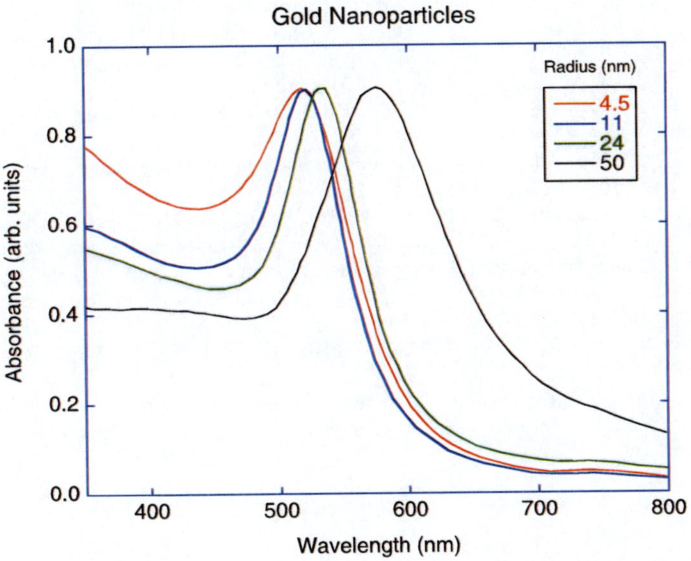

Fig. 2.21 Experimental absorbance of gold nanoparticles in solution reprinted with permission from [9] as a function of particle radius

Mie theory and (2.170) and shown in Fig. 2.20. A particle with a radius of 40 nm shows a strong absorption resonance with enhanced scattering. Smaller particles have broader resonances because of increased surface scattering of the free electrons, and smaller efficiencies because they move into the Rayleigh limit. The particles begin to move out of the Raleigh limit and into the Mie regime as the particle radii become larger than about 50 nm. Experimental normalized absorbance of gold nanoparticles in solution is shown in Fig. 2.21 from [9] for diameters from 9 to 99 nm. The resonance shifts to longer wavelengths with increasing size.

Selected Bibliography

van de Hulst, H.C.: Light Scattering by Small Particles. Dover, New York (1981) (The old classic on light scattering)

Bohren, C.F., Huffman, D.R.: Absorption and Scattering of Light by Small Particles. Wiley, New York (1998) (This is more up-to-date, and includes computer codes for scattering)

Berne, B.J., Pecora, R.: Dynamic Light Scattering: With Applications to Chemistry, Biology, and Physics. Dover, New York (2000) (This is a treasure trove of dynamic light scattering, but with many older references)

References

1. Zhao, M., Cho, W., Regnier, F., Nolte, D.: Differential phase-contrast BioCD biosensor. Appl. Opt. **46**, 6196–6209 (2007)
2. Felidj, N., Aubard, J., Levi, G.: Discrete dipole approximation for ultraviolet-visible extinction spectra simulation of silver and gold colloids. J. Chem. Phys. **111**(3), 1195–1208 (1999)
3. Gouesbet, G., Grehan, G.: Generalized Lorenz–Mie theory for assemblies of spheres and aggregates. J. Opt. A: Pure Appl. Opt. **1**(6), 706–712 (1999)
4. Brus, L.: Electronic wave-functions in semiconductor clusters – experiment and theory. J. Phys. Chem. **90**(12), 2555–2560 (1986)
5. Tomasulo, A., Ramakrishna, M.V.: Quantum confinement effects in semiconductor clusters II. J. Chem. Phys. **105**(9), 3612–3626 (1996)
6. Dabbousi, B.O., RodriguezViejo, J., Mikulec, F.V., Heine, J.R., Mattoussi, H., Ober, R., Jensen, K.F., Bawendi, M.G.: (CdSe)ZnS core-shell quantum dots: synthesis and characterization of a size series of highly luminescent nanocrystallites. J. Phys. Chem. B **101**(46), 9463–9475 (1997)
7. Etchegoin, P.G., Le Ru, E.C., Meyer, M.: An analytic model for the optical properties of gold, vol. 125, p. 164705, 2006
8. Johnson, P.B., Christy, R.W.: Optical-constants of noble-metals. Phys. Rev. B **6**(12), 4370–4379 (1972)
9. Link, S., El-Sayed, M.A.: Spectral properties and relaxation dynamics of surface plasmon electronic oscillations in gold and silver nanodots and nanorods. J. Phys. Chem. B **103**(40), 8410–8426 (1999)

Chapter 3
Speckle and Spatial Coherence

Speckle is a ubiquitous phenomenon that arises during any quasi-monochromatic interrogation of biological specimens, whether by spatial scanning of focused beams or broad-field illumination, and whether using laser sources or low-coherence illumination. It is a direct manifestation of spatial interference among multiple waves that have been scattered or refracted from a complex target sample. Speckle is detrimental to high-resolution optical imaging, and many techniques are applied to mitigate its effects. However, speckle is a form of self interference and hence is intrinsically an interferometric phenomenon, and due to its interferometric sensitivity, speckle carries information from the smallest scales of the target (organelles, membranes, vesicles, etc.) to provide statistical information on subcellular structure and dynamics.

Dynamic speckle is speckle that is changing in time due to an active target. If speckle is strongly fluctuating, it provides a means for natural speckle reduction simply by time averaging, which makes dynamic speckle useful for its own reduction [1]. But dynamic speckle can be used more directly. For instance, dynamic speckle can be used to monitor blood flow [2–10] and can assess the health of living tissue [11–14].

This chapter provides the mathematical and physical foundations for understanding the origins of speckle. Speckle is part of statistical optics, which provides a formalism for identifying what information can be extracted from a speckle field. A quantity known as speckle contrast is particularly important in biological and biomedical imaging applications because it can be used to quantify fluctuating properties of biological systems. Speckle holography is a powerful technique for selecting and studying subsets of random speckle fields, such as by gating information arising from a selected depth inside biological tissue by using low coherence.

Speckle is directly related to concepts of spatial coherence. Spatial coherence causes the properties of the optical field to be correlated over characteristic length scales. A key question about speckle is how much information it can provide about a sample. In some cases it provides *no* information: fully developed speckle carries no structural information and cannot be used for imaging to resolve details of the target.

D.D. Nolte, *Optical Interferometry for Biology and Medicine*, Bioanalysis 1,
DOI 10.1007/978-1-4614-0890-1_3, © Springer Science+Business Media, LLC 2012

However, in cases when the speckle is not fully developed, and the spatial statistics drift (nonstationary statistics), speckle can provide some spatial information. Alternatively, even fully developed speckle can have strong temporal fluctuations that provide significant dynamic information about the activity of the target.

3.1 Random Fields

In the analysis of the statistical properties of speckle, it is helpful to consider a random configuration of N discrete random sources that contribute to a total field. Each specific configuration of N sources is called an instance, and a group of many of these instances is called an ensemble. Each instance is randomly configured within the statistical constraints of the physical system. The goal is to find how the configuration-to-configuration variability depends on the number N of scatterers. Many of the physical trends of speckle, and how speckle behaves in biological imaging systems, can be uncovered in this way. This discrete approach is also the starting point for a continuum approach to speckle, which is applicable to large probe or coherence volumes.

The total field from N discrete sources (Fig. 3.1) is

$$E_{\mathrm{T}} = \sum_{i=1}^{N} E_i \mathrm{e}^{\mathrm{i}\phi_1} \tag{3.1}$$

in which the amplitudes E_i of the fields are taken to be real-valued, and the phases ϕ_i span the unit circle modulo 2π. Each of these are stochastic variables that are drawn from a statistical probability distribution. This representation can be understood as a

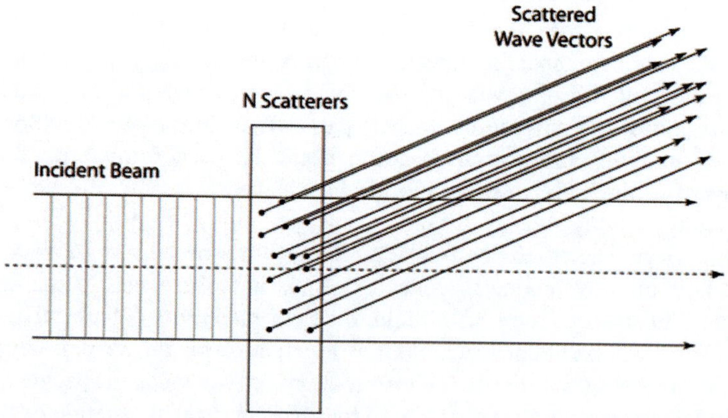

Fig. 3.1 A configuration of N points scattering partial waves to the far field. A configuration of N sources is one instance of an ensemble of many random configurations of N scatterers

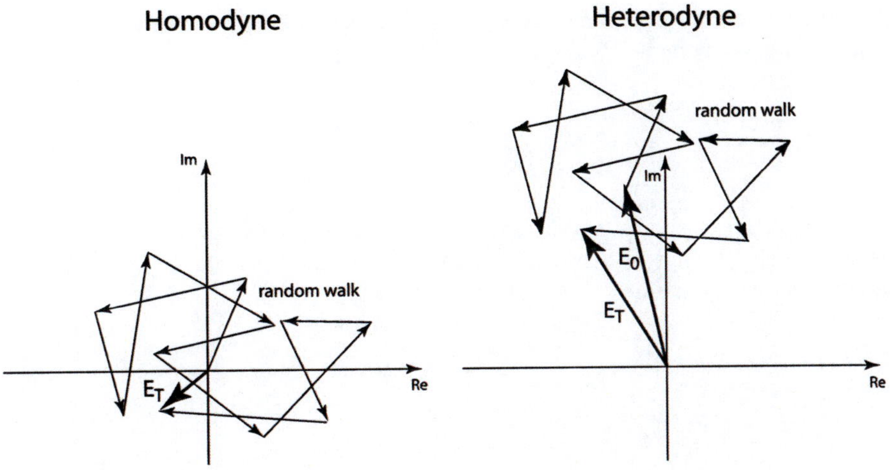

$$\langle E_T \rangle_{ensemble} = 0 \qquad \langle E_T \rangle_{ensemble} = E_0$$

$$\langle E_T^2 \rangle_{ensemble} = N|\delta E|^2 \qquad \langle E_T^2 \rangle_{ensemble} = |E_0|^2 + N|\delta E|^2$$

Fig. 3.2 A random walk of phasors on the complex plane define a Gaussian diffusion process. The homodyne condition occurs when there is no background or reference wave. The heterodyne condition adds a constant reference phasor E_0 to the diffuse field

random walk in the complex plane, as shown in Fig. 3.2. Each source contributes a phasor of length E_i and orientation defined by ϕ_i. In the special case when the phasor length obeys a Gaussian probability distribution with variance δE^2, and when the phase is uniformly distributed on 2π, this process describes Gaussian diffusion. The mean value of the field across an ensemble is

$$\langle E_{\mathrm{T}} \rangle \equiv 0 \tag{3.2}$$

because of the completely random orientations of the phase angles ϕ_i. However, the variance in the field is nonzero and depends on the number of particles and on the variance of the probability distribution.

When only the scattered fields are considered (without a reference wave), this corresponds to a homodyne condition. The intensity from a single instance of N sources is

$$\begin{aligned} I_{\mathrm{Hom}} &= |E_{\mathrm{Hom}}|^2 = \left| \sum_{i=1}^{N} E_i \mathrm{e}^{\mathrm{i}\phi_1} \right|^2 \\ &= \sum_{i=1}^{N} |E_i|^2 + \sum_{i \neq j} E_i E_j \cos(\phi_i - \phi_j) \end{aligned} \tag{3.3}$$

In the ensemble average, the second term averages to zero because of the random phases in the exponent. Therefore,

$$\langle I_{\mathrm{Hom}} \rangle = \sum_{i=1}^{N} |E_i|^2 = N\delta E^2 \tag{3.4}$$

This simple result shows that the average intensity depends linearly on the number N of sources, which is the result of an *incoherent* sum. The configuration-to-configuration fluctuations in the average intensity are

$$\Delta I^2 = \langle I_{\mathrm{Hom}}^2 \rangle - \langle I_{\mathrm{Hom}} \rangle^2 \tag{3.5}$$

The expression for I_{Hom}^2 is

$$\begin{aligned} I_{\mathrm{Hom}}^2 &= \left(\sum_{i=1}^{N} I_i + \sum_{i \neq j} E_i E_j \cos(\phi_i - \phi_j) \right)^2 \\ &= \sum_{i,j=1}^{N} I_i I_j + \text{cross terms} \end{aligned}$$

$$\langle I_{\mathrm{Hom}}^2 \rangle = 2N^2 \delta E^4 \tag{3.6}$$

because the cross-terms average to zero and the factor of 2 comes from adding two random Gaussian distributions. The variance of the intensity is

$$\begin{aligned} \Delta I^2 &= 2N^2 \delta E^4 - N^2 \delta E^4 \\ &= N^2 \delta E^4 \\ &= \langle I_{\mathrm{Hom}} \rangle^2 \end{aligned} \tag{3.7}$$

This is the important result that, for a Gaussian diffusion process on the complex plane, the fluctuations in the homodyne intensity are equal to the average intensity

$$\Delta I_{\mathrm{Hom}} = \langle I_{\mathrm{Hom}} \rangle \tag{3.8}$$

These results are easily extended to the case when the random fields are added to a constant (reference) field for a heterodyne condition

$$E_{\mathrm{Het}} = E_0 \mathrm{e}^{\mathrm{i}\phi_0} + \sum_{i=1}^{N} E_i \mathrm{e}^{\mathrm{i}\phi_1} \tag{3.9}$$

where ϕ_0 is the reference phase. The intensity of the configuration is

$$
\begin{aligned}
I_{\text{Het}} = |E_{\text{Het}}|^2 &= \left| E_0 e^{i\phi_0} + \sum_{i=1}^{N} E_i e^{i\phi_1} \right|^2 \\
&= I_0 + 2E_0 \sum_{i=1}^{N} E_i \cos(\phi_1 - \phi_0) + \sum_{i=1}^{N} E_i^2 + 2\sum_{i \neq j} E_i E_j \cos(\phi_i - \phi_j)
\end{aligned}
\quad (3.10)
$$

The ensemble average is

$$
\langle I_{\text{Het}} \rangle = I_0 + N\delta E^2 = I_0 + \langle I_{\text{Hom}} \rangle \quad (3.11)
$$

When the magnitude E_0 is much larger than δE then

$$
\begin{aligned}
\Delta I_{\text{Het}}^2 &= I_0^2 + 2NE_0^2 \delta E^2 - I_0^2 \\
&= 2NI_0 \delta E^2
\end{aligned}
\quad (3.12)
$$

for which the heterodyne intensity fluctuations are

$$
\Delta I_{\text{Het}} = \sqrt{2NI_0}\delta E \quad (3.13)
$$

In this case with a strong reference wave, the intensity fluctuations scale as the square root of the number of sources.

3.2 Dynamic Light Scattering (DLS)

Any motion of (or in) a scattering target causes dynamic light scattering (DLS). This motion can be uniform, such as directed drift of organelles along cytoskeletal tracks in cells, random, such as Brownian motion of small particles, and active, such as the undulations of cell membranes driven by the cytoskeleton. The motions of the target components produce time-varying changes in the phase of the scattered partial waves. Because scattering targets are composed of many components, each of which may be executing uniform or random motions, the phase drift of all the partial waves causes time-varying interferences that lead to intensity fluctuations.

The goal of DLS experiments is to separate the fluctuating light intensities into contributions from: (1) different types of scattering objects within the target (mitochondria, nucleus, membrane, etc.), and (2) separate types of motion (drift, diffusion or active). This separation is made possible, in part, by the angular dependence of light scattering that depends on the sizes and shapes of the scattering components, and in part by the different fluctuation spectral signatures of random vs. directed motions.

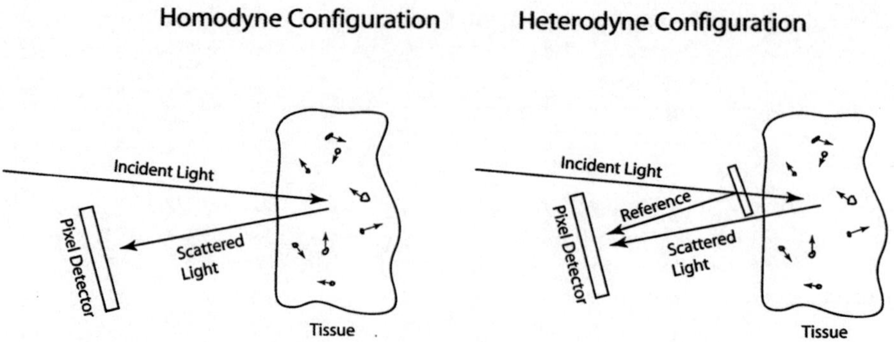

Fig. 3.3 Homodyne vs. heterodyne detection configurations for dynamic light scattering. For short-coherence interferometry, the reference must be path-matched to the signal light

DLS is caused by the change in the optical phase of the scattered light as the particle moves. For instance, the field scattered from a single moving particle is

$$E(t) = E_s \exp(-\mathrm{i}(\mathbf{k}_f - \mathbf{k}_i) \cdot \mathbf{r}(t)) \tag{3.14}$$

As the position $r(t)$ of the particle changes, the phase of the scattered light changes. The difference in phase from one position to the next is

$$\begin{aligned} \Delta\phi &= -(\mathbf{k}_f - \mathbf{k}_i) \cdot (\mathbf{r}(t_2) - \mathbf{r}(t_1)) \\ &= -(\mathbf{k}_f - \mathbf{k}_i) \cdot \Delta\mathbf{r} \\ &= -\mathbf{q} \cdot \Delta\mathbf{r} \end{aligned} \tag{3.15}$$

where $\mathbf{q} = (\mathbf{k}_f - \mathbf{k}_i)$ is the scattering vector. The direction of the initial k-vector defines $\theta = 0$, and the magnitude of q for a scattering angle θ is

$$q = 2k \sin\frac{\theta}{2} = k\sqrt{2(1 - \cos\,\theta)} \tag{3.16}$$

For a collection of N particles, the fluctuating scattered field is

$$E(t) = \sum_{n=1}^{N} E_n^s \exp(-\mathrm{i}\mathbf{q} \cdot \mathbf{r}_n(0)) \exp(-\mathrm{i}\mathbf{q} \cdot \Delta\mathbf{r}_n(t)) \tag{3.17}$$

where $r_n(0)$ sets the instantaneous phase and $\Delta r(t)$ are the fluctuations.

As for static speckle, the two general scattering configurations for DLS are homodyne and heterodyne configurations. In homodyne, only the scattered light is detected, without any contribution from the original unscattered illumination field. In heterodyne detection, part of the original illumination field interferes with the scattered light. These configurations are shown in Fig. 3.3. This distinction between heterodyne and homodyne is important, because it is the difference between detecting field-based fluctuations vs. intensity-based fluctuations, respectively. Because of the squared relationship between field and intensity, the statistics of homodyne detection are different than the statistics of heterodyne detection.

3.2.1 Heterodyne: Field-Based Detection

In field-based detection, the mixing of the scattered field with a reference field produces the net field

$$E(t) = E_0 + \sum_{n=1}^{N} E_n \exp(-iqx_n(0)) \exp(-iq\Delta x_n(t)) \tag{3.18}$$

where the fields are assumed to be detected through a polarizer, and the coordinate x is defined parallel to the scattering vector q. Time autocorrelation functions provide a direct way to characterize the properties of dynamic speckle. To construct the time autocorrelation function, the product of the field with itself is defined by

$$\begin{aligned} E^*(0)E(t) = & \left[E_0 + \sum_{n=1}^{N} E_n \exp(iqx_n(0))\right] \\ & \times \left[E_0 + \sum_{n=1}^{N} E_n \exp(-iqx_n(0)) \exp(-iq\Delta x_n(t))\right] \\ = & E_0^2 + E_0 \sum_{n=1}^{N} E_n \exp(iqx_n(0)) + E_0 \sum_{n=1}^{N} E_n \exp(-iqx_n(0)) \exp(-iq\Delta x_n(t)) \\ & + \sum_{n=1}^{N} E_n^2 \exp(-iq\Delta x_n(t)) + \text{cross terms} \end{aligned} \tag{3.19}$$

If the particles are randomly positioned, then in the limit of large N the random initial phases average to zero to yield

$$E^*(0)E(t) = E_0^2 + \sum_{n=1}^{N} E_n^2 \exp(-iq\Delta x_n(t)) \tag{3.20}$$

The autocorrelation is obtained by taking an ensemble average of this quantity, or by integrating over all time. (Ensemble averages and time averages are equivalent under stationary statistics.) The stochastic sum is evaluated using a probability distribution as

$$\langle E^*(0)E(t)\rangle = \sum_n P(\Delta x_n(t))E^*(0)E(t) \tag{3.21}$$

where $P(\Delta x_i(t))$ is the probability functional of the displacement Δx. This probability functional is determined by the physical process, such as diffusion or drift or active transport. The autocorrelation is

$$\langle E^*(0)E(t)\rangle = E_0^2 + \sum_{n=1}^{N} E_n^2 P(\Delta x_n(t)) \exp(-iq\Delta x_n(t)) \tag{3.22}$$

and the stochastic sum is equivalent to an integral over the probability distribution

$$\begin{aligned}\langle E^*(0)E(t)\rangle &= E_0^2 + NI_s \int_{-\infty}^{\infty} P(\Delta x) \exp(-iq\Delta x) \\ &= I_0 + NI_s \,\mathrm{FT}(P(\Delta x))\end{aligned} \tag{3.23}$$

where the autocorrelation is proportional to the Fourier transform of the probability functional.

Different types of motion have different probability functionals:

Diffusion: For diffusion, the probability functional is

$$\begin{aligned}P_\mathrm{D}(\Delta x_i(t)) &= \frac{1}{\sqrt{4\pi Dt}} \exp(-(\Delta x_i(t))^2/(4Dt)) \\ &= \frac{1}{\sqrt{2\pi\Delta x_\mathrm{rms}^2}} \exp(-(\Delta x_i(t))^2/2(\Delta x_\mathrm{rms}^2))\end{aligned} \tag{3.24}$$

which gives the diffusion autocorrelation function

$$A_\mathrm{D}^\mathrm{E}(\tau) = \langle E^*(t)E(t+\tau)\rangle - I_0 = NI_s \exp(-q^2 D\tau) \tag{3.25}$$

RMS velocity: For a Gaussian distribution of velocities, the probability functional is

$$P_\mathrm{v}(\Delta x_i) = \frac{1}{\sqrt{2\pi}v_\mathrm{rms}} \exp(-v_i^2/2v_\mathrm{rms}^2) = \frac{1}{\sqrt{2\pi}v_\mathrm{rms}} \exp(-\Delta x_i{}^2/2v_\mathrm{rms}^2 t^2) \tag{3.26}$$

which gives the autocorrelation function

$$A_\mathrm{v}^\mathrm{E}(\tau) = NI_s \exp(-q^2 v_\mathrm{rms}^2 \tau^2/2) \tag{3.27}$$

Drift: For an average drift in a specific direction, the probability functional is

$$P_\mathrm{v}(\Delta x_i) = \frac{1}{\sqrt{2\pi}v_\mathrm{rms}} \exp(-(v_i - v_0)^2/2v_\mathrm{rms}^2) \tag{3.28}$$

which gives the drift autocorrelation function

$$A_{\mathrm{v}}^{\mathrm{E}}(\tau) = NI_{\mathrm{s}} \exp(-q^2 v_{\mathrm{rms}}^2 \tau^2/2) \cos(qv_0\tau) \tag{3.29}$$

where the cosine term is equivalent to a Doppler beat frequency. For all three cases the motion is co-parallel to the scattering vector q, which gives all these dynamic-light scattering correlation functions one-dimensional character, even though the actual motion is in three dimensions. This is because the light scattering is only sensitive to longitudinal motions and not transverse.

3.2.2 Homodyne: Intensity-Based Detection

For homodyne detection, there is no reference field, in which case the field is

$$E(t) = \sum_{n=1}^{N} E_{\mathrm{s}} \exp(-iqx_n(0)) \exp(-iq\Delta x_n(t)) \tag{3.30}$$

and the intensity is

$$\begin{aligned} I(t) = E^*E &= \left| \sum_{n=1}^{N} E_{\mathrm{s}} \exp(-iqx_n(0)) \exp(-iq\Delta x_n(t)) \right|^2 \\ &= NI_{\mathrm{s}} + I_{\mathrm{s}} \sum_{i \neq j} \exp(-iqx_i(0)) \exp(-iq\Delta x_i(t)) \exp(iqx_j(0)) \exp(iq\Delta x_j(t)) \end{aligned} \tag{3.31}$$

The intensity autocorrelation is constructed from the product of the intensity with itself, neglecting terms of random phase in the large N limit, to be

$$\begin{aligned} I^*(0)I(t) &= \left[NI_{\mathrm{s}} + I_{\mathrm{s}} \sum_{i \neq j} \exp(iqx_i(0)) \exp(-iqx_j(0)) \right] \\ &\quad \times \left[NI_{\mathrm{s}} + I_{\mathrm{s}} \sum_{i \neq j} \exp(-iqx_i(0)) \exp(-iq\Delta x_i(t)) \exp(iqx_j(0)) \exp(iq\Delta x_j(t)) \right] \\ &= N^2 I_{\mathrm{s}}^2 + NI_{\mathrm{s}}^2 \sum_{i \neq j} \exp(-iq\Delta x_i(t)) \exp(iq\Delta x_j(t)) \end{aligned} \tag{3.32}$$

and averaged over the Gaussian joint probability distribution to give

$$\begin{aligned} \langle I^*(0)I(t) \rangle &= N^2 I_{\mathrm{s}}^2 + NI_{\mathrm{s}}^2 \sum_{i \neq j} P_{ij} \mathrm{e}^{iq(\Delta x_i(t))} \mathrm{e}^{-iq(\Delta x_j(t))} \\ &= N^2 I_{\mathrm{s}}^2 [1 + \exp(-q^2 2\Delta x(t)^2)] \end{aligned} \tag{3.33}$$

This is a general expression for a Gaussian random process, and is related to the heterodyne autocorrelation function through the Siegert relation (for single-mode detection)

$$A^{\mathrm{I}}(\tau) = N^2 I_{\mathrm{s}}^2 + |A^{\mathrm{E}}(\tau)|^2 \tag{3.34}$$

The homodyne results are therefore

Diffusion:

$$\langle I^*(0)I(t)\rangle = N^2 I_{\mathrm{s}}^2 [1 + \exp(-q^2 2Dt)] \tag{3.35}$$

Maxwell velocity:

$$\langle I^*(0)I(t)\rangle = N^2 I_{\mathrm{s}}^2 [1 + \exp(-q^2 v_{\mathrm{rms}}^2 t^2)] \tag{3.36}$$

Drift:

$$\langle I^*(0)I(t)\rangle = N^2 I_{\mathrm{s}}^2 [1 + \exp(-q^2 v_{\mathrm{rms}}^2 t^2)\cos^2(qv_0 t)] \tag{3.37}$$

These simple expressions for the homodyne and heterodyne autocorrelation functions each hold only under specific conditions. In all cases, the scattering density is assumed to be dilute and without multiple scattering. The effects of multiple scattering on DLS are discussed in Chap. 10 on the optical properties of tissues which are optically thick. In addition, each expression is for a uniform type of motion, but real biological systems are always a mixture of many types of scatterers and many types of motion. For these reasons, the experimental autocorrelation functions extracted from living systems tend to have less distinct signatures.

3.2.3 Fluctuation Power Spectra: Wiener-Khinchin Theorem

The Wiener-Khinchin theorem makes a direction connection between autocorrelation functions and spectral power density. The Wiener-Khinchin theorem states

$$S(\omega) = F(\omega)F^*(\omega) = \mathrm{FT}\left(\int_{-\infty}^{\infty} f(t)f(t+\tau)\mathrm{d}t\right) = \mathrm{FT}(A(\tau)) \tag{3.38}$$

Therefore, any information contained within the autocorrelation function is also contained within the spectral power density. Because many random processes have exponentially decaying correlations without specific frequency content, autocorrelation functions provide a natural way to characterize a dissipative system without any additional need to consider the fluctuation power spectrum. But fluctuation spectra do provide a different perspective on fluctuating systems. For instance,

Table 3.1 Autocorrelations and spectral densities for DLS

	$g^{(1)}(\tau)$	$g^{(2)}(\tau)$	$S^{(1)}(\omega)$
Diffusion	$e^{-q^2 D\tau}$	$1 + e^{-q^2 2D\tau}$	$\frac{1}{\pi}\left(\frac{1}{1+\omega^2/q^2 D}\right)$
Maxwell velocity	$\exp(-q^2 v_{rms}^2 \tau^2/2)$	$1 + \exp(-q^2 v_{rms}^2 \tau^2)$	$\frac{1}{2\pi}\exp(-\omega^2/2q^2 v_{rms}^2)$
Maxwell with drift	$\exp(-q^2 v_{rms}^2 \tau^2/2)$ $\cos(qv_0\tau)$	$1 + \exp(-q^2 v_{rms}^2 \tau^2)$ $\cos^2(qv_0\tau)$	$\frac{1}{2\pi}\exp(-(\omega - qv_0)^2/2q^2 v_{rms}^2)$

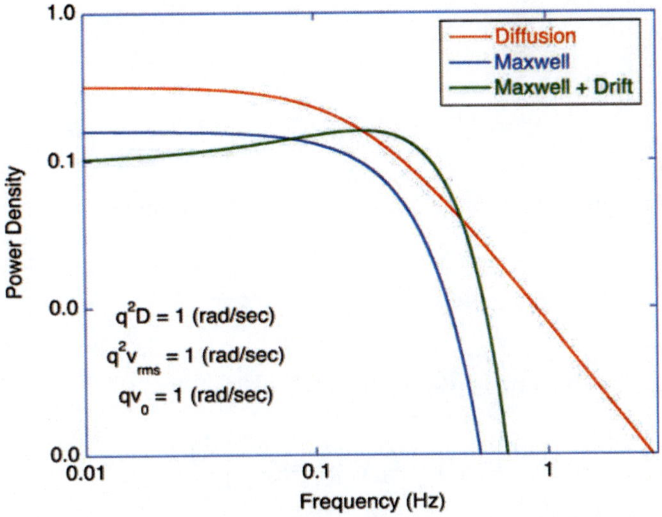

Fig. 3.4 Spectral power densities for diffusion, Maxwell velocity and Maxwell plus drift. Diffusion is characterized by a $1/\omega^2$ roll-off with a knee frequency, while the Maxwell densities have a sharp drop-off. The peak in the drift curve is at the Doppler frequency

spectral analysis of complex systems that have many simultaneous processes and several characteristic times sometimes can help isolate different contributions. And for the specific case of systems with drift motions, the Doppler frequency is captured directly in the spectrum. For these reasons, fluctuation spectra are useful tools for studying biological systems, and can be used to isolate different types of motion in living tissue, as in holographic tissue dynamics spectroscopy (TDS) described in Chap. 12. A summary of the autocorrelation functions and power spectra for diffusion and drift is given in Table 3.1. The characteristic shapes of the spectral power density are shown in Fig. 3.4 for heterodyne spectra. The diffusion spectral shape has a $1/\omega^2$ roll-off above the knee frequency. The Maxwell distribution has a much sharper drop-off. The peak in the drift curve is at the Doppler frequency.

3.3 Statistical Optics

Speckle properties can be obtained more generally by using principles of statistical optics. At the heart of statistical methods are probability distributions. For instance $p_I(I)$ represents the probability of observing an intensity I in a speckle intensity pattern, while $p_E(E)$ is the probability to detect the field amplitude E (assumed here to be a real-valued positive amplitude). These probability distributions are related through the basic relation

$$I = |E|^2 \tag{3.39}$$

from which

$$p_I(I) = p_E(E)\left|\frac{dE}{dI}\right| = \frac{1}{2\sqrt{I}}p_E(\sqrt{I}) \tag{3.40}$$

For instance, consider a Gaussian-distributed field amplitude

$$p_E(E) = \frac{E}{\sigma_E^2}\exp\left(-\frac{E^2}{2\sigma_E^2}\right) \tag{3.41}$$

When this is inserted into (3.40) the intensity distribution is

$$\begin{aligned} p_I(I) &= \frac{\sqrt{I}}{\sigma_E^2}\exp\left(-\frac{I}{2\sigma_E^2}\right)\frac{1}{2\sqrt{I}} \\ &= \frac{1}{2\sigma_E^2}\exp\left(-\frac{I}{2\sigma_E^2}\right) \end{aligned} \tag{3.42}$$

The moments of this distribution are

$$\langle I^q\rangle = (2\sigma_E^2)^q q! \tag{3.43}$$

and the first moment is

$$\langle I\rangle = (2\sigma_E^2) \tag{3.44}$$

Therefore

$$p_{I,Hom}(I) = \frac{1}{\langle I\rangle}\exp\left(-\frac{I}{\langle I\rangle}\right) \tag{3.45}$$

which is called fully developed speckle with the important property

$$\sigma_I = \langle I \rangle \tag{3.46}$$

such that the standard deviation of the intensity is equal to the mean intensity. The contrast of a speckle field is defined as

$$C = \frac{\sigma_I}{\langle I \rangle} \tag{3.47}$$

and hence fully developed speckle has unity contrast. For finite N (number of scatterers), the contrast is

$$C = \sqrt{1 - \frac{1}{N}} \tag{3.48}$$

These results hold for a homodyne condition under Gaussian statistics.

For a heterodyne condition with a reference intensity I_r, the intensity distribution is [15]

$$p_{I,\mathrm{Het}}(I) = \frac{1}{\langle I_s \rangle} \exp\left(-\frac{I + I_r}{\langle I_s \rangle}\right) I_0\left(2\frac{\sqrt{II_r}}{\langle I_s \rangle}\right) \tag{3.49}$$

in which $I_0(x)$ is a modified Bessel function of the first kind order zero. This distribution has the properties

$$\langle I \rangle = \left(1 + \frac{I_r}{\langle I_s \rangle}\right) \langle I_s \rangle$$

$$\sigma_I = \langle I_s \rangle \sqrt{1 + 2\frac{I_r}{\langle I_s \rangle}}$$

$$C = \frac{\sqrt{1 + 2\frac{I_r}{\langle I_s \rangle}}}{\left(1 + \frac{I_r}{\langle I_s \rangle}\right)} \tag{3.50}$$

The speckle contrast in the presence of a coherent reference is plotted as a function of the reference-to-speckle ratio in Fig. 3.5. The speckle contrast survives up to high reference levels, enabling high-background applications of speckle interferometry and holography.

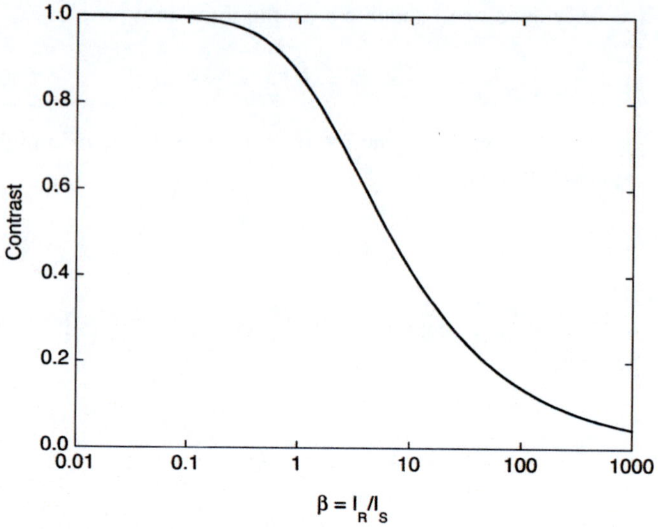

Fig. 3.5 Speckle contrast in the presence of a coherent reference intensity. The contrast falls slowly (square-root) as a function of beam ratio, enabling speckle to survive up to high reference backgrounds

3.4 Spatial Coherence

The probability functions of (3.40) and (3.49) for speckle patterns are distributed spatially, which manifests itself as an average speckle size that depends on the properties of the optical system. Speckle size is roughly synonymous with spatial coherence length because it is the characteristic length over which the phase is relatively constant at the observation plane. Speckle size depends on the optical system and so is not an intrinsic property, but it usually depends on only a few simple properties such as wavelength and distances and illuminated areas, and therefore can be easily estimated in most cases.

3.4.1 Autocorrelation Function and Power Spectrum

The chief analytical tools of spatial coherence are the autocorrelation function and its associated power spectral density. The quantitative behavior of these functions provides many of the statistical properties of speckle patterns. Consider an emission plane (a rough plane illuminated by a temporally coherent light source) and an observation plane separated by the distance z (Fig. 3.6). The emission plane has isotropic (or at least high-angle) emission from a finite area. The complex field amplitude just to the right of the emission plane is $f(x', y')$.

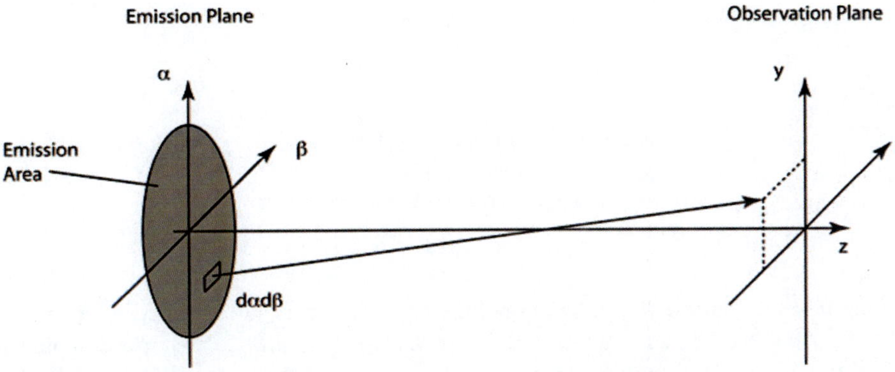

Fig. 3.6 Emission and detection planes for formation of speckle patterns

As we saw in (2.69) the amplitude at the observation plane for free-space propagation in the paraxial approximation is

$$E(x,y) = \frac{1}{\mathrm{i}\lambda z} \mathrm{e}^{\mathrm{i}kz} \mathrm{e}^{\mathrm{i}\frac{k}{2z}(x^2+y^2)} \iint f(x',y') \mathrm{e}^{\mathrm{i}\frac{k}{2z}(\alpha^2+\beta^2)} \mathrm{e}^{-\mathrm{i}\frac{k}{z}(xx'+yy')}\, \mathrm{d}x'\, \mathrm{d}y' \tag{3.51}$$

which expresses a Fourier-transform relation (within a quadratic phase factor) between $f(x',y')$ and $E(x,y)$. The first-order speckle autocorrelation function we seek is

$$\begin{aligned} g^{(1)}(\Delta x, \Delta y) &= \frac{\iint E(x,y)E^*(x-\Delta x, y-\Delta y)\, \mathrm{d}x\, \mathrm{d}y}{\iint E(x,y)E^*(x,y)\, \mathrm{d}x\, \mathrm{d}y} \\ &= \frac{\iint E(x,y)E^*(x-\Delta x, y-\Delta y)\, \mathrm{d}x\, \mathrm{d}y}{\iint I(x,y)\, \mathrm{d}x\, \mathrm{d}y} \end{aligned} \tag{3.52}$$

This correlation can be related to the amplitude $f(x',y')$ using the convolution theorem

Convolution theorem:

$$\mathrm{FT}(A * B) = \mathrm{FT}(A)\mathrm{FT}(B) \tag{3.53}$$

to give

$$g^{(1)}(\Delta x, \Delta y) = \frac{\mathrm{FT}^{-1}(|f(x',y')|^2)}{\iint I(x,y)\, \mathrm{d}x\, \mathrm{d}y} \tag{3.54}$$

The condition for (3.51) and (3.54) to hold is the paraxial approximation for small scattering angles. Within the paraxial approximation, these results apply equally well for the near and the far fields, and hence also for Fourier-transform lenses.

Using the intensity at the emission plane as

$$I(x',y') = |f(x',y')|^2 \tag{3.55}$$

then the first-order normalized amplitude correlation coefficient becomes

$$g^{(1)}(\Delta x, \Delta y) = \frac{\iint I(x',y')\mathrm{e}^{-\mathrm{i}(2\pi/\lambda z)(\Delta x x' + \Delta y y')}\,\mathrm{d}x'\,\mathrm{d}y'}{\iint I(x',y')\,\mathrm{d}x'\,\mathrm{d}y'} \tag{3.56}$$

This shows the simple relation between the first-order correlation function of spatial coherence at the detection plane and the intensity distribution at the emission plane. The top integral is simply a Fourier transform, and the bottom integral is the normalization. The result for $g^{(1)}$ has Fourier spatial frequencies given by

$$\nu_x = \frac{\Delta x}{\lambda z} \quad \nu_y = \frac{\Delta y}{\lambda z} \tag{3.57}$$

But the correlation function is not in a frequency domain, but rather is in the space-domain. The arguments Δx and Δy determine how much of a fluctuation remains at a lateral distance Δx or Δy away. As the displacements increase, the residual fluctuation tends to decrease.

The second-order (intensity) normalized correlation function is defined as

$$g^{(2)}(\Delta x, \Delta y) = \frac{\iint I(x,y)I(x-\Delta x, y-\Delta y)\,\mathrm{d}x\,\mathrm{d}y}{\iint I^2(x,y)\,\mathrm{d}x\,\mathrm{d}y} \tag{3.58}$$

which is related directly to the first-order correlation through the Siegert relation

$$g^{(2)}(\nu) = 1 + |g^{(1)}(\Delta x, \Delta y)|^2 \tag{3.59}$$

under the condition of full temporal coherence.

Intensity autocorrelation functions are related, through a Fourier transform, to power spectra by the Wiener-Khinchin theorem

$$S_\mathrm{I}(\nu_x, \nu_y) = \langle I\rangle^2 \iint g^{(2)}(\Delta x, \Delta y)\mathrm{e}^{-\mathrm{i}2\pi(\nu_x\Delta x + \nu_y\Delta y)}\,\mathrm{d}\Delta x\,\mathrm{d}\Delta y \tag{3.60}$$

The resulting power spectral density is expressed in units of intensity squared per unit spatial frequency squared

$$\mathrm{d}I^2 = S_\mathrm{I}(\nu_x, \nu_y)\,\mathrm{d}\nu_x\,\mathrm{d}\nu_y \tag{3.61}$$

The results of this section are general for any value z as long as the paraxial approximation is valid. One special case is when a Fourier-transform lens is placed a distance f from the emission plane, and the observation is on the Fourier plane a distance f from the lens. In this case $z = f$ for all of the formula of this section.

Example: Gaussian Beam Illumination

The most common example (and the most easy to calculate analytically) is illumination by a Gaussian beam.

$$I(x',y') = S_0 \exp(-(x'^2 + y'^2)/w_0^2) \tag{3.62}$$

The first-order (amplitude) correlation function is

$$\begin{aligned} g^{(1)}(\Delta x, \Delta y) &= \frac{\iint I(x',y') \mathrm{e}^{-\mathrm{i}(2\pi/\lambda z)(\Delta x x' + \Delta y y')}\, \mathrm{d}x'\, \mathrm{d}y'}{\iint I(x',y')\, \mathrm{d}x'\, \mathrm{d}y'} \\ &= \frac{\pi w_0^2 \mathrm{e}^{-(1/4)(2\pi w_0/\lambda z)^2(\Delta x^2 + \Delta y^2)}}{\pi w_0^2} \\ &= \exp\left(-\frac{1}{4}\left(\frac{2\pi w_0}{\lambda z}\right)^2 (\Delta x^2 + \Delta y^2)\right) \end{aligned} \tag{3.63}$$

with a second-order (intensity) correlation

$$g^{(2)}(\Delta x, \Delta y) = 1 + \exp\left(-\frac{1}{2}\left(\frac{2\pi w_0}{\lambda z}\right)^2 (\Delta x^2 + \Delta y^2)\right) \tag{3.64}$$

which has a lateral coherence diameter

$$d_c = \frac{\lambda z}{2\pi w_0} \tag{3.65}$$

The power spectrum is

$$\begin{aligned} P_I(\nu_x, \nu_y) &= I_0^2 \iint g^{(2)}(\Delta x, \Delta y) \mathrm{e}^{-\mathrm{i}2\pi(\nu_x \Delta x + \nu_y \Delta y)}\, \mathrm{d}\Delta x\, \mathrm{d}\Delta y \\ &= I_0^2 \left[\delta(\nu_x, \nu_y) + 2\pi \left(\frac{\lambda z}{2\pi w_0}\right)^2 \exp\left(-\frac{1}{2}\left(\frac{\lambda z}{w_0}\right)^2\right)(\nu_x^2 + \nu_y^2)\right] \end{aligned} \tag{3.66}$$

with a coherence frequency given by

$$\upsilon_c = \frac{w_0}{\lambda z}, \quad k_c = \frac{2\pi w_0}{\lambda z} \tag{3.67}$$

which is related to the inverse of (3.65) by a factor of 2π.

Examples of Gaussian speckle are shown in Fig. 3.7 for changing illumination radius $w_0 = 64$, 32, 16, and 8 μm for $z = 1$ cm and $\lambda = 500$ nm with a field-of-view of 5 mm. The second-order coherence function $g^{(2)}$ is shown in Fig. 3.8 for changing illumination radius.

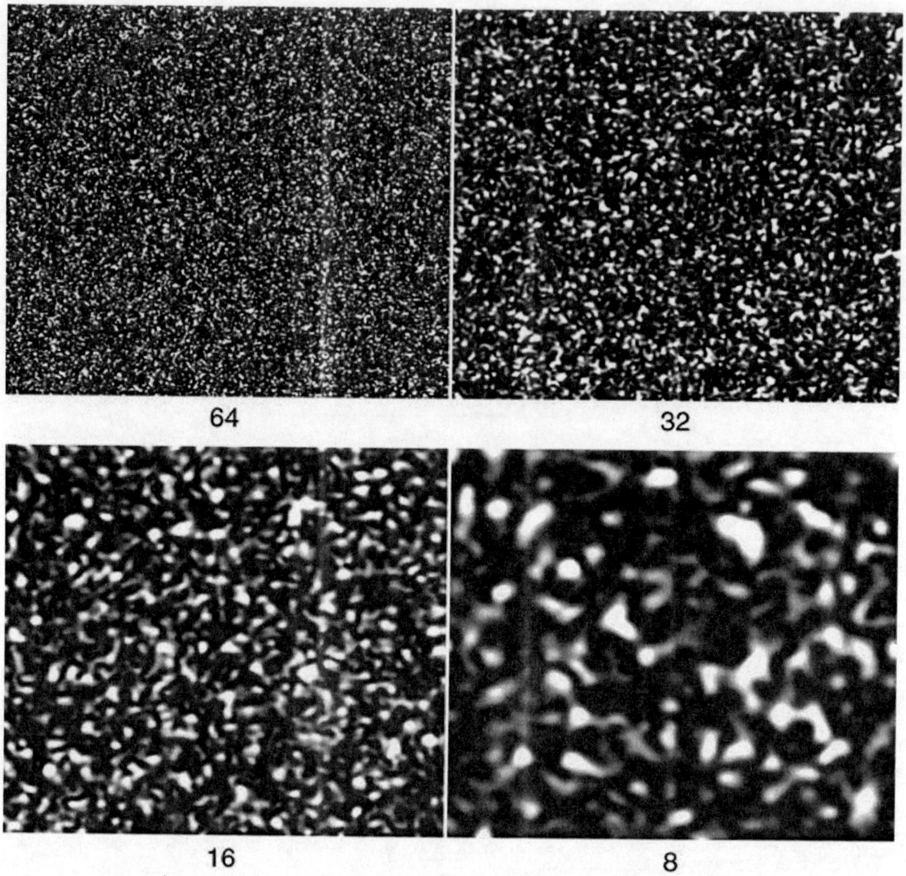

Fig. 3.7 Speckle intensities for Gaussian illumination of a random phase screen for changing illumination radius $w_0 = 64$, 32, 16, and 8 microns for $f = 1$ cm and $\lambda = 500$ nm with a field-of-view of 5 mm

3.4.2 Coherence Area

Diffraction from a screen or target is principally a two-dimensional process, and speckles are generally non-isotropic. Therefore, a one-dimensional measure of spatial coherence length can be artificial, while a two-dimensional coherence area can be a better measure for spatial coherence. To arrive at a coherence area, start with the normalized covariance function

$$c_I(\Delta x, \Delta y) = g^{(2)}(\Delta x, \Delta y) - 1 \tag{3.68}$$

and define the equivalent area of the normalized covariance function as

$$A_c = \iint c_I(\Delta x, \Delta y)\,\mathrm{d}\Delta x\,\mathrm{d}\Delta y = \iint |g^{(1)}(\Delta x, \Delta y)|^2\,\mathrm{d}\Delta x\,\mathrm{d}\Delta y \tag{3.69}$$

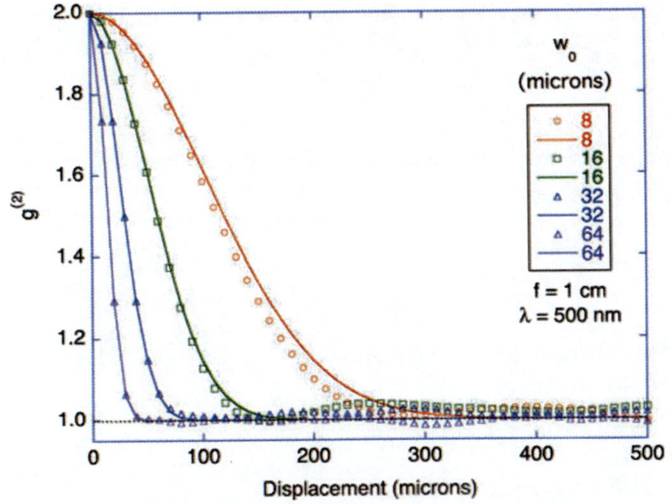

Fig. 3.8 Second-order correlation functions for the speckle fields in Fig. 3.7 compared with the theory of (3.64)

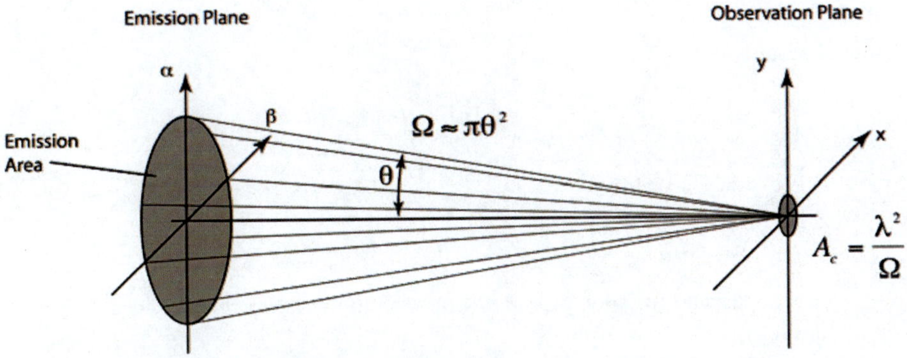

Fig. 3.9 The coherence area on the observation plane is λ^2 divided by the solid angle subtended by the emission area seen from the observation plane

For an illuminated area A with constant intensity, this yields

$$A_c = \frac{(\lambda z)^2}{A} = \frac{\lambda^2}{\Omega_s} \tag{3.70}$$

where Ω_s is the solid angle subtended by the illumination area as seen from the observation point (Fig. 3.9). This assumes that the angular spread of the light scattered from the illumination area is very broad. For a uniformly illuminated circular aperture of diameter D, the coherence area is

$$A_c = \frac{4(\lambda z)^2}{\pi D^2} \tag{3.71}$$

and the diameter of the coherence area on the observation plane is

$$d_c = \frac{4}{\pi}\frac{\lambda z}{D} \tag{3.72}$$

More generally, for an illumination that is not uniform, or an illumination area that is not symmetric

$$A_c = (\lambda z)^2 \frac{\iint I^2(\alpha,\beta)\,d\alpha\,d\beta}{\left[\iint I(\alpha,\beta)\,d\alpha\,d\beta\right]^2} \tag{3.73}$$

For a Fourier-transform lens, the equations are the same, but with the focal length f substituted in for z.

3.4.2.1 Example: Gaussian Beam

For the Gaussian intensity distribution

$$I(x',y') = S_0 \exp(-(x'^2 + y'^2)/w_0^2) \tag{3.74}$$

The coherence area is

$$\begin{aligned} A_c &= (\lambda z)^2 \frac{\iint I^2(\alpha,\beta)\,d\alpha\,d\beta}{\left[\iint I(\alpha,\beta)\,d\alpha\,d\beta\right]^2} \\ &= \frac{(\lambda z)^2}{2\pi w_0^2} = \frac{1}{2\pi}\left(\frac{z}{w_0}\right)^2 \lambda^2 \end{aligned} \tag{3.75}$$

Larger distances and smaller pinholes improve the coherence area in a coherent optical system.

3.5 Speckle Holography

Speckle holography involves the interference of a coherent reference wave with a speckle field. To start, it is simplest to consider light that is temporally coherent but has spatial diversity (limited spatial coherence). Off-axis holography with a plane-wave reference uses a finite crossing angle between the k-vector of the reference wave and the optic axis of the speckle. In this case, each speckle will have interference fringes. Although the phases of the fringes are different in each speckle, the fringe periodicity is defined by the angle between the optic axis of the imaging optics and the optic axis of the reference beam. The grating vector in the recording plane is

$$\vec{K} = \pm(\vec{k}_s - \vec{k}_r) \tag{3.76}$$

The fringe spacing is

$$\Lambda = \frac{\lambda}{\sin\theta} \tag{3.77}$$

where θ is the angle between the beam axes.

An example of a speckle hologram is shown in Fig. 3.10a, with interference fringes from a coherent reference wave tilted at an azimuthal angle of 45°. The fringes have the same orientations and periodicity from one speckle to another. However, because the phase drifts from speckle to speckle, there are many dislocations in the overall fringe pattern. The hologram in Fig. 3.10a is a Fourier-domain hologram on the Fourier plane of the imaging system. The Fourier-transform reconstruction of the hologram is shown in Fig. 3.10b. The two side-bands are evident along with the zero-order. An example of an in-line speckle hologram is shown in Fig. 3.11. The reference wave used for recording and reconstruction in this case is a spherical wave.

The highest practical resolution in speckle holography is achieved when there are approximately three fringes per speckle. This allows high resolution to be retained when the hologram is demodulated. However, in the case of digital holography, there is also the limit set by the size of the digital pixel. In general, there should be approximately three pixels per fringe. Nyquist sampling requires two pixels per fringe, but to avoid aliasing problems, three or four pixels are more practical. Therefore, there are about ten pixels per speckle. Whether the hologram is in the Fourier domain or the image domain, this limits the spatial resolution of the reconstructed image.

3.6 Caustics

Biological tissue is macroscopically heterogeneous on length scales that are larger than a wavelength of light. This causes refractive aberrations of light that can be best understood from the point of view of geometric or ray optics. For instance, the nuclei inside cells, and cells themselves, may act as microlenses that refract light and focus or defocus it, much as ripples on the surface of a swimming pool refract sunlight and cast the characteristic bright lines of focused light on the swimming pool floor. These bright lines are called caustics, and are concentrations of rays by the refractive surface. The same effect occurs in biological tissue, leading to refractive aberrations of propagating light. This effect is sometimes called the shower-glass effect, in which imaging resolution degrades linearly with increasing depth of propagation through tissue. In the limit of the far field (in the Fourier domain) these refractive aberrations produce caustics.

Caustics share many properties in common with speckle, but are not specifically coherent phenomena. They are the ray-optics consequences of macroscopic random refractive heterogeneities, just as speckle is the diffraction optics consequence of

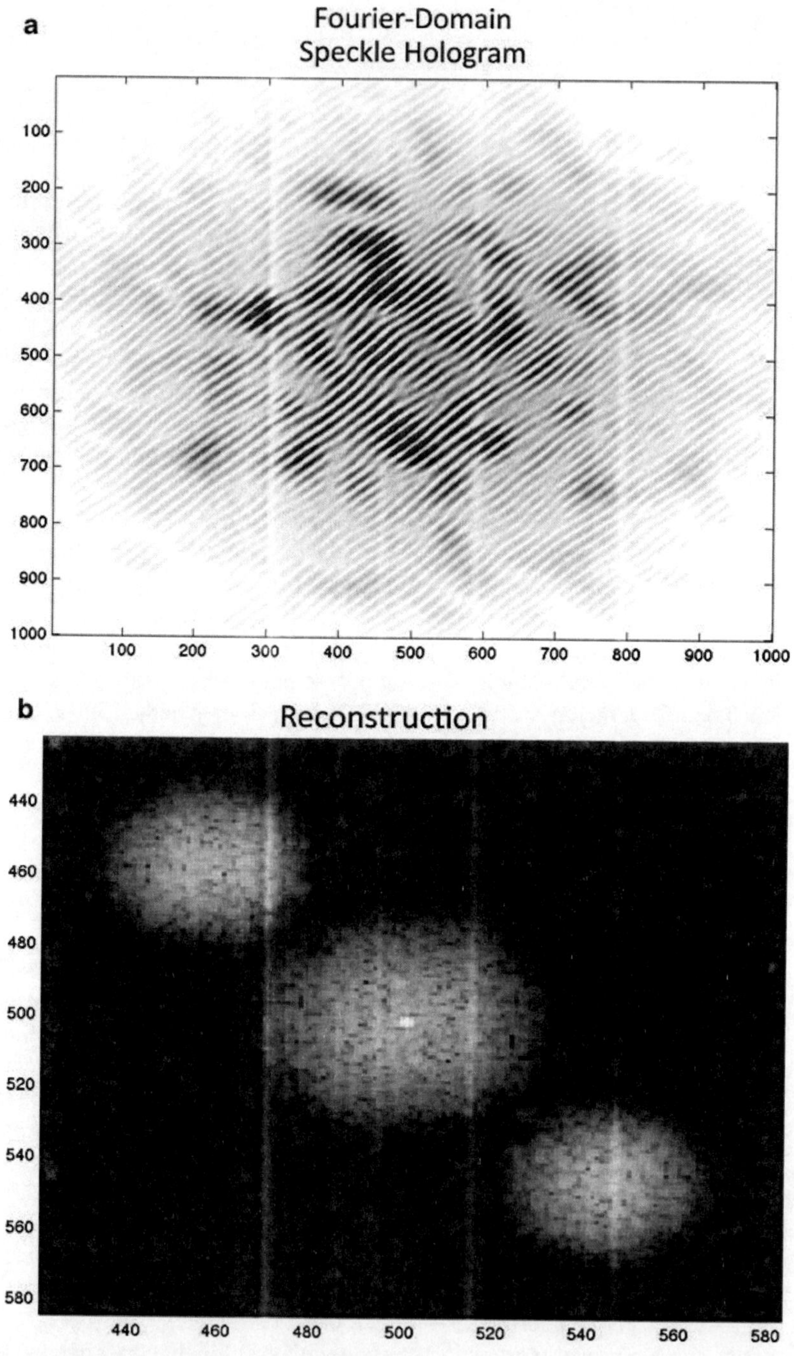

Fig. 3.10 (**a**) Fourier-domain speckle with coherent interference fringes from a reference wave (black is high intensity). Each speckle has the same fringe spacing, but different relative phase. (**b**) Fourier-transform of the Fourier-domain speckle hologram, showing the two sidebands plus the zero-order

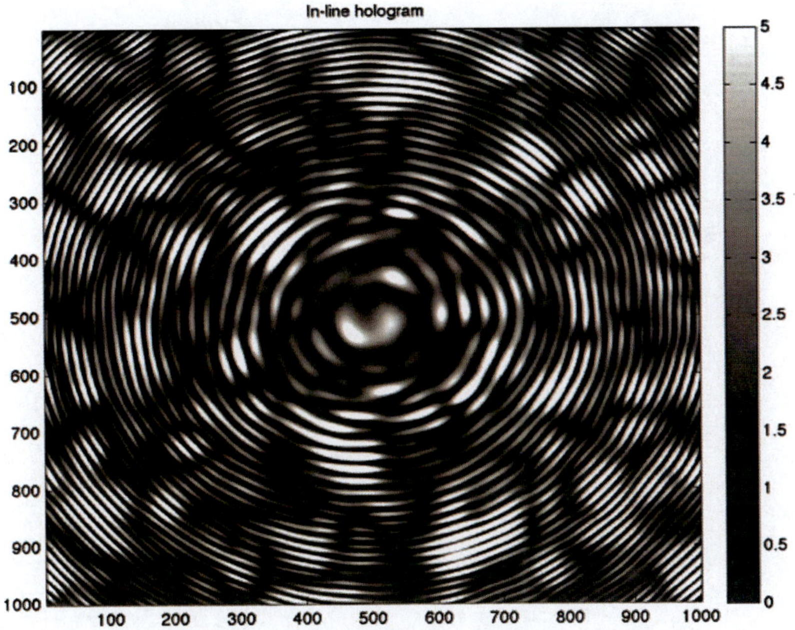

Fig. 3.11 Example of an in-line speckle hologram. The recording and reconstruction are performed using a spherical reference wave

microscopic refractive heterogeneities. Like speckle, they degrade imaging resolution by causing large intensity variances. In dynamic media, time-dependent caustics are the ray-optics analog of dynamic speckle from DLS. When cellular tissue is undergoing cell death, cells tend to ball up and tissue surfaces can ruffle. These lead to time-dependent Fourier-domain caustics that participate together with DLS speckle.

Caustics are understood mathematically as the envelope functions of multiple rays that converge in the Fourier domain (angular deflection). These are points of mathematical stationarity, in which the ray density is invariant to first order in deviations in the refracting surface. The basic geometry is shown in Fig. 3.12 for a ray incident on a nonplanar surface emerging into a less-dense medium. From Snell's law

$$\sin\theta_2' = n\sin\theta_1$$

$$\theta_2' \approx n\theta_1 \tag{3.78}$$

where n is the relative index (ratio), and the small-angle approximation has been made. The incident angle θ_1 is simply related to the slope of the interface $\mathrm{d}h/\mathrm{d}x$ as

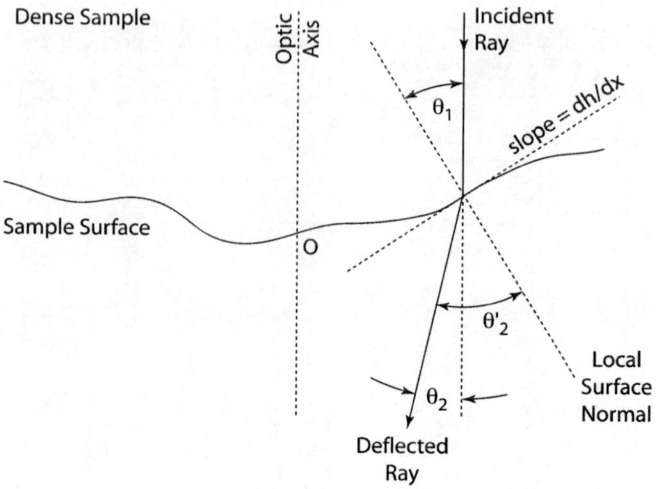

Fig. 3.12 Geometry for ray refraction at a nonplanar surface between an optically denser sample and the surrounding medium. The deflection angle θ_2 is a function of the slope of the surface $\mathrm{d}h/\mathrm{d}x$

$$\begin{aligned}\tan\theta_1 &= \mathrm{d}h/\mathrm{d}x \\ \theta_1 &\approx \mathrm{d}h/\mathrm{d}x\end{aligned} \tag{3.79}$$

where the small-angle approximation is used again. The angular deflection relative to the optic axis is then

$$\begin{aligned}\theta_2 &\approx \theta_2' - \mathrm{d}h/\mathrm{d}x \\ &= n\theta_1 - \mathrm{d}h/\mathrm{d}x \\ &= (n-1)\,\mathrm{d}h/\mathrm{d}x\end{aligned} \tag{3.80}$$

which is equal to the optical path difference through the sample.

In two dimensions, the optical path difference can be replaced with a general potential

$$\Phi(x,y) = [n(x,y) - 1]h(x,y) \tag{3.81}$$

and the two orthogonal angular deflections (measured in the far field on a Fourier plane) are

$$\theta_x = \frac{\partial\Phi(x,y)}{\partial x} \qquad \theta_y = \frac{\partial\Phi(x,y)}{\partial y} \tag{3.82}$$

These angles describe the deflection of the rays across the sample surface.

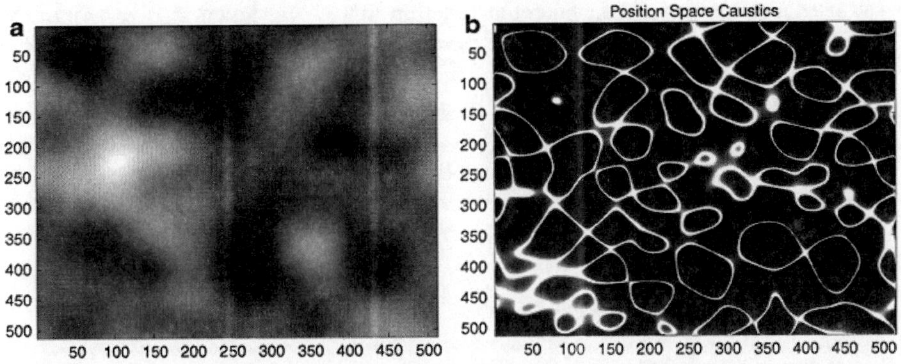

Fig. 3.13 The potential function $\Phi(x, y)$ in (**a**) and the corresponding Jacobian density in (**b**)

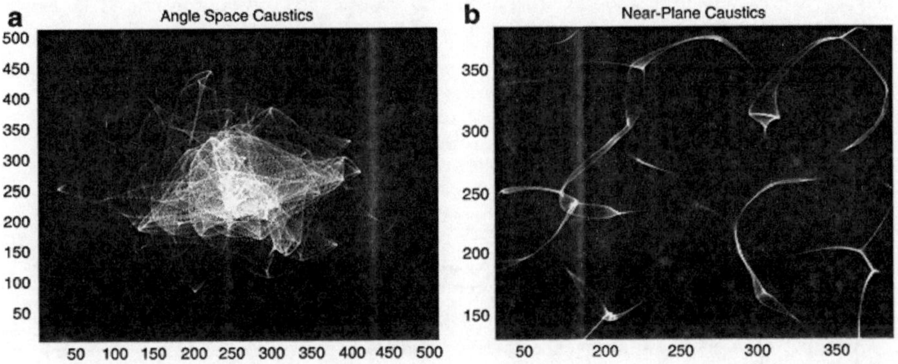

Fig. 3.14 (**a**) Angular caustics and (**b**) near-screen caustics

Caustics are lines of stationarity, meaning that the density of rays is independent of first-order changes in the refracting sample. The condition of stationarity is defined by the Jacobian of the transformation from (x, y) to (θ_x, θ_y), with

$$J = \begin{vmatrix} \dfrac{\partial\theta_x}{\partial x} \dfrac{\partial\theta_x}{\partial y} \\ \\ \dfrac{\partial\theta_y}{\partial x} \dfrac{\partial\theta_y}{\partial y} \end{vmatrix} = 0 \text{ or } J = \begin{vmatrix} \dfrac{\partial^2\Phi}{\partial x\partial x} \dfrac{\partial^2\Phi}{\partial y\partial x} \\ \\ \dfrac{\partial^2\Phi}{\partial x\partial y} \dfrac{\partial^2\Phi}{\partial y\partial y} \end{vmatrix} = 0 \tag{3.83}$$

When this condition is satisfied, the envelope function bounding groups of collected rays is stationary to perturbations in the inhomogeneous sample.

An example of a random potential function $\Phi(x, y)$ is shown in Fig. 3.13a. The associated Jacobian density $(1/J(x,y))$ is shown in Fig. 3.13b. These regions on the wave front contribute to the caustics in the far field. The far-field caustic, when the random surface is illuminated by a plane wave, is shown in Fig. 3.14a. When the observation screen is not at infinity, but is at some distance Z from the random surface, then near-field caustics arise, as in Fig. 3.14b These types of caustics are commonly seen on the bottom of a swimming pool on a sunny day.

Selected Bibliography

Goodman, J.W.: Speckle phenomena in optics, Roberts (2006) (This is a recent book with great detail on speckle statistics)

Dainty, J.C.: Laser speckle and related phenomena. 2nd edl edn. Springer, Berlin (1984) (The original classic on speckle phenomena)

Nye, J. F.: Natural focusing and fine structure of light: caustics and wave dislocations. IOP, Bristol (1999) (One of the few accessible texts on caustics)

References

1. Briers, J.D.: Speckle fluctuations and biomedical optics: implications and applications. Opt. Eng. **32**(2), 277–283 (1993)
2. Dunn, A.K., Bolay, T., Moskowitz, M.A., Boas, D.A.: Dynamic imaging of cerebral blood flow using laser speckle. J. Cereb. Blood Flow Metab. **21**(3), 195–201 (2001)
3. Yuan, S., Devor, A., Boas, D.A., Dunn, A.K.: Determination of optimal exposure time for imaging of blood flow changes with laser speckle contrast imaging. Appl. Opt. **44**(10), 1823–1830 (2005)
4. Li, P.C., Ni, S.L., Zhang, L., Zeng, S.Q., Luo, Q.M.: Imaging cerebral blood flow through the intact rat skull with temporal laser speckle imaging. Opt. Lett. **31**(12), 1824–1826 (2006)
5. Paul, J.S., Al Nashash, H., Luft, A.R., Le, T.M.: Statistical mapping of speckle autocorrelation for visualization of hyperaemic responses to cortical stimulation. Ann. Biomed. Eng. **34**(7), 1107–1118 (2006)
6. Briers, J.D.: Laser speckle contrast imaging for measuring blood flow. Opt. Appl. **37**(1–2), 139–152 (2007)
7. Wang, Z., Hughes, S., Dayasundara, S., Menon, R.S.: Theoretical and experimental optimization of laser speckle contrast imaging for high specificity to brain microcirculation. J. Cereb. Blood Flow Metab. **27**(2), 258–269 (2007)
8. Cheng, H.Y., Yan, Y.M., Duong, T.Q.: Temporal statistical analysis of laser speckle images and its application to retinal blood-flow imaging. Opt. Express **16**(14), 10214–10219 (2008)
9. Duncan, D.D., Kirkpatrick, S.J.: Can laser speckle flowmetry be made a quantitative tool? J. Opt. Soc. Am. A Opt. Image Sci. Vis. **25**(8), 2088–2094 (2008)
10. Parthasarathy, A.B., Tom, W.J., Gopal, A., Zhang, X.J., Dunn, A.K.: Robust flow measurement with multi-exposure speckle imaging. Opt. Express **16**(3), 1975–1989 (2008)
11. Yu, P., Peng, L., Mustata, M., Turek, J.J., Melloch, M.R., Nolte, D.D.: Time-dependent speckle in holographic optical coherence imaging and the state of health of tumor tissue. Opt. Lett. **29**(1), 68–70 (2004)

12. Jeong, K., Turek, J.J., Nolte, D.D.: Functional imaging by dynamic speckle in digital holographic optical coherence imaging. Coher. Domain Opt. Meth. Opt. Coher. Tomogr. Biomed. XI **6429**, 6429V1–6429V9 (2007)
13. Jeong, K., Turek, J.J., Nolte, D.D.: Imaging motility contrast in digital holography of tissue response to cytoskeletal anti-cancer drugs. Op0t. Express **15**, 14057 (2007)
14. Jeong, K., Turek, J.J., Melloch, M.R., Nolte, D.D.: Functional imaging in photorefractive tissue speckle holography. Opt. Commun. **281**(7), 1860–1869 (2008)
15. Goodman, J.W.: Speckle Phenomena in Optics. Roberts, Colorado (2006)

Chapter 4
Surface Optics

Many optical biosensors detect molecules that either are attached to a surface, or are in close proximity to one. Therefore, understanding the optical properties of surfaces, and how these properties affect molecular interferometry, is a central topic of this book. Surfaces are planes of dielectric discontinuity that split the amplitude of waves into transmitted and reflected partial waves (Fig. 4.1). Surfaces impose electromagnetic boundary conditions that produce constructive or destructive interference of the incident and reflected waves. On the wrong type of surface (unity reflection with a π phase shift, as for a metal surface), a thin protein film can be entirely invisible to an incident plane wave (but not a surface wave), because the electric field strength at the molecular layer is canceled by destructive interference between the incident and reflected waves. In this situation, a light wave can pass right through the layer without ever polarizing it, and hence never sensing any change in the optical path length. This is perhaps the oldest and simplest form of optical cloaking [1]. Conversely, on the right type of surface (a partial reflection with a $\pi/2$ phase shift) a dilute dusting of biological molecules can produce easily measurable signals in the far field, making this type of surface ideal for sensitive detection of molecules [2].

4.1 Reflection from Planar Surfaces

When incident fields arrive at an interface at an oblique angle, as in Fig. 4.2, the continuity equations take the form

$$(\ldots)_i e^{i(\vec{k}_i \cdot \vec{r} - \omega t)} + (\ldots)_r e^{i(\vec{k}_r \cdot \vec{r} - \omega t)} = (\ldots)_t e^{i(\vec{k}_t \cdot \vec{r} - \omega t)} \tag{4.1}$$

D.D. Nolte, *Optical Interferometry for Biology and Medicine*, Bioanalysis 1,
DOI 10.1007/978-1-4614-0890-1_4,

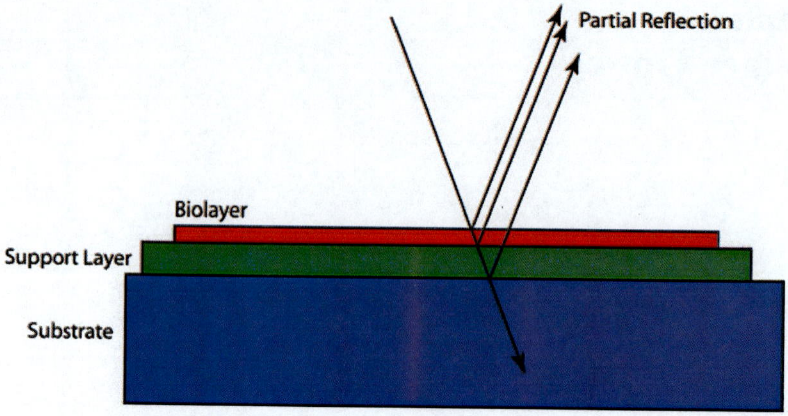

Fig. 4.1 Many optical biosensors use light to probe thin molecular biolayers on solid dielectric supports. The incident wave is split into numerous partial waves with relative phase differences that interfere in the far field where the light is detected. The biolayer shifts these phases and modifies the detected intensity

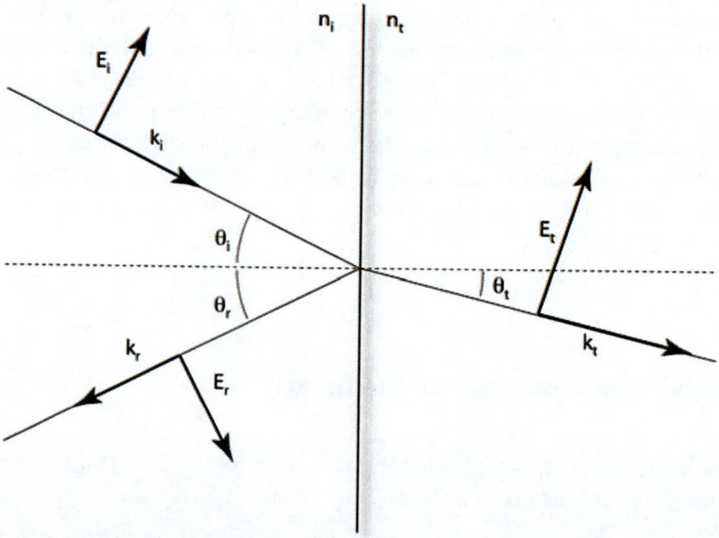

Fig. 4.2 Oblique incidence of an electromagnetic wave on an interface between two materials with refractive indices n_i and n_t. The "plane of incidence" is the plane that contains the *k*-vectors (the flat page in this case). This figure shows transverse-magnetic (TM) polarization because the magnetic fields are perpendicular to the plane of incidence. This polarization is also known as p-polarized light because the electric field is in the plane of incidence

where the coefficients in parentheses are polarization components of the fields. This general boundary condition must hold for all x, y and z and t, yielding the three-way equality

$$\vec{k}_i \cdot \vec{r} = \vec{k}_r \cdot \vec{r} = \vec{k}_t \cdot \vec{r} \tag{4.2}$$

These expression are statements of conservation of transverse photon momentum

Conservation of Transverse Momentum:

$$\hbar k_{\|}^{i} = \hbar k_{\|}^{r} = \hbar k_{\|}^{t} \tag{4.3}$$

The simple relations of (4.3) lead to the well-known Snell's Law

Snell's Law:

$$\frac{\sin\theta_i}{\sin\theta_t} = \frac{n_t}{n_i} \tag{4.4}$$

and to the law of reflection

Law of Reflection:

$$\theta_r = \theta_i \tag{4.5}$$

When the four Maxwell equations are converted to the four electromagnetic boundary conditions at the interface, the resulting Fresnel equations for the reflection and transmission coefficients are

The Fresnel Equations:

$$\begin{aligned} r_\perp &= \frac{n_i\cos\theta_i - n_t\cos\theta_t}{n_i\cos\theta_i + n_t\cos\theta_t} & r_\| &= \frac{n_t\cos\theta_i - n_i\cos\theta_t}{n_i\cos\theta_t + n_t\cos\theta_i} \\ t_\perp &= \frac{2n_i\cos\theta_i}{n_i\cos\theta_i + n_t\cos\theta_t} & t_\| &= \frac{2n_i\cos\theta_i}{n_i\cos\theta_t + n_t\cos\theta_i} \end{aligned} \tag{4.6}$$

where the "perpendicular" or "parallel" signs relate to the electric field orientation relative to the plane of incidence. When the electric field is perpendicular to the plane of incidence, then this is known as TE-polarization, or s-polarization. When the electric field is parallel to the plane of incidence, then this is known as TM-polarization, or p-polarization. Alternative expressions for the Fresnel equations entirely in terms of angles are

Alternate Fresnel Equations:

$$\begin{aligned} r_\perp &= -\frac{\sin(\theta_i - \theta_t)}{\sin(\theta_i + \theta_t)} & r_\| &= +\frac{\tan(\theta_i - \theta_t)}{\tan(\theta_i + \theta_t)} \\ t_\perp &= +\frac{2\sin\theta_t\cos\theta_i}{\sin(\theta_i + \theta_t)} & t_\| &= +\frac{2\sin\theta_t\cos\theta_i}{\sin(\theta_i + \theta_t)\cos(\theta_i - \theta_t)} \end{aligned} \tag{4.7}$$

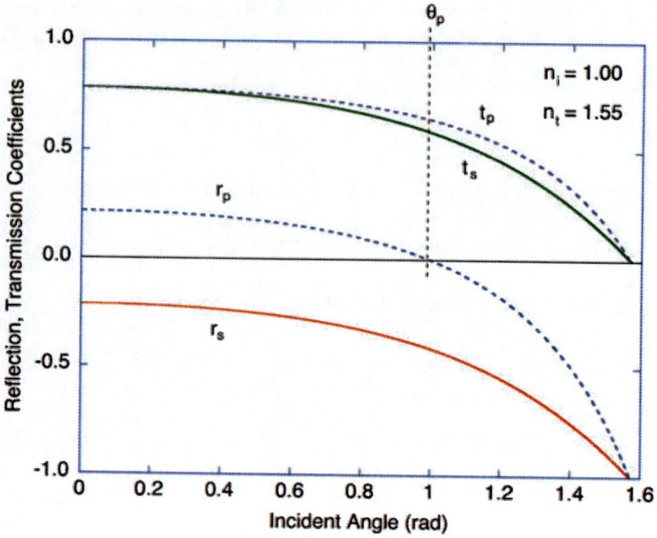

Fig. 4.3 Reflection and transmission coefficients for a wave incident on glass from air. The p-polarized light has a Brewster angle at which the reflection vanishes

The dependence of these coefficients on angle are shown in Figs. 4.3–4.5 for air and glass ($n = 1.55$). In Fig. 4.3, the incident medium is air. The p-polarized light has a special angle θ_p, known as the Brewster angle, at which the reflectance goes to zero. If unpolarized light is incident at this angle, then purely s-polarized light is reflected. Note that at normal incidence ($\theta = 0°$), the reflection coefficients have opposite signs. This is an artifact of the sign convention chosen in Fig. 4.2. Although this choice of sign is not "physical," this convention must be maintained in subsequent derivations, for instance of Mie scattering near a surface.

When the incident medium is glass and the wave is incident on air, then total internal reflection (TIR) occurs above a critical angle for which the reflectance is unity and the transmittance is zero. In the regime of total internal reflection, fields still exist on the air side of the interface, with real and imaginary components that introduce a relative phase shift that is important for waveguides. But they are not propagating modes, and are confined near the interface. There is also an internal Brewster angle for p-polarized light.

For interferometric derivations and applications, the complex reflection and transmission coefficients are the relevant quantities, because they contain phase information. Furthermore, the sign conventions that are implicit in the derivation of the Fresnel equations must be consistent with the phase conventions chosen for optical path lengths in interferometric devices, and with the phase conventions chosen for scattering and diffraction. For instance, throughout this book, the plane wave convention is chosen as $\psi \sim \exp(i\vec{k}\cdot\vec{r} - i\omega t)$ for all derivations. The reader should be aware that this convention is not standardized, and that many textbooks choose to place the minus sign in front of the space term rather than the time term.

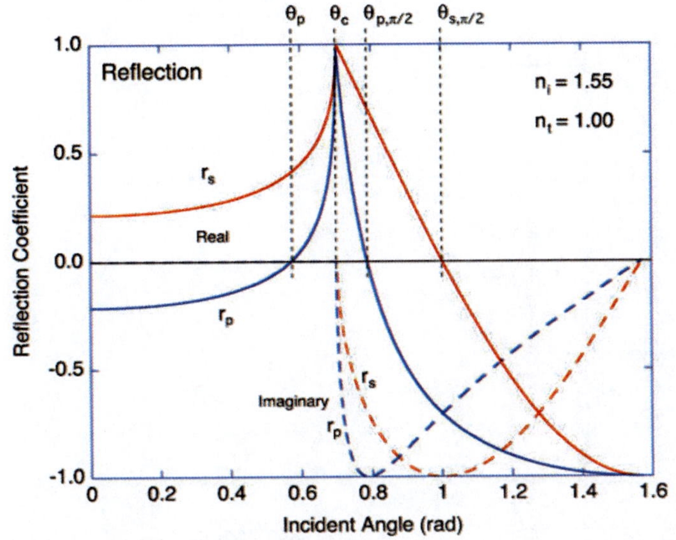

Fig. 4.4 Reflection coefficients for a wave incident on air from glass. The reflectance goes to unity at and beyond the critical angle for total-internal reflection. There is also an internal Brewster angle for p-polarized light

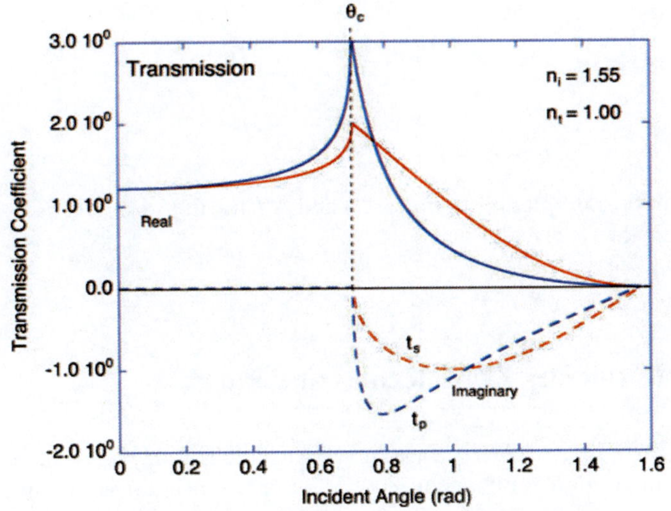

Fig. 4.5 Transmission coefficients for a wave incident on air from glass

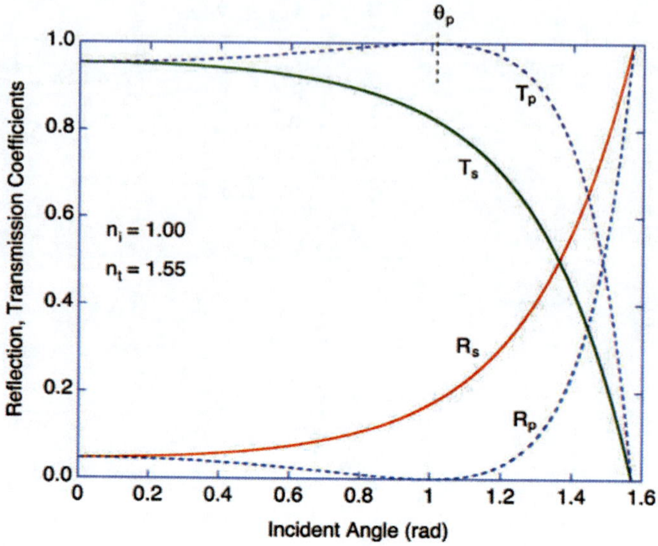

Fig. 4.6 Reflectance and transmittance for a wave incident on glass from air

For simple intensity measurements of single-interface transmittance and reflectance, the phase does not matter, and reflectance and transmittance are related directly to the reflection and transmission coefficients through the expressions

$$\begin{aligned} R_\perp &= r_\perp^2 \\ R_\parallel &= r_\parallel^2 \\ T_\perp &= \left(\frac{n_t \cos\theta_t}{n_i \cos\theta_i}\right) t_\perp^2 \\ T_\parallel &= \left(\frac{n_t \cos\theta_t}{n_i \cos\theta_i}\right) t_\parallel^2 \end{aligned} \tag{4.8}$$

These quantities are plotted in Figs. 4.6 and 4.7 for the cases of Figs. 4.3–4.5 for glass and air interfaces.

4.2 Reflectometry of Molecules and Particles

The reflection of light off of a surface carrying a molecular layer is an interferometric process. In the limit of a dilute coverage, the light reflected from the surface can be separated into a homogeneous reflection from the ideal surface plus light scattered from the molecules. These fields interfere in the far field to generate a resultant

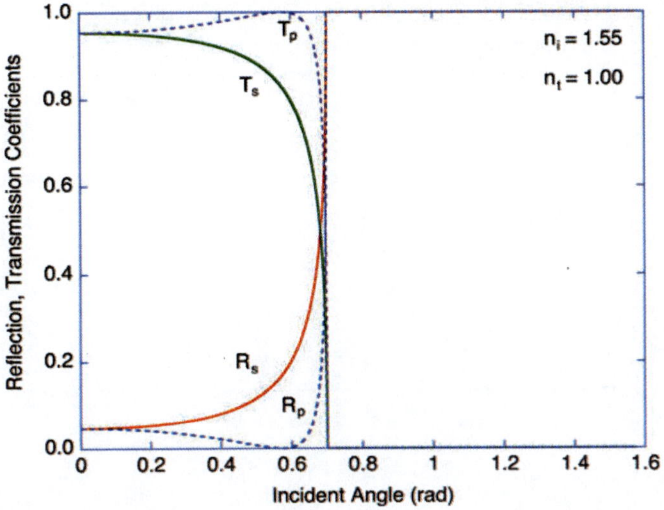

Fig. 4.7 Reflectance and transmittance for a wave incident on air from glass

amplitude and phase related to the density of molecules on the surface. In the opposite limit of a thin homogeneous layer, interference effects occur through the partial waves reflected from each interface. Either way, molecular films on surfaces produce interference effects that can be used to sense the presence or the change in density of molecules in many types of optical biosensors.

4.2.1 Molecules on Surfaces

To understand the optical properties of a dilute dispersion of molecules on a surface, it helps first to consider a single molecule scattering light in close proximity to a mirror with complex reflectance $r(\theta)$. The scattering configuration is shown in Fig. 4.8. The net scattering amplitude (neglecting optical path length associated with the distance of the molecule from the surface) is

$$f_{net} = f_b + f_f r + r f_f + r f_b r \tag{4.9}$$

that constitutes four partial waves scattered from the molecule: a backscattered wave, a forward scattered wave that is reflected, a reflected wave that is forward scattered and a reflected wave that is backscattered and reflected again.

If $r = -1$ (a perfectly conducting mirror with a nodal surface), then $f_{net} = f_b - f_f + f_b - f_f$, and if the scattering is furthermore isotropic ($f_b = f_f$) then the net scattering vanishes, and the molecule is invisible. On the other hand

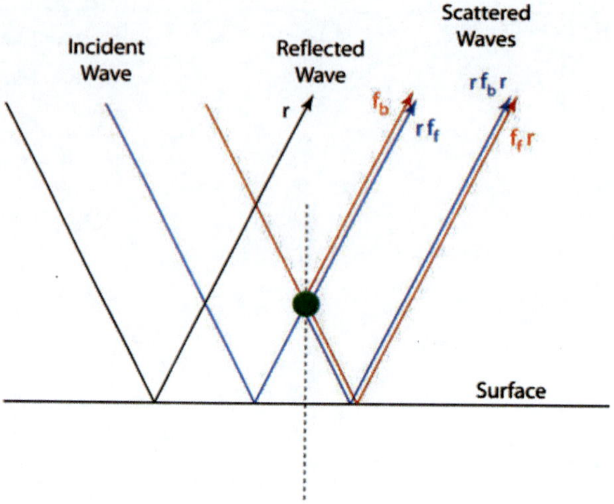

Fig. 4.8 Scattering by a molecule on a surface generates four partial scattered waves. The phase of the reflection coefficient contributes to the relative phases of the partial waves that add and interfere. These waves collectively interfere in the far field with the reflected unscattered wave

if $r = +1$ (an antinodal surface), then $f_{net} = f_b + f_f + f_b + f_f$, and if the scattering is isotropic, then the net scattered field is 4 times larger than for the same molecule scattering in free space.

For the more general case of a dielectric surface, the net scattering amplitude is

$$f_{net} = f_b(1 + r^2) + f_f 2r \tag{4.10}$$

which, for isotropic scattering, becomes

$$f_{net} = f_{scat}(1 + r)^2 \tag{4.11}$$

In the far field, this scattered field is added to the unscattered field reflected from the surface. The unscattered field has an additional $\pi/2$ phase acquired from the diffraction integration over the continuous incident field (see Chap. 2). The effect on the far field is then

$$\begin{aligned} E_{far} &= irE_0 + E_0 f_{scat}(1 + r)^2 \\ &= irE_0\left(1 - i\,\frac{f_{scat}(1 + r)^2}{r}\right) \end{aligned} \tag{4.12}$$

with a phase contribution

$$\phi_{reflect} = atan\left[\frac{\mathrm{Im}\,(ir + f_{scat}(1 + r)^2)}{\mathrm{Re}\,(ir + f_{scat}(1 + r)^2)}\right] \tag{4.13}$$

The ultimate goal of surface interferometry, as for any type of interferometry, is to transduce a phase change into a detectable intensity change. Here the goal is to detect a change in intensity in the far field that is related to the molecular scattering at the surface. The intensity in the far field for $r = r_1 + ir_2$ is

$$\begin{aligned} I_{\text{far}} &= |E_{\text{far}}|^2 = |irE_0 + E_0 f_{\text{scat}}(1+r)^2|^2 \\ &= |r|^2 - 2f_{\text{scat}} r_2 (1 - |r|^2) \\ &= R - 2f_{\text{scat}} r_2 (1 - R) \end{aligned} \tag{4.14}$$

with a relative change in the reflected intensity

$$\frac{\Delta I}{I} = 2f_{\text{scat}} r_2 \frac{1-R}{R} \tag{4.15}$$

which depends linearly on the scattering f_{scat} from the molecule. These results show that the intensity modulation caused by the molecular scattering vanishes when the reflectance equals unity, even though the phase modulation may be a maximum for the case of $r = r_1 = +1$. The maximum intensity modulation $\Delta I = 0.77 f_{\text{scat}}$ is attained for

$$\begin{aligned} |r|^2 &= 0.58^2 = 0.34 \qquad && r_1 = 0 \\ \phi &= \tfrac{\pi}{2} = 1.57 && r_2 = 0.58 \end{aligned} \tag{4.16}$$

The phase condition for maximum intensity modulation is equal to $\pi/2$, with a 34% surface reflectance. The $\pi/2$ phase shift on reflection places the reflected field in phase quadrature with the light scattered from the molecule. This condition of phase quadrature causes the scattered field to be detected directly as an intensity in the far field where the light is detected. Therefore, for molecular detection on surfaces, using either laser scanning or direct microscopy, a phase quadrature condition is needed.

Silicon is a common substrate to use in a biosensor. However, in the case of bare silicon the electric field at the surface is very small because $r_1 = -0.57$ and the intensity modulation nearly vanishes for dilute molecular scattering. This is why molecules on bare silicon surfaces are nearly invisible in normal-incidence reflectance. However, when a thermal oxide is placed on silicon, choosing a thickness of 120 nm and a wavelength of 633 nm, then $r_1 = 0.013$ and $r_2 = 0.33$ with a phase of 1.5 rad (near phase quadrature) with $|r| = 0.33$ and $R = 0.11$, yielding an intensity modulation $\Delta I = 0.59 f_{\text{scat}}$, which approaches the maximum value. This is why thermal oxide on silicon is a nearly ideal surface from which direct detection in reflection is made possible [3–5].

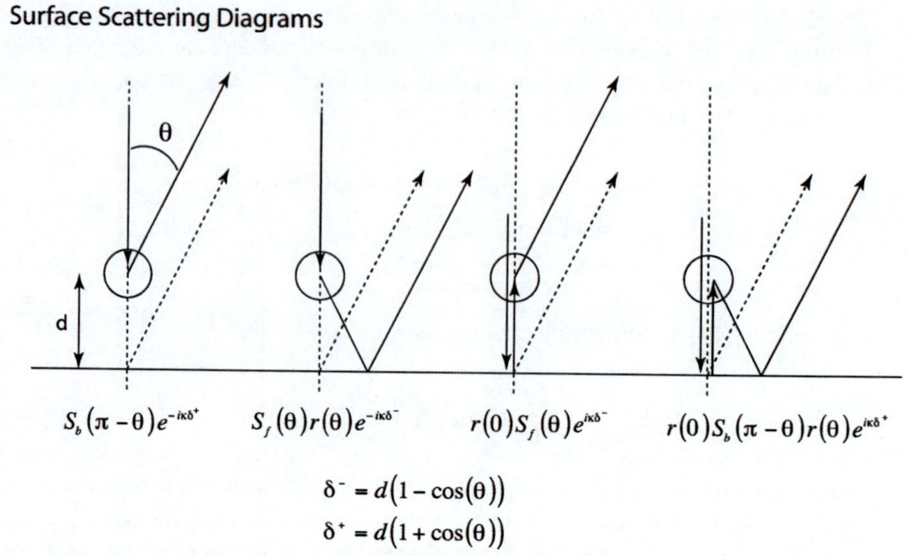

Fig. 4.9 Surface scattering and reflection processes that contribute to the total scattering S. The angle θ is assumed small with near-normal incident light

4.2.2 *Particles on Surfaces*

Just as molecules on surfaces give rise to image dipoles in light scattering, particles on surfaces give rise to similar effects. This particle scattering model is called the double-interaction (DI) model [6, 7]. It takes the same form as molecular scattering, but replaces f_b and f_f with Mie scattering coefficients $S(\pi - \theta)$ and $S(\theta)$, respectively (Fig. 4.9). It also introduces a "shadow" function $F(ka)$ which is a phenomenological way to include the partial screening of fields at the surface (Fig. 4.9).

When the Mie scatterer is placed on a surface the scattered fields reflect from the surface, leading to four terms in the total scattered field. These are

$$S(\pi-\theta)\mathrm{e}^{-ik\delta^+} \quad S(\theta)r(\theta)F(ka)\mathrm{e}^{-ik\delta^-} \quad r(0)S(\theta)F(ka)\mathrm{e}^{ik\delta^-} \quad r(0)S(\pi-\theta)r(\theta)F^2(ka)\mathrm{e}^{ik\delta^+} \tag{4.17}$$

where the path difference is

$$\begin{aligned}\delta^- &= d(1-\cos\theta)\\ \delta^+ &= d(1+\cos\theta)\end{aligned} \tag{4.18}$$

and the center of the nanoparticle is a distance d above the reflecting surface. The function $F(ka)$ is a shadow function that is unity in the case of point-scatterers like molecules, but can vanish for large opaque particles that screen the field at the surface. The forward and back scattering coefficients are complex quantities

$$\begin{aligned} S(\pi-\theta) &= S_\mathrm{b} \\ S(\theta) &= S_\mathrm{f} \end{aligned} \tag{4.19}$$

With the complex reflection coefficient

$$r = r_0 \mathrm{e}^{\mathrm{i}\phi} \tag{4.20}$$

the scattering amplitude is

$$S = (S_\mathrm{b})(\mathrm{e}^{-ik\delta^+} + F^2(ka)r_0^2\mathrm{e}^{\mathrm{i}2\phi}\mathrm{e}^{ik\delta^+}) + (S_\mathrm{f})r_0\mathrm{e}^{\mathrm{i}\phi}F(ka)(\mathrm{e}^{-ik\delta^-} + \mathrm{e}^{ik\delta^-}) \tag{4.21}$$

which represents the total scattering.

For illumination by a Gaussian beam, the total field in the far field is now

$$\begin{aligned} E = \frac{k\sqrt{S_0}}{f}\Bigg[&\frac{S}{-ik^2}\exp\left(-\frac{(x'^2+y'^2)}{2w_0^2}\right)\exp(-ikx'x/f)\exp(-iky'y/f) \\ &-irw_0^2\exp\left(-\frac{2\pi^2 w_0^2}{(f\lambda)^2}(x^2+y^2)\right)\Bigg] \end{aligned} \tag{4.22}$$

where (x', y') are the coordinates of the particle and (x, y) are the coordinates on the Fourier plane.

When there is a surface density N_A of spherical scatterers occupying a volume fraction

$$f_\mathrm{V} = \frac{2\pi}{3}N_A a^2 \tag{4.23}$$

the integral over the surface particle density weighted by the Gaussian incident field is

$$N_A\int_0^\infty \mathrm{e}^{-r'^2/2w_0^2}2\pi r'\mathrm{d}r' = 2\pi N_A w_0^2 \tag{4.24}$$

and the total far field in the nanoparticle double-interaction model becomes

$$E = \frac{kw_0^2\sqrt{S_0}}{f}\left[3f_\mathrm{V}\frac{S}{-ik^2a^2} - ir\right]\exp\left(-\frac{2\pi^2 w_0^2}{(f\lambda)^2}(x^2+y^2)\right) \tag{4.25}$$

for an illuminating beam with beam radius w_0.

4.3 Surface Films

The preceding discussion is for highly dilute molecules or particles on a surface. They do not constitute a thin film, but are isolated weak scatterers. In this limit, they do not modify the waves in the substrate, and hence are in the first Born Approximation (they do not affect the incident wave). However, if the surface density grows large enough, then the optical scattering takes on the properties of a thin film on the surface. This film can be assigned a refractive index and a thickness as in effective medium theories. Once a refractive index and thickness are assumed, these in turn can be used as the properties of a solid dielectric film on a dielectric substrate.

Light incident on a film on a substrate is split at each interface into a series of partial reflections. These all add coherently to yield the resultant reflected and transmitted fields (Fig. 4.10).

The nth partial reflection amplitude is

$$r_n = t_{01}t_{10}r_{12}(r_{10}r_{12})^n e^{i(n+1)\delta} \tag{4.26}$$

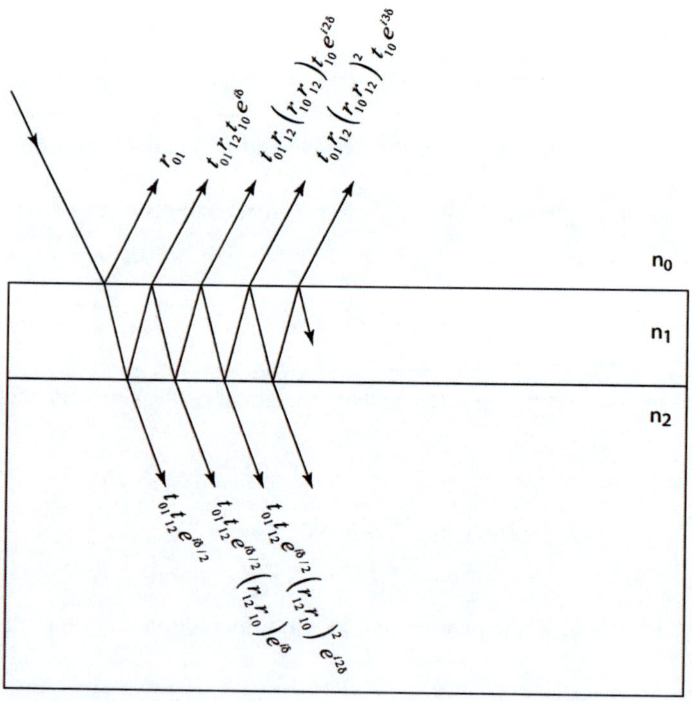

Fig. 4.10 Partial reflections for a single layer on a substrate. The infinite sum of the partial waves leads to the transmitted and reflected fields

where

$$\delta = \frac{4\pi n_1 d_1 \cos\theta_1}{\lambda} \tag{4.27}$$

The total reflection is the sum of all the partial reflections

$$\begin{aligned} r = r_{01} &+ t_{01}t_{10}r_{12}\mathrm{e}^{\mathrm{i}\delta} + t_{01}t_{10}r_{12}(r_{10}r_{12})\mathrm{e}^{\mathrm{i}2\delta} + t_{01}t_{10}r_{12}(r_{10}r_{12})^2\mathrm{e}^{\mathrm{i}3\delta} + \cdots \\ &+ t_{01}t_{10}r_{12}(r_{10}r_{12})^n\mathrm{e}^{\mathrm{i}(n+1)\delta} + \cdots \end{aligned} \tag{4.28}$$

Likewise, the nth partial transmission amplitude is

$$t_n = t_{01}t_{12}(r_{12}r_{10})^n\mathrm{e}^{\mathrm{i}(n+1/2)\delta} \tag{4.29}$$

and the total transmission is the sum of all the partial transmissions

$$\begin{aligned} t = t_{01}t_{12}\mathrm{e}^{\mathrm{i}\delta/2} &+ t_{01}t_{12}(r_{12}r_{10})\mathrm{e}^{\mathrm{i}3\delta/2} + t_{01}t_{12}(r_{12}r_{10})^2\mathrm{e}^{\mathrm{i}5\delta/2} + \cdots \\ &+ t_{01}t_{12}(r_{12}r_{10})^n\mathrm{e}^{\mathrm{i}(n+1/2)\delta} + \cdots \end{aligned} \tag{4.30}$$

The infinite series for reflection converges to

$$r = r_{01} + \frac{t_{01}t_{10}r_{12}\mathrm{e}^{\mathrm{i}\delta}}{1 - r_{10}r_{12}\mathrm{e}^{\mathrm{i}\delta}} \tag{4.31}$$

and for transmission to

$$t = \frac{t_{01}t_{12}\mathrm{e}^{\mathrm{i}\delta/2}}{1 - r_{10}r_{12}\mathrm{e}^{\mathrm{i}\delta}} \tag{4.32}$$

By using the relations from the Stoke’s treatment of reversibility

$$r_{01} = -r_{10} \tag{4.33}$$

and no absorption

$$1 = (t_{01}t_{10} - r_{01}r_{10}) \tag{4.34}$$

these become

$$r = \frac{r_{01} + r_{12}\mathrm{e}^{\mathrm{i}\delta}}{1 + r_{01}r_{12}\mathrm{e}^{\mathrm{i}\delta}} \tag{4.35}$$

and

$$t = \frac{t_{01}t_{12}\mathrm{e}^{\mathrm{i}\delta/2}}{1 + r_{01}r_{12}\mathrm{e}^{\mathrm{i}\delta}} \tag{4.36}$$

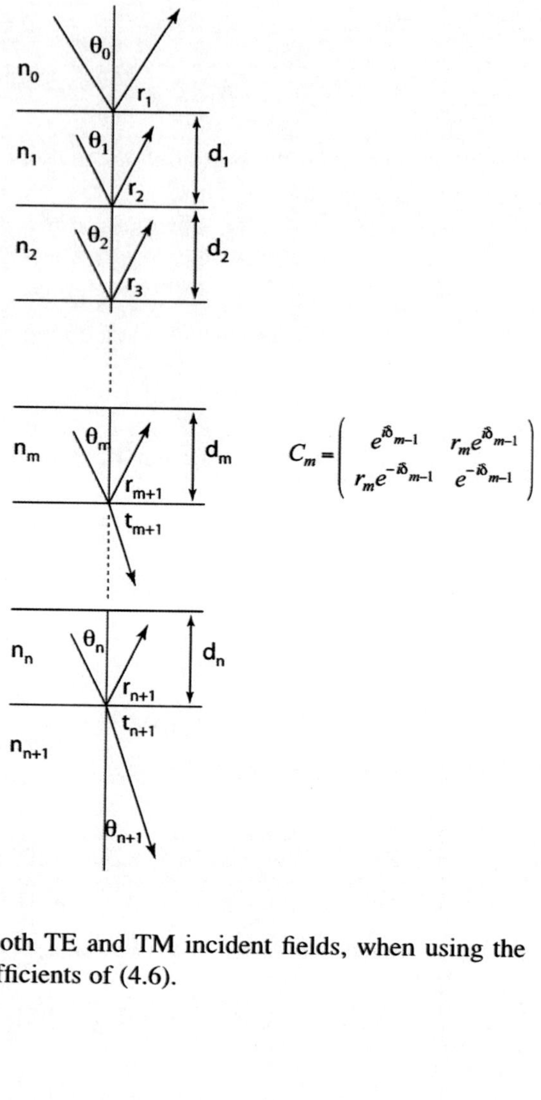

Fig. 4.11 A n-layer stack on a substrate. Each interface has an associated matrix. The subsequent propagation is described by an ordered product of the transfer matrices

These expressions are valid for both TE and TM incident fields, when using the appropriate Fresnel reflection coefficients of (4.6).

4.3.1 *Transfer Matrix*

Multiple layers on a surface are encountered frequently in biosensor applications. The reflection and transmission coefficients of light incident on a multilayer media are derived from the transfer matrix approach, in which each interface of the multilayer stack has a matrix associated with it.

The configuration of a multilayer stack is shown in Fig. 4.11. The incident medium has index n_0 and the substrate has index n_{n+1}. This configuration has n layers. The mth interface (overlying the mth layer) has the associated matrix

$$C_m = \begin{pmatrix} \mathrm{e}^{\mathrm{i}\delta_{m-1}} & r_m \mathrm{e}^{\mathrm{i}\delta_{m-1}} \\ r_m \mathrm{e}^{-\mathrm{i}\delta_{m-1}} & \mathrm{e}^{-\mathrm{i}\delta_{m-1}} \end{pmatrix} \tag{4.37}$$

where the r_m are the Fresnel reflection coefficients of (4.6) for the appropriate incident polarization. The phase terms are

$$\delta_m = \frac{2\pi n_m \cos\theta_m}{\lambda} d_m \tag{4.38}$$

where λ is the vacuum wavelength, θ_m is the angle in the mth layer, and the sum is over the thicknesses of m layers. After passing through all interfaces, the resultant matrix is the ordered product of the C_m matrices

$$C' = C_1 C_2 \ldots C_{n+1} = \begin{pmatrix} c_{\mathrm{in}} & b \\ c_{\mathrm{ref}} & d \end{pmatrix} \tag{4.39}$$

The reflection coefficient is

$$r = \frac{c_{\mathrm{ref}}}{c_{\mathrm{in}}} \tag{4.40}$$

and the transmission coefficient is

$$t = \frac{t_1 t_2 \ldots t_{n+1}}{c_{\mathrm{in}}} \tag{4.41}$$

where the t_m are the Fresnel transmission coefficients of (4.6) for the appropriate incident polarization.

4.3.2 *Biolayers on a Substrate*

One of the most common configurations for a label-free biosensor is to immobilize or capture a thin biofilm onto a surface. In the case of a DNA array, DNA molecules are immobilized onto the surface, which then capture an additional amount of target DNA molecules out of a sample. Similarly, in the case of a protein immunoassay, antibodies are immobilized on a surface which then capture target molecules and bind them to the surface. The change in the surface reflection amplitudes caused by the immobilized or captured molecules can be measured directly in reflectance measurements.

In principle, the added biolayer affects all the partial waves in all of the underlying layers of the substrate, which may be composed of a multilayer dielectric stack. However, in the limit of a thin layer (for instance a monolayer of protein), the resulting reflection coefficient can be described simply as a function of the original reflection coefficient. This simplifies the analysis because the underlying multilayer structure contributes once to the original reflection coefficient, which is expressed simply in terms of an amplitude and a phase $r = |r|e^{i\phi}$.

For a thin biolayer of thickness d on a (possibly multilayered) substrate, the reflection coefficient is

$$r_{\text{biolayer}} = \frac{r_{\text{p}} + r_{\text{ps}} e^{i\delta}}{1 + r_{\text{p}} r_{\text{ps}} e^{i\delta}} \tag{4.42}$$

where $r_{\text{p}} = r_{\text{mp}} = (n_{\text{m}} - n_{\text{p}})/(n_{\text{m}} + n_{\text{p}})$ is the reflection coefficient of the ambient medium interface with the protein layer, and r_{ps} is the reflection coefficient at the interface between the protein layer and the substrate. The phase term in the exponential is

$$\delta = \frac{4\pi n_{\text{p}} d \cos\theta_{\text{p}}}{\lambda} \tag{4.43}$$

We are most interested in the expression for the reflection coefficient when the protein layer modifies the original reflection relative to the ambient medium, i.e., in terms of $r_0 = r_{\text{ms}}$ instead of r_{ps}. This is achieved using the basic identity

$$r_{ij} = \frac{r_{ik} + r_{kj}}{1 + r_{ik} r_{kj}} \tag{4.44}$$

to give

$$r_{\text{ps}} = \frac{r_{\text{pm}} + r_{\text{ms}}}{1 + r_{\text{pm}} r_{\text{ms}}} \tag{4.45}$$

Therefore,

$$r_{\text{biolayer}} = \frac{r_{\text{mp}}(1 - r_{\text{mp}} r_{\text{ms}}) + (r_{\text{ms}} - r_{\text{mp}}) e^{i\delta}}{(1 - r_{\text{mp}} r_{\text{ms}}) + r_{\text{mp}}(r_{\text{ms}} - r_{\text{mp}}) e^{i\delta}} \tag{4.46}$$

This reflection coefficient is referenced relative to the top of the protein layer. However, in experiments, a convenient reference level is the top of the substrate. Therefore, the difference in phase between a portion of the substrate carrying a biolayer and a portion that is free of the biolayer is

$$\Delta r_{\text{biolayer}} = \frac{r_{\text{p}}(1 - r_{\text{p}} r_0) + (r_0 - r_{\text{p}}) e^{i\delta}}{(1 - r_{\text{p}} r_0) + r_{\text{p}}(r_0 - r_{\text{p}}) e^{i\delta}} - r_0 e^{i\delta_0} \tag{4.47}$$

where

$$\delta_0 = \frac{4\pi n_{\text{m}} d \cos\theta_{\text{m}}}{\lambda} \tag{4.48}$$

is the equivalent phase shift of the medium across the thickness d. The relative quantity in (4.47) is what is measured in imaging biosensors and also in diffraction biosensors.

In most situations, the biolayer thickness has molecular dimensions of nanometers for which d is very small. The change in reflection can then be expanded to give

$$\Delta r_{\text{biolayer}} = i\frac{4\pi d}{\lambda}\left[n_{\text{p}}\cos\theta_{\text{p}}\frac{(r_0 - r_{\text{p}})(1 - r_0 r_{\text{p}})}{(1 - r_{\text{p}}^2)} - n_{\text{m}}\cos\theta_{\text{m}} r_0\right] \tag{4.49}$$

Under the condition of s-polarization, (4.49) is further simplified to [2]

Thin Film Reflection:

$$r_{\text{biolayer}} = r_0 + \frac{(1 + r_0)^2}{n_{\text{m}}\cos\theta_{\text{m}}}\frac{\pi i}{\lambda}(n_{\text{m}}^2 - n_{\text{p}}^2)d_{\text{p}} \tag{4.50}$$

This form can be recognized as the first Born approximation where the wave reflecting from the original surface is unaffected by the layer, and the layer simply contributes an additional scattered field.

The maximum change in imaging (intensity) contrast caused by the immobilized biolayer comes by maximizing the relation

$$\text{Im}\left[(1 + r_0)^2 r_0^*\right] = \text{Im}\,(r_0)(R - 1) \tag{4.51}$$

that reaches extrema of ∓ 0.385 at $r = \pm 0.577i$ ($|r|$ is always less than 1). This factor is plotted in Fig. 4.12 as a function of the amplitude of the reflection coefficient and the phase of the reflection. The maximum occurs when the phase of the original reflection is $\pi/2$, which provides a quadrature condition and produces the maximum imaging contrast of a biolayer on a substrate. It is important to note that these conditions are met closely by thermal oxide on silicon [4].

The reflection coefficient of the surface carrying the biolayer can be reexpressed as

$$\begin{aligned} r_{\text{biolayer}} &= r_0\left(1 + \frac{(1 + r_0)^2}{r_0 n_{\text{m}}\cos\theta_{\text{m}}}\frac{\pi i}{\lambda}(n_{\text{m}}^2 - n_{\text{p}}^2)d\right) \\ &\approx r_0\exp(i\Delta\phi_{\text{b}}) \end{aligned} \tag{4.52}$$

where the shift in phase induced by the biolayer is

Thin Film Phase:

$$\Delta\phi_{\text{b}} = \frac{(1 + r_0)^2}{2r_0}\frac{k_0}{n_{\text{m}}\cos\theta_{\text{m}}}(n_{\text{m}}^2 - n_{\text{p}}^2)d \tag{4.53}$$

This expression replaces the overly simplistic approximation $\Delta\phi = 2k_0(n - n_{\text{m}})d$ often assumed for the phase shift imposed by a thin layer on a reflecting surface.

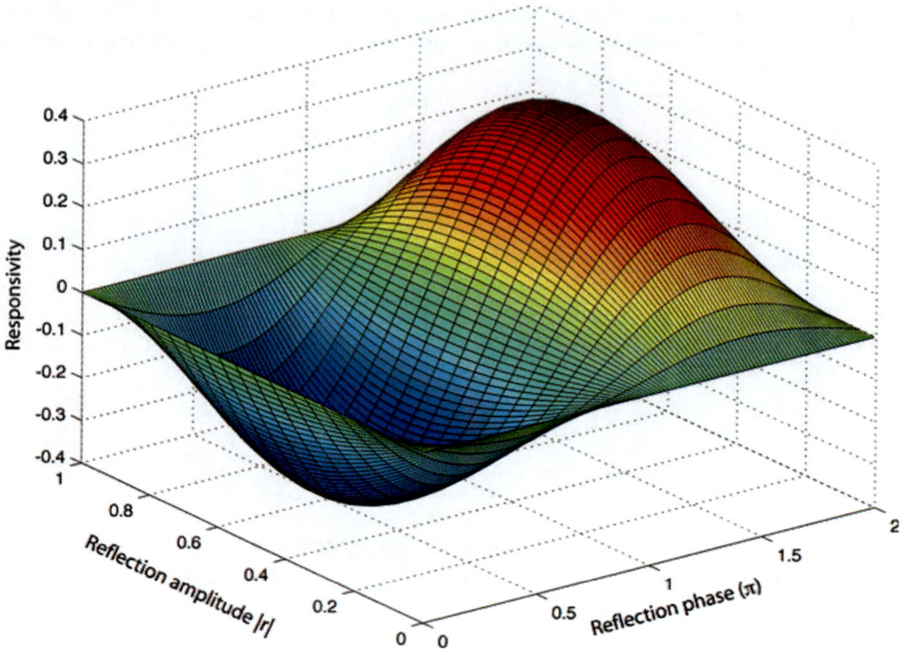

Fig. 4.12 The imaging contrast factor of (4.51) plotted vs. the amplitude of the reflection coefficient and the phase of the ambient reflection. The factor vanishes for both unity and zero reflectance (from [2])

4.4 Surface Plasmons

Gold layers on surfaces support surface plasmons. These are propagating waves that travel along the surface of the gold supported by charge oscillations in the free-electrons of the gold metal. Surface plasmons were the source of the plasmon resonance of gold nanoparticles in Chap. 2, and similar effects occur when plasmons are launched on the surface of planar gold sheets. Biosensors based on surface plasmon absorption are among the most common optical biosensors for measuring binding kinetics from solution [8, 9].

4.4.1 Planar Gold Films

From Maxwell's equations, for TM-polarization, the relationship between E_x and H_y is

$$E_x = -i\frac{1}{\omega\varepsilon_0\varepsilon}\frac{\partial H_y}{\partial z} \tag{4.54}$$

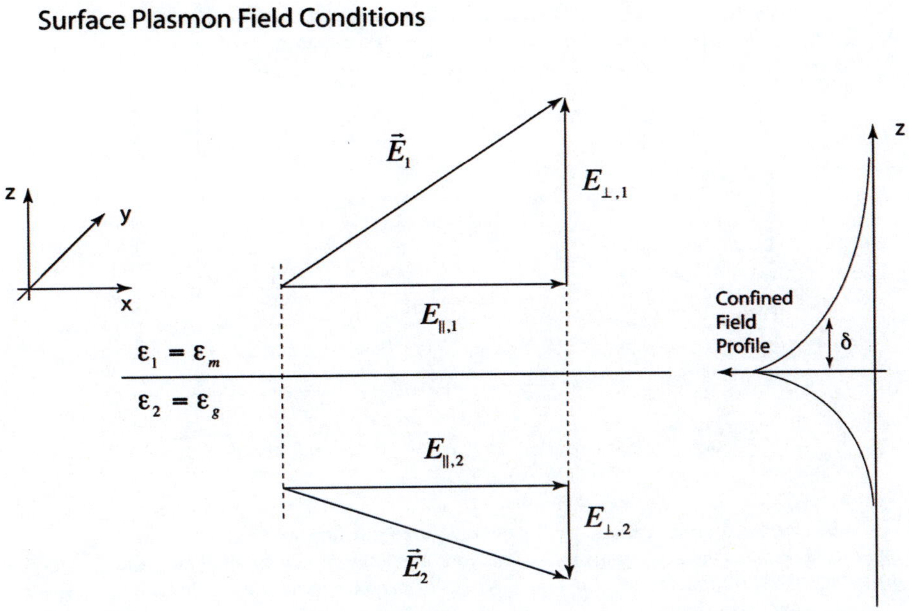

Fig. 4.13 Plasmon field conditions at the boundary of a metal and a dielectric. Only TM incident light polarization can generate surface plasmons, confined at the interface, that support longitudinal electric fields caused by charge oscillations on the metal surface

A surface wave localized at the interface has the property of free propagation along the interface, but exponential decay away from the interface. The field solution in this case has the form

$$\begin{aligned} H_y(z) &= A\mathrm{e}^{\mathrm{i}k_x x}\mathrm{e}^{-k_{z1}z}, \quad z > 0 \\ H_y(z) &= A\mathrm{e}^{\mathrm{i}k_x x}\mathrm{e}^{k_{z2}z}, \quad z < 0 \end{aligned} \tag{4.55}$$

When this field is applied to (4.54) at the interface between two materials with dielectric constants ε_m and ε_g, and the continuity of the transverse electric field at the interface is used (Fig. 4.13), the following condition is obtained

$$\frac{k_{z1}}{\varepsilon_m} + \frac{k_{z2}}{\varepsilon_g} = 0 \tag{4.56}$$

Because k_{zi} is positive for each material, this solution can only be satisfied when the real parts of the dielectric constant have opposite signs. Hence, localized surface waves only exist when one of the dielectric constants are negative, which is satisfied by nearly free electron metals at frequencies below the plasma frequency of the metal. Equation (4.56) is combined with the definition for the propagation constant in each material,

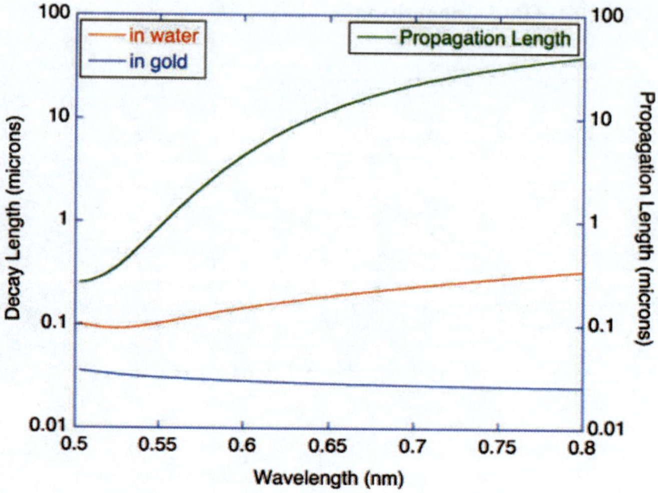

Fig. 4.14 Propagation length and field penetration lengths into water and gold at a gold–water interface. Plasmons propagate only tens of microns along the surface before being attenuated. The decay into the surrounding media is typically tens of nanometers into the gold, and up to several hundred nanometers into the water

$$k_x^2 - k_{zm,g}^2 = \varepsilon_{m,g} \left(\frac{\omega}{c}\right)^2 \tag{4.57}$$

and is solved for the longitudinal component of the k-vector

$$k_x = \frac{\omega}{c} \sqrt{\frac{\varepsilon_m(\omega)\varepsilon_g(\omega)}{\varepsilon_m(\omega) + \varepsilon_g(\omega)}} \tag{4.58}$$

which is in general complex, consisting of a propagating part modulated by an attenuation. The transverse k-vector is

$$k_{zm,g} = \pm i \frac{\omega}{c} \frac{\varepsilon_{m,g}}{\sqrt{\varepsilon_{m,g} + \varepsilon_{g,m}}} \tag{4.59}$$

which represents the decaying field into each medium (gold and substrate).

The decay lengths for plasmons on a gold–water interface are plotted in Fig. 4.14 as a function of wavelength. The decay length in the direction of propagation varies from a few to tens of microns. The decay into the half spaces is much smaller. The decay of the fields into the gold occurs in a short range of tens of nanometers, and into the water is about 100 nm.

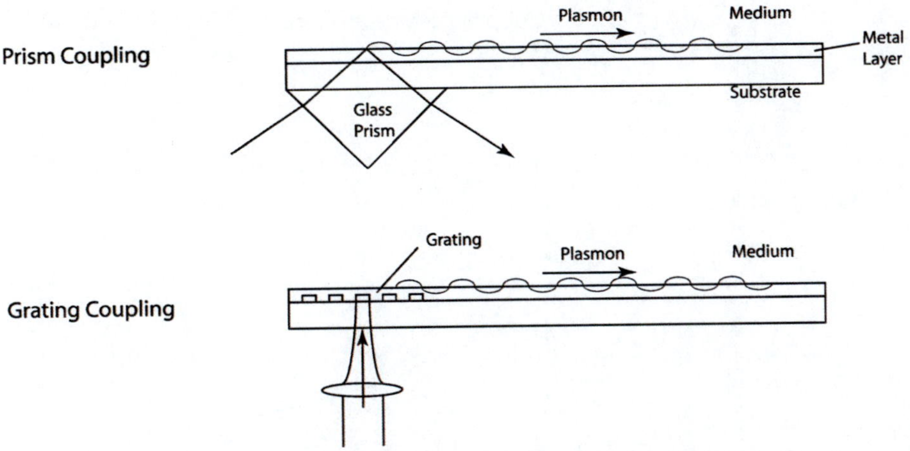

Fig. 4.15 Plasmon polariton coupling for Kreitschmann (*top*) and grating (*bottom*). Note the figure is not to scale: the plasmon propagation distance is only tens of microns

4.4.2 *Plasmon Polariton Coupling*

Surface plasmons are excited by optical waves when the transverse component of the light wave *k*-vector matches the *k*-vector of the surface plasmon. The most common configuration for the coupling is through the use of a glass prism in the Kreitschmann configuration, shown in Fig. 4.15. However, it is also possible to match *k*-vectors by using a diffraction grating relief in the metal film.

The *k*-vector of the surface plasmon at the sample–gold interface, as a function of frequency, is given by (4.58) and the transverse *k*-vector for light is

$$k_t = \omega n_{\mathrm{p}} \sin \theta / c \tag{4.60}$$

where n_{p} is the refractive index of the substrate (prism). The coupling condition is therefore

$$n_{\mathrm{p}} \sin \theta = \sqrt{\frac{\varepsilon_{\mathrm{s}}(\omega)\varepsilon_{\mathrm{Au}}(\omega)}{\varepsilon_{\mathrm{s}}(\omega) + \varepsilon_{\mathrm{Au}}(\omega)}} \tag{4.61}$$

When this equation is satisfied, then light couples into the surface plasmons. This is detected as a decrease of the reflectance of the light from the gold surface, either as a function of angle, or as a function of wavelength.

The plasmon *k*-vector is compared with the transverse *k*-vector of light in glass in Fig. 4.16. When the real-part of the plasmon *k*-vector matches the *k*-vector of light in the substrate, then the conditions are satisfied for launching a surface plasmon. Because of dissipation in the gold, the plasmon coupling resonance is not extremely narrow. The reflectance of TM-polarized light as a function of angle is shown in Fig. 4.17 with and without a 3 nm biolayer. The presence of the biolayer shifts the resonance condition. The amount of shift is linearly proportional to the protein thickness, which can be calibrated accurately because of the sharp resonance condition.

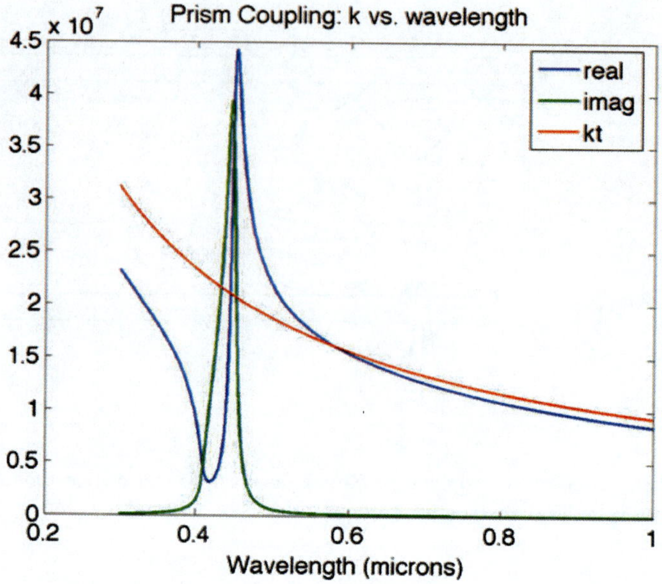

Fig. 4.16 Transverse k-vector for the optical wave in the glass prism incident at an angle of 75°, compared with the real and imaginary parts of the surface plasmon k-vector. The sample on the gold is water with $n = 1.33$. The light couples to the surface plasmons at a wavelength just less than 600 nm for this condition

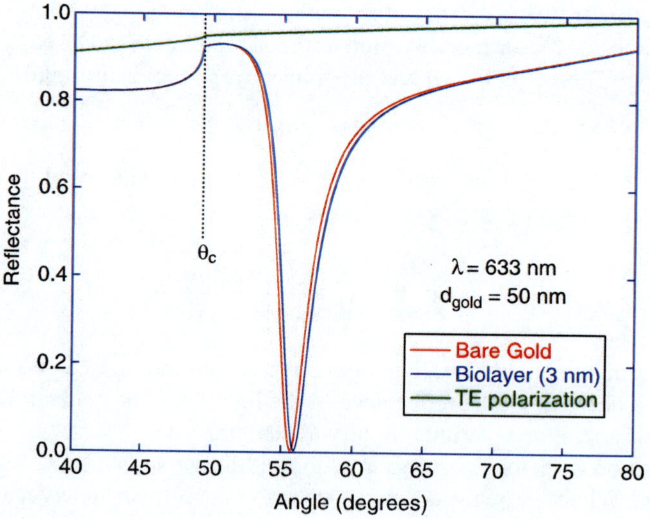

Fig. 4.17 Reflectance for a 3 nm biofilm in water on gold on Schott glass for TM polarization (with and without film) and for TE polarization

Selected Bibliography

Heavens, O.S.: Optical Properties of Thin Solid Films. Dover, New York (1991) (An old classic on thin films)

Maier, S.A.: Plasmonics: Fundamentals and Applications. Springer, New York (2007) (A good modern review of plasmonics)

References

1. Ergin, T., Stenger, N., Brenner, P., Pendry, J.B., Wegener, M.: Three-dimensional invisibility cloak at optical wavelengths. Science **328**, 337–339 (2010)
2. Wang, X., Zhao, M., Nolte, D.D.: Common-path interferometric detection of protein on the BioCD. Appl. Opt. **46**, 7836–7849 (2007)
3. Jenison, R., La, H., Haeberli, A., Ostroff, R., Polisky, B.: Silicon-based biosensors for rapid detection of protein or nucleic acid targets. Clin. Chem. **47**, 1894–1900 (2001)
4. Zhao, M., Wang, X., Nolte, D.D.: Molecular interferometric imaging. Opt. Express **16**, 7102–7118 (2008)
5. Ozkumur, E., Yalcin, A., Cretich, M., Lopez, C.A., Bergstein, D.A., Goldberg, B.B., Chiari, M., Unlu, M.S.: Quantification of DNA and protein adsorption by optical phase shift. Biosens. Bioelectron. **25**, 167–172 (2009)
6. de la Pena, J.L., Gonzalez, F., Saiz, J.M., Moreno, F., Valle, P.J.: Application of a double-interaction model to the backscattering from particulate surfaces. Opt. Eng. **38**, 1017–1023 (1999)
7. Nebeker, B.M., de la Pena, J.L., Hirleman, E.D.: Comparisons of the discrete-dipole approximation and modified double interaction model methods to predict light scattering from small features on surfaces. J. Quant. Spectrosc. Radiat. Transf. **70**, 749–759 (2001)
8. Homola, J.: Surface Plasmon Resonance Based Sensors. Springer, Berlin (2006)
9. Schasfoort, R.B.M.: Handbook of Surface Plasmon Resonance. Royal Society of Chemistry, Cambridge (2008)

Part II
Molecular Interferometry and Biosensors

Chapter 5
Interferometric Thin-Film Optical Biosensors

When molecules accumulate out of solution onto a dielectric substrate that is illuminated by an optical field, the reflected field is altered by the accumulating surface-bound molecules. If the molecular surface attachment is specific, meaning that only predefined target molecules are bound, while other nontarget molecules do not bind, then by monitoring the reflection change, the target molecule concentration in the solution can be obtained. This is the basic process of all surface-capture affinity optical biosensors: specific molecular recognition, followed by transduction of the captured molecules to a measurable optical signal. The primary goal of all affinity biosensors is to achieve the highest specificity (ratio of specific to nonspecific captured molecules) at the highest sensitivity (smallest detectable molecular signal).

Because of their simplicity, reflectometric thin-film sensors are among the most popular biosensors. Optical beams are incident from an overlying half-space, interact with the surface-bound molecules, and are reflected and detected. However, in this class of biosensor the optical path length through the molecular layer is extremely small. For instance, the optical phase change for passage through 1nm layer of protein is only about 10 or 20 mrad. Therefore, these biosensors must achieve extremely low-noise operation with high-precision referencing to achieve high sensitivity. They also should maximize the signal change (reflectivity change) per molecule, called the responsivity. Most of the design effort of thin-film optical biosensors is put into the pursuit of maximized responsivity. In all cases of maximized response, interferometric principles play either an explicit or an implicit role.

This chapter presents a perspective on thin-film interferometric biosensors, describing their operational principles and their typical performance. These sensors include colorimetric biosensors based on color changes, ellipsometric biosensors based on polarization changes, molecular interferometric imaging, and the related BioCD approaches.

D.D. Nolte, *Optical Interferometry for Biology and Medicine*, Bioanalysis 1,
DOI 10.1007/978-1-4614-0890-1_5,

5.1 Label-Free Optical Biosensors and Direct Detection

One of the driving forces for label-free biosensors is their potential to perform high-throughput detection in which hundreds or thousands of assays could be performed simultaneously – a high degree of multiplexing. A multiplexed assay is one that seeks many target analytes (from tens to hundreds) in a single biological sample. High multiplexing is possible with label-free approaches because the molecular interactions are detected directly and do not require additional labels. For instance, in conventional multiplexed fluorescence assays, one requires as many different fluorescence labels as target analytes. This has limited fluorescent multiplexing to a few dozen analytes at a time, although that number continually increases.

One example where highly multiplexed label-free assays could provide value is in screens of blood proteins. There are thousands of proteins in the blood, with a concentration range that spans 12 orders of magnitude. Most of these proteins are not intrinsic to the blood, but arise from all parts and functions of the body. They are the faint trace of the overall health of the individual. A plot of a partial inventory of 70 important blood proteins is given in Fig. 5.1. The most abundant proteins are hemoglobin and albumin at mg/mL concentrations, followed by immunoglobulins at μg/mL concentrations. Beyond those, most of the proteins lie on an exponential trend of concentration as a function of their concentration rank [1]. Important cytokines, such as TNF and IL-4, are present at sub-picomolar concentrations.

The challenge to direct detection is the absence of intrinsic selectivity for many of the approaches. The most common label-free sensors are called "mass-sensitive"

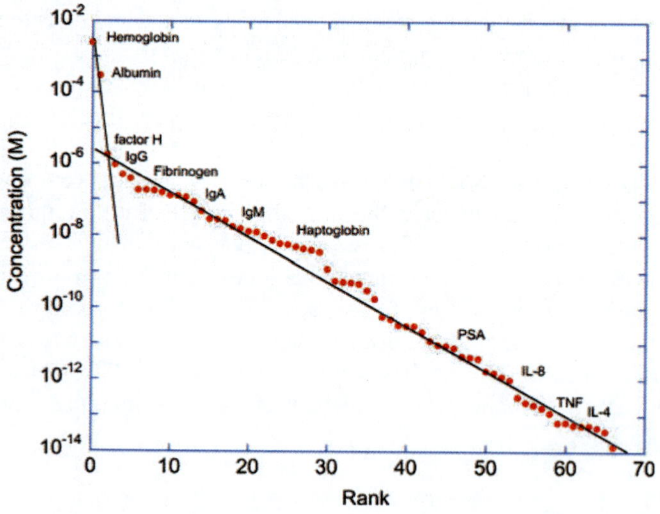

Fig. 5.1 Plot of protein concentration in blood as a function of its rank of abundance. The most abundant blood proteins are immunoglobulins and albumin. The main trend exhibits an exponential dependence on rank that spans eight orders of magnitude in concentration

because they detect only the presence of target analyte. The nomenclature "mass-sensitive" arises not because they detect mass (although quartz-crystal balances do detect mass density), but because dipole density of biological molecules is closely proportional to molecular weight. Of the mass-detection platforms, the most common is surface-plasmon resonance, which senses the effective index of accumulating molecules on the surface of the plasmonic gold sensor. Other common types of mass-sensitive detectors are the interferometric sensors described in this and the next several chapters. The problem with direct detection is that any molecule that attaches to a recognition site is measured, whether it is the desired target or not. Such nonspecific binding is the fundamental bottleneck to high-sensitivity multiplexed label-free assays. In the case of DNA, the high specificity of the recognition molecule can suppress nonspecific signal. However, in the case of protein assays, the specificity of antibodies is not as high. Even simple chemical residues act as nonspecific mass. An additional challenge for label-free protein assays relative to gene chips is the lack of amplification that is readily achieved for DNA using the PCR reaction.

These two challenges – nonspecific binding and lack of amplification – limit highly multiplexed label-free protein assays to the range of 100 pg/mL or higher. This level of sensitivity still leaves the door open for about a hundred abundant blood proteins. However, it is not yet clear whether these abundant molecules are high-specificity biomarkers for important diseases. For this reason, most interferometric biosensors are used today for chemical evaluation of reagents and recognition molecules like antibodies or aptamers, measuring association and dissociation rates for chemical reactions. In this application, the absence of tag molecules attached to the molecules of interest ensure a higher relevance of the measured kinetic rates than might be obtained using fluorescent tags or biotin–avidin labeling.

5.2 Ellipsometric Biosensors

Ellipsometry is an established experimental approach to measure the optical properties of thin films. Ellipsometry measures the change in polarization, upon reflection from a surface, of obliquely incident light. As a biological film accumulates on the substrate surface, the polarization of the reflected light shifts according to the amount of biomolecule accumulation.

5.2.1 Experimental Ellipsometry on Biolayers

The key property to extract from ellipsometric experiments is the ratio of p-polarized to s-polarized reflection coefficients. The most common polarization-control experiment that performs ellipsometry is shown in Fig. 5.2, called the PCSA (Polarizer-Compensator-Sample-Analyzer) configuration. It consists of a polarizer (P), compensator (C), sample (S), and analyzing polarizer (A) in

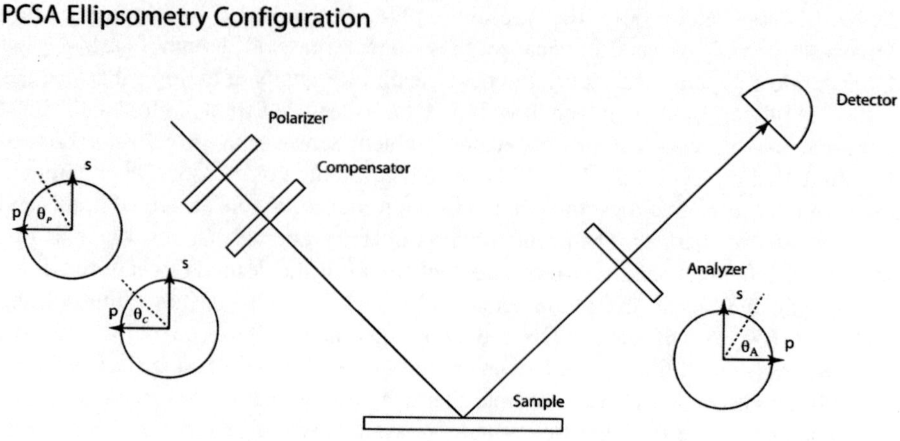

Fig. 5.2 Polarizer-compensator-sample-analyzer (PCSA) configuration for ellipsometry measurements. The compensator angle is usually set to $\theta_C = \pi/4$. The polarizer angle θ_P and the analyzer angle θ_A are changed to null the reflectance

sequence. In null ellipsometry the compensator angle θ_C is usually set to $+\pi/4$, and the angles θ_P and θ_A are adjusted to null the intensity at the detector.

The detected intensity is defined as [2]

$$I = K|L|^2 \tag{5.1}$$

where K is a proportional constant, and the function L is (when the compensator angle is fixed at $\theta_C = \pi/4$)

$$\begin{aligned} L = \frac{R_s}{\sqrt{2}} &\left[\tan \Psi e^{i\Delta} \cos \theta_A \{\cos(\theta_P - \pi/4) + i \sin(\theta_P - \pi/4)\}\right. \\ &\left. + \sin \theta_A \{\cos(\theta_P - \pi/4) - i \sin(\theta_P - \pi/4)\}\right] \end{aligned} \tag{5.2}$$

The ratio of p-polarized to s-polarized intesities is expressed as

$$\frac{\rho_P}{\rho_s} = \tan \Psi \exp(i\Delta) \tag{5.3}$$

When the detected intensity is nulled by adjusting θ_A and θ_P, the quantities in (5.3) are

$$\Psi = \theta_A^{null} \Delta = 3\pi/2 - 2\theta_P^{null} \tag{5.4}$$

The off-null intensity is then

$$I = K\frac{|R_s|^2}{8\cos^2\theta_A^{\text{null}}}\left[\sin^2\left(\theta_A - \theta_A^{\text{null}}\right) + \sin 2\theta_A \sin 2\theta_A^{\text{null}}\sin^2\left(\theta_P - \theta_P^{\text{null}}\right)\right] \tag{5.5}$$

This equation shows the quadratic dependence of the intensity on the deviation from the null angles.

In off-null ellipsometry, the aim is to measure the time-dependent change in θ_A^{null} and θ_P^{null} as molecules accumulate on the surface. Because surface coverage cannot be easily characterized by a thickness, a more apt surface coverage characteristic is the mass density. This can be related back to film thickness d and index through the relation

$$\Gamma = \frac{d(n_f - n_m)}{dn/dc} \tag{5.6}$$

in units of g/mm^2, where [3]

$$dn/dc = 0.18 \text{ mL/g} \tag{5.7}$$

for globular proteins, and h_f is the film thickness in units of meters. For instance, for a 1 nm film, the surface mass concentration for a typical protein is $\Gamma = 1$ ng/mm^2. In terms of the surface mass density, the changes in the null angles are

$$d\theta_A^{\text{null}} = K_A\Gamma \qquad d\theta_P^{\text{null}} = K_P\Gamma \tag{5.8}$$

calculated using an appropriate multilayer numerical model, and (5.5) becomes

$$I = K\frac{|R_s|^2}{8\cos^2\theta_A^{\text{null}}}\left[K_A^2 + \sin 2\theta_A^0 \sin 2\theta_A^{\text{null}} K_P^2\right]\Gamma^2 \tag{5.9}$$

where θ^0 is the null analyzer angle on a bare surface. This equation again shows the quadratic dependence of the intensity on the surface mass density.

High multiplexing ability is possible with an imaging extension of the PCSA single-channel configuration by using expanded parallel illumination and a CCD camera [4]. A schematic is shown in Fig. 5.3. The system operates in the off-null mode by monitoring real-time changes to the spatially resolved reflectance ratio. The system is generally nulled on a reference surface, possibly being simply an area of the surrounding substrate that is clear of binding. The spatial resolution of the system was 5 μm and the field of view was 5 mm.

The field of view of imaging ellipsometry is restricted to several millimeters, limiting the number of parallel assay measurements that can be performed. A broad-area laser scanning system operates under the same principles and with the same advantages as imaging ellipsometry, but can scan areas several centimeters in size [5].

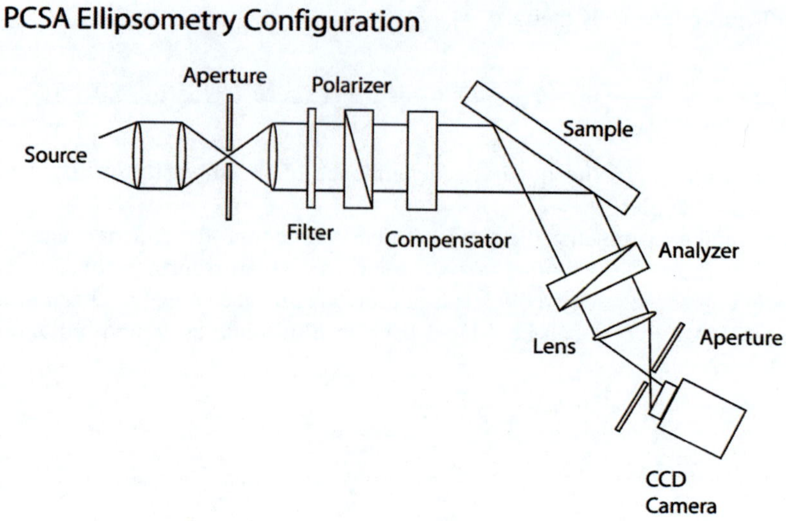

Fig. 5.3 Imaging ellipsometer PCSA configuration with a 5 mm field of view and 5 μm spatial resolution. With permission from Jin et al. [4]

5.2.2 *Interferometric Ellipsometry on Biolayers*

When the biolayer rests on a separate dielectric layer on a substrate, interference effects occur between the top and bottom surfaces of the extra layer. Under the appropriate angles and wavelengths, antireflection conditions can be established for one polarization or the other. At the antireflection condition, a biolayer lifts the null reflectance with a quadratic dependence on the biolayer thickness. Despite the small intensity modulation that results, contrast can be high because of the null reflectance.

One configuration with this condition is a thermal oxide on silicon for high incident angle with s-polarization. The reflectance at $\lambda = 633$ nm for a 260 nm thermal oxide ($n = 1.46$) on silicon ($n = 3.8678 + 0.0188i$) in a water medium ($n = 1.33$) is shown in Fig. 5.4a as a function of incident angle. Near grazing incidence around 85°, the s-polarization has a reflectance null. The change in reflectance for a 1 nm biolayer is shown in Fig. 5.4b at about 0.25% of the incident intensity. However, the contrast is high because of the null reflectance. This principle has been used to develop a binding assay called arrayed imaging reflectometry (AIR) [6].

The grazing angle of incidence in AIR is difficult to maintain with relaxed bandwidths and beam collimations. However, a related effect called Brewster Angle Straddling Interferometry (BASI) occurs for p-polarization, but at a lower angle and with a broader operating window, using an extremely thin layer of thermal oxide on silicon [7]. A simulation of the reflectance is shown in Fig. 5.5a

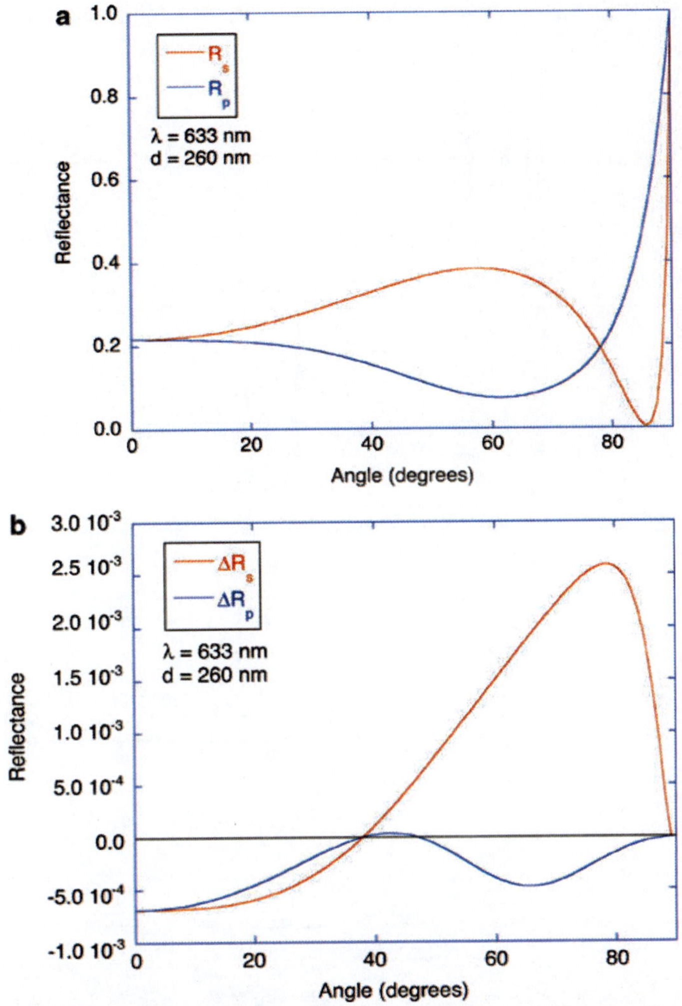

Fig. 5.4 A reflectance null for s-polarization occurs at grazing incidence from water on a 260 nm thermal oxide on silicon. (**a**) The reflectance vs. angle and polarization. (**b**) The reflectance change caused by 1 nm protein layer

with the reflectance change caused by 1 nm of protein shown in Fig. 5.5b. In this configuration the thermal oxide thickness is only 2 nm. The p-polarized light has a reflectance null at the Brewster angle caused by balanced destructive interference between the top and bottom interface of the oxide. The results are shown for a water medium. Although the reflectance change is extremely small because of the extremely weak electric field condition at the biolayer, the null condition can produce high contrast and a broader range of performance.

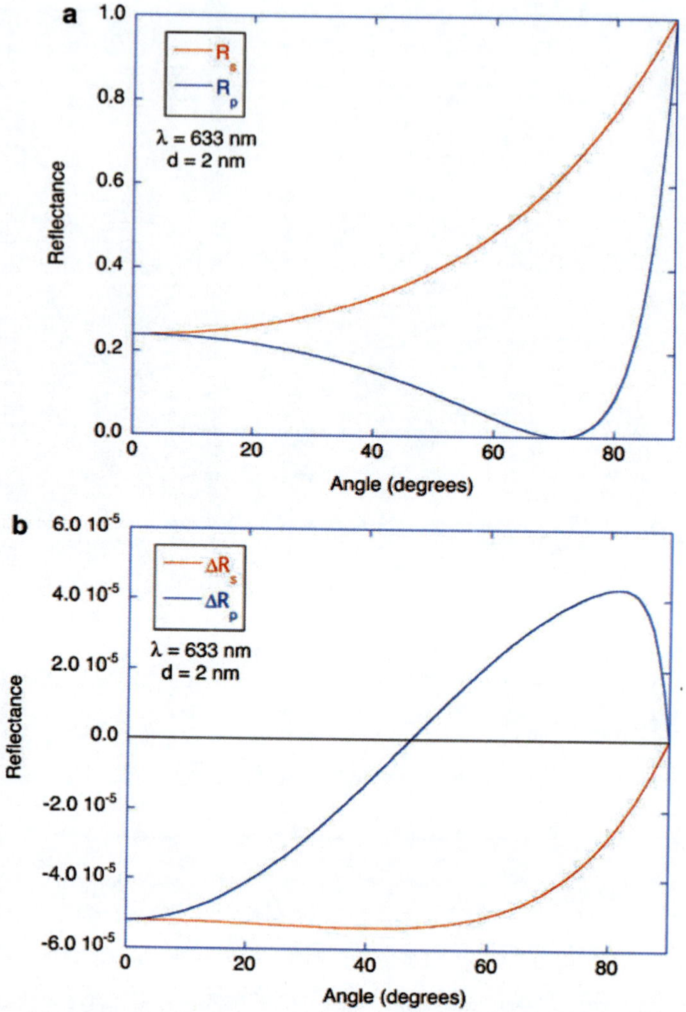

Fig. 5.5 Brewster angle straddling interferometry (BASI) for 2 nm thermal oxide on silicon. The p-polarized light has a reflectance null at the Brewster angle. (**a**) The p- and s-polarized reflectance from water onto thermal oxide on silicon. (**b**) The change in reflectance caused by 1 nm of protein

5.3 Thin-Film Colorimetric Biosensors

Anyone who has owned a good pair of binoculars or an expensive camera has probably experienced colorimetric detection of biolayers. All good commercial optics have antireflection-coated primary lenses. These appear as a deep blue or purple when viewed in natural lighting. If you inadvertently put your finger on the

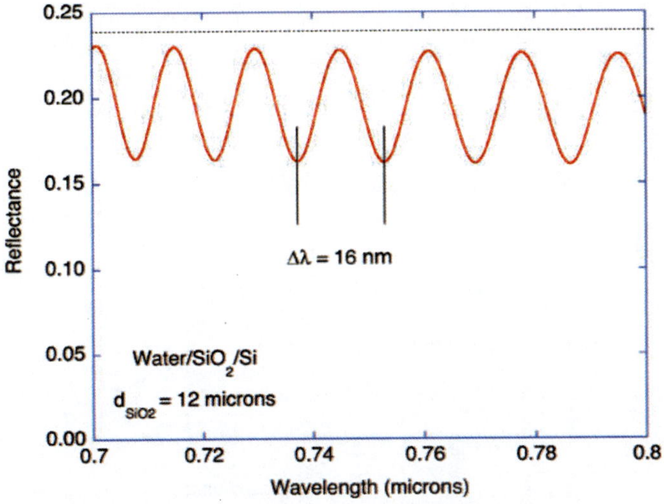

Fig. 5.6 Spectral interference fringes from a 12 μm oxide on silicon in water

coating, it leaves a multicolored smudge that is clearly visible, despite the thinness of the biological film. This same principle can be applied intentionally to thin film biosensors to detect very thin biolayers. As a layer of molecules accumulates on specially designed surfaces, the color of the reflected light changes, forming the basic transduction for colorimetric biosensors. By using a spectrometer, very small color changes can be measured, corresponding to molecularly thin layers.

From (4.42) in Chap. 4, the equation for the reflectance of a film on a substrate is

$$r = \frac{r_{01} + r_{12}e^{i\delta}}{1 + r_{01}r_{12}e^{i\delta}} \tag{5.10}$$

For example, a 12-μm thick thermal oxide grown on silicon has the spectral reflectance shown in Fig. 5.6. The free spectral range in this case is 16 nm. The regular periodicity of the reflectance as a function of wavelength provides a multi-channel measurement for subtle shifts in the reflectance spectral fringes. For example, a thickness change of 3 nm in the oxide layer causes the fringes to shift, as shown in Fig. 5.7a. The shift is only 0.3 nm in wavelength for a 3 nm shift in layer thickness. However, it can be measured accurately. The height responsivity is shown in Fig. 5.7b.

Colorimetric biosensors typically operate in two limits of the supporting layer thickness. For relatively thin layers around 50 nm thick, the phrase "colorimetric" is literal, meaning that the color of the printed capture spot actually changes color to the eye [8]. The color-changing readout is somewhat like strip biosensors (pregnancy tests or glucose tests) that are the most commercially successful biosensors. In the opposite extreme, thick layers are used to generate multiple spectral fringes, as in Fig. 5.6. This type of sensor uses a spectrometer with multiple detection channels to accurately measure the shift in fringes [9].

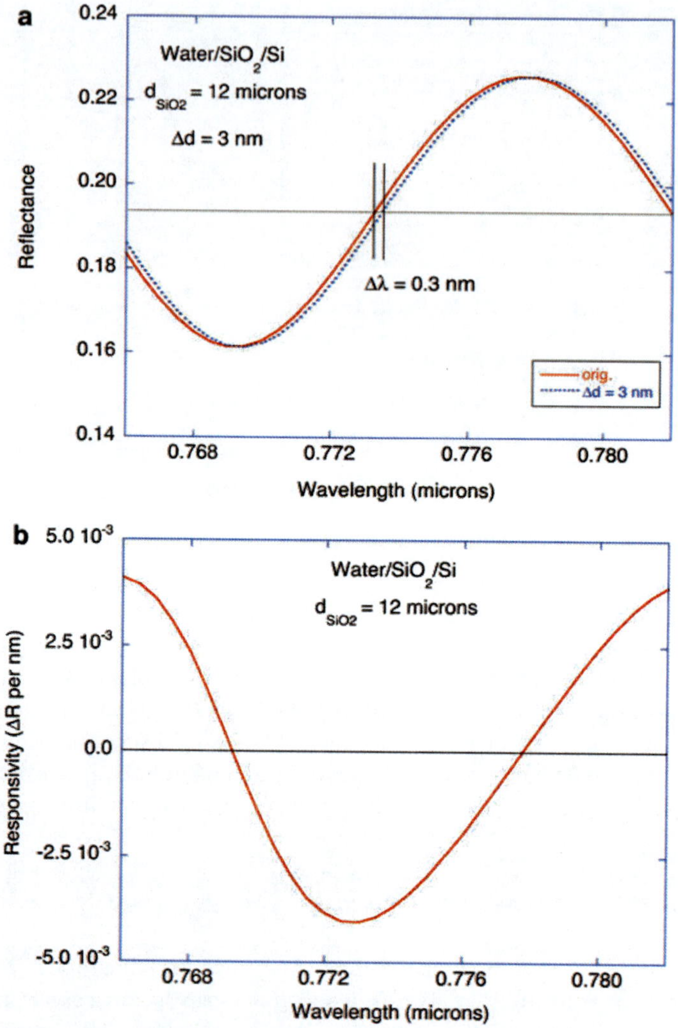

Fig. 5.7 (**a**) Shift in the spectral interference fringe for a thickness change of 3 nm. The equivalent wavelength shift of the fringes is 0.3 nm. (**b**) Height responsivity on 12 μm oxide on silicon

5.4 Molecular Interferometric Imaging

A simpler alternative to ellipsometric or colorimetric measurement of biomolecular layers is direct molecular interferometric imaging (MI2) in which a biolayer resides on an interferometric surface that has a phase quadrature condition [10]. In this case, the dipole density on the surface is transduced directly to an intensity that can be imaged. There are trade-offs between molecular interferometric imaging and ellipsometry. For instance, interferometric imaging has much simpler optics with higher molecular responsivity, but has much higher background. Molecular interferometric imaging also requires an interferometric surface that provides the phase quadrature condition, but this is easily supplied by appropriate thicknesses of thermal oxide on silicon.

5.4.1 *In-line Quadrature*

When a thin biolayer is added to a substrate that has an original reflection coefficient r_0, the new reflection coefficient r is related to r_0 at normal incidence through (4.50)

$$r = r_0 + (1 + r_0)^2 \frac{\pi i}{\lambda} \frac{(n_p^2 - n_m{}^2)}{n_m} d \tag{5.11}$$

where n_p is the refractive index of the protein and d is the protein thickness. The details of the substrate structure are not relevant for the extraction of the optical properties of the biolayer – only the reflection coefficient is needed to extract its properties. The condition for largest absolute change in the reflectance is

$$r_0 = \pm \frac{i}{\sqrt{3}} \tag{5.12}$$

which is purely imaginary ($\pi/2$ phase shift) and establishes the condition of phase quadrature for direct phase-to-intensity transduction of the molecular film.

One of the simplest (and most ideal) approaches to establish this reflection condition is the use of a single thermal oxide (SiO_2) layer on silicon. Silicon and thermal oxide are among the world's most perfect materials, partly because of their use in the semiconductor and integrated circuit industries. The surface of silicon can be made nearly molecularly flat and provides an ideal surface for molecular imaging. The reflection coefficient and reflectance of thermal oxide on silicon are shown in Fig. 5.8 as a function of oxide thickness at a wavelength of 633 nm. The reflection coefficient at zero thermal oxide thickness is purely real and negative (a nodal surface with a minimum in the surface electric field). The reflection coefficient becomes purely imaginary (and is close to the optimal value of $r = 0.577$) for a thickness around 54 nm. This thickness has the special value

$$d_{\text{quad}} = \frac{\lambda}{8 n_{\text{SiO}_2}} \tag{5.13}$$

that is an eighth wavelength of the probe light. Upon double pass through the oxide layer by reflection from the oxide/silicon interface, the resulting phase shift is $\pi/2$, which establishes the quadrature condition for direct phase-to-intensity transduction. Another quadrature condition occurs at 3/8 of a wavelength around 162 nm. The maximum surface field occurs at the antinodal thickness of a quarter wavelength around 108 nm.

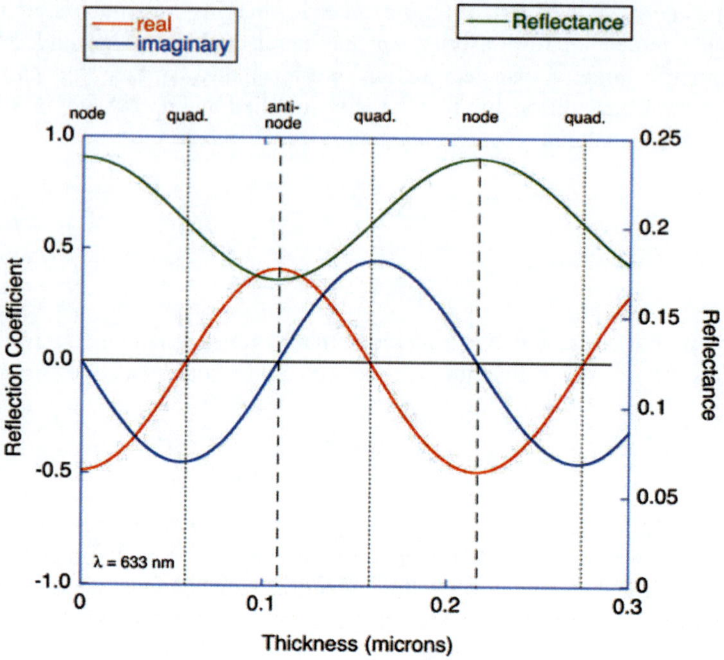

Fig. 5.8 Reflection coefficient and reflectance for thermal oxide on silicon in water as a function of oxide thickness at a wavelength of $\lambda = 633$ nm. The nodal, quadrature, and antinodal surface field conditions are marked. The maximum intensity sensitivity to a surface biolayer occurs at the quadrature conditions. The refractive index values are $n_{SiO_2} = 1.46$ and $n_{Si} = 3.87 + i0.02$

The geometry of the quadrature condition is shown in Fig. 5.9 when the supporting layer has an eighth wavelength thickness. The phase difference between light reflected from the top of the oxide layer relative to light reflected at the silicon interface is $\pi/2$. When a surface biolayer is present, the phase difference it causes is transduced to an intensity (reflectance) difference that can be imaged directly through a microscope.

The relative reflectance responsivity per nanometer of protein is given by

$$C(\lambda) = \frac{1}{R}\frac{dR}{d(h_f)} = \frac{\mathbb{R}_h}{R} = \frac{\Delta R(h_f)}{R}\frac{1}{h_f} \tag{5.14}$$

assuming linear dependence of reflectance on protein thickness, where ΔR is the reflectance change caused by the biolayer. The relative responsivity is normalized by the reflectance and is larger for fixed ΔR at smaller reflectance R. The relative responsivity is shown in Fig. 5.10 at three wavelengths as a function of oxide thickness in air. The responsivity is approximately 2.5% reflectance modulation per nanometer of protein at the maximum values. The thicknesses where these maximum relative responsivities occur are shifted from the quadrature condition toward the antinode position where the reflectance is a minimum because of the

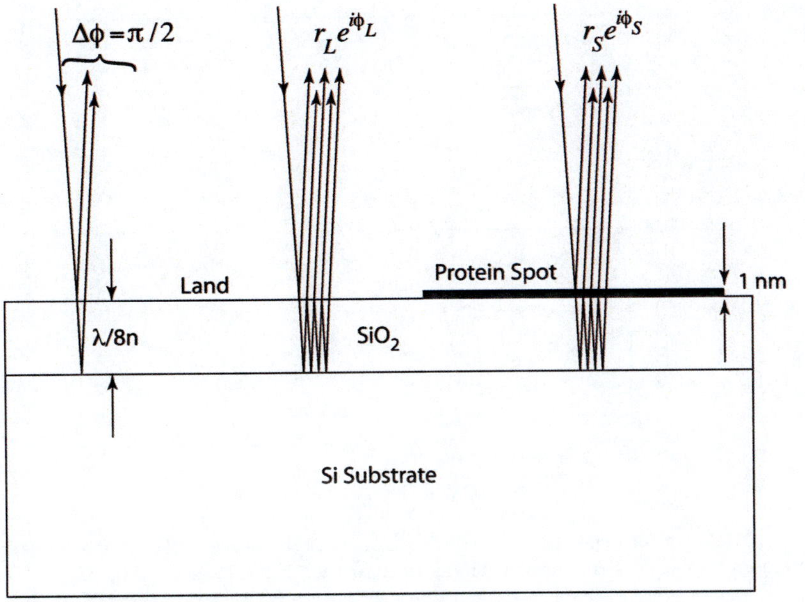

Fig. 5.9 The phase quadrature condition that converts the phase shift caused by a surface biolayer directly into intensity that can be imaged through a microscope

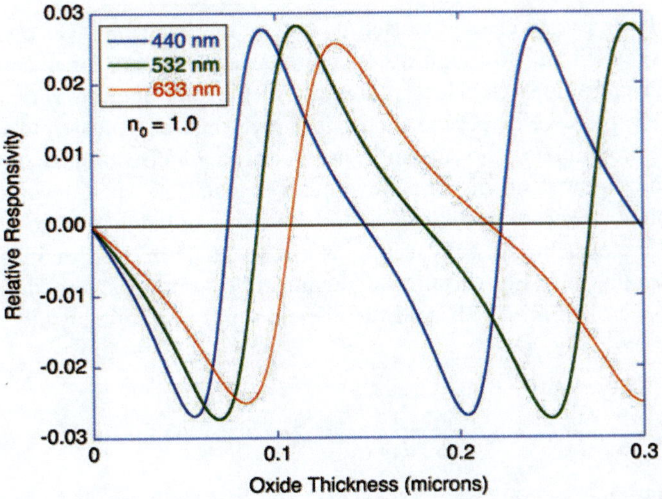

Fig. 5.10 Relative responsivity in air (normalized by reflectance) in response to 1 nm of protein

normalization (see Fig. 5.10). If the overlying medium is water, then the responsivity decreases because of the smaller index contrast between protein ($n = 1.46$) and water ($n = 1.33$). The comparison of in-air vs. in-water responsivity is shown in Fig. 5.11. The responsivity is smaller for detection in water by about a factor of 5.

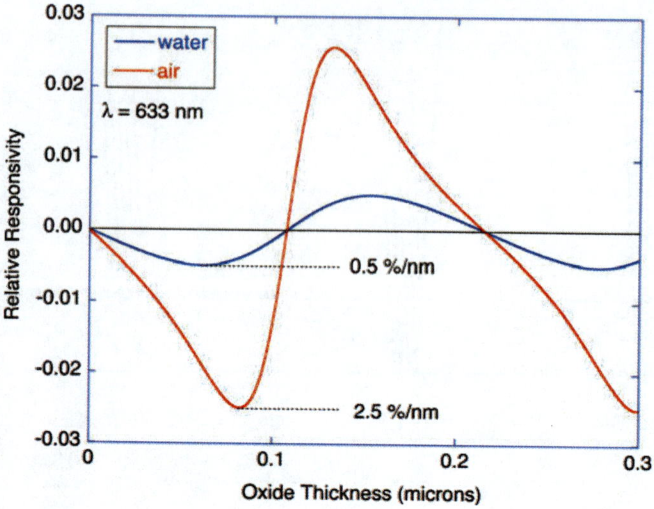

Fig. 5.11 Relative responsivity of protein on thermal oxide measured in air and in water. The water-based responsivity is approximately 5 times smaller than for detection in air

5.4.2 Image Shearing and Molecular Sensitivity

Although the intensity change caused by an immobilized biolayer on the surface can be imaged directly through a conventional microscope, even under the best illumination conditions (Koehler illumination) the illumination varies across the field of view by several percent. This is larger than a typical intensity change caused by a protein layer. For instance, in Fig. 5.12a the direct image of two protein spots exhibits almost no discernable protein signal. The spatial variation of the illumination is a form of spatial 1/*f* noise (slow drift in intensity across the field of view). Fortunately, this spatial variation is usually static in time, making it possible to normalize the intensity variation by acquiring two images, one shifted relative to the other, as in Fig. 5.12b, and differencing and normalizing the two images through

$$\frac{\Delta I}{I} = 2 \times \left(\frac{I_2 - I_1}{I_2 + I_1}\right) = \mathbb{R}_d(\lambda)(h(x + \Delta x) - h(x)) \tag{5.15}$$

where $h(x)$ is the "height" of the wafer topology, Δx is the spatial shift between the two images, and $\mathbb{R}_d(\lambda)$ is the responsivity per nm. This procedure, applied to the direct image in (a), is shown in (c) for the two protein spots (IgG antibody spots approximately 100 μm in diameter). The surface between the spots, known as "land," is extremely flat and uniform, and the protein spots appear with extremely high contrast. The two spots show up as double images, one positive and one negative (in color, one is red and the other is blue). The two antibody spots in

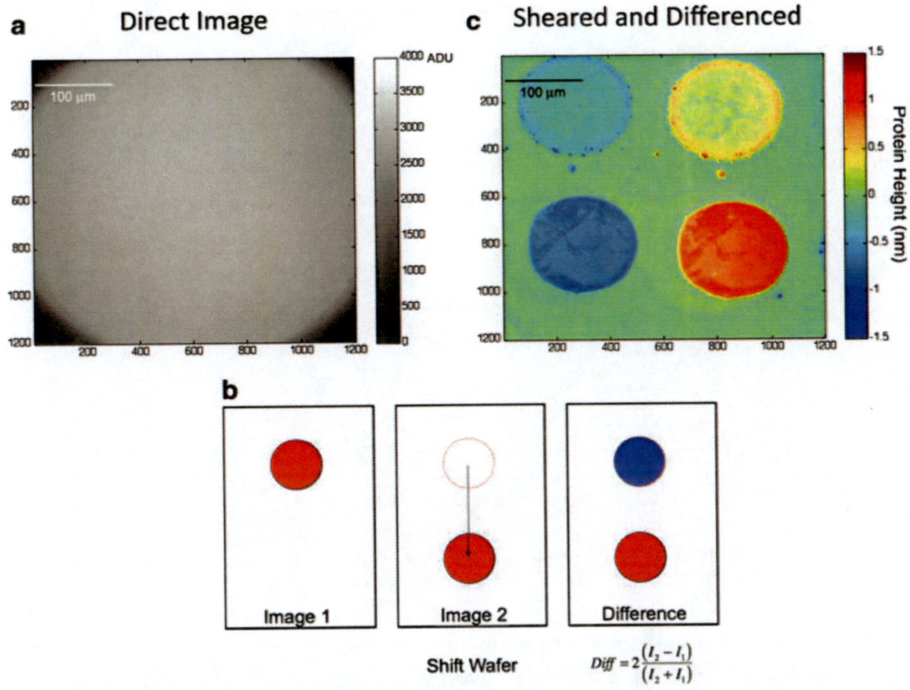

Fig. 5.12 The shearing procedure for intensity normalization in molecular interferometric imaging. Two protein spots in the direct image in (**a**) are not visible. By acquiring two images, with a spatial shift of the silicon wafer in between, and differencing and normalizing as in (**b**) produces the image in (**c**) of two antibody spots on the wafer

this example did not print evenly. One has an average height of 1 nm, while the other has an average spot height of 0.4 nm, with a clear "dry-down" ring that is common for protein spotting.

The image acquisition is performed with a charged coupled device (CCD) that can have shot-noise-limited performance. The ultimate surface height sensitivity is set by the shot noise of the background. Molecular interferometric imaging is a high background application in which the signal to be extracted is a small (only about a percent) spatial intensity modulation on a high-intensity background. The shot-noise-limited height resolution is set when the *S/N* ratio equals 1. The surface height sensitivity is then

$$\delta h_{\text{shot}} = \frac{1}{\mathbb{R}_d(\lambda)} \sqrt{\frac{1}{N}} \tag{5.16}$$

where N is the product of the number of pixels and number of repetitive samplings. The value for $\mathbb{R}_d(\lambda)$ from Fig. 5.10 is about 2% per nm. For a 12-bit camera the bit depth is 4,096 bits, yielding a height resolution for 1,000 averages of about 20 pm. The experimental height repeatability of successive images as a function of the number of averaged frames is shown in Fig. 5.13 for three different magnifications.

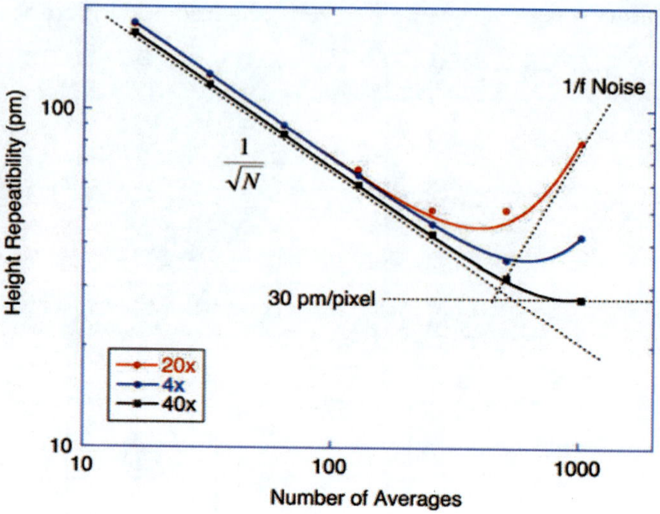

Fig. 5.13 Height repeatability for MI2 as a function of imaging magnification and number of frame averages. Long-term drift sets in above about 1,000 frames

The 40× magnification has the best height repeatability because it has the smallest spatial 1/*f* noise. For large numbers of averages, temporal 1/*f* noise (long-term drift) begins to dominate. The best performance is around 30 pm/pixel for 40× with the 12-bit camera operating at several frames per second.

The height sensitivity of molecular interferometric imaging can be converted to a molecular sensitivity. The number of molecules per pixel corresponding to a height resolution δh is

$$N_{\text{molec}} = \frac{\delta h A_{\text{res}} \rho_{\text{m}}}{m_w} \frac{1}{ff} \tag{5.17}$$

where A_{res} is the resolution area of the optical imaging, and ff is the CCD fill factor (typically equal to 1/2 for interline chips). For a protein density $\rho_{\text{m}} = 1.0\ \text{g/cm}^3$ and a 150 kDa molecule (antibody molecule) at 40× magnification (0.4 μm resolution) the number of antibody molecules detectable per resolution area in the shot-noise-limit is approximately 50 molecules. Therefore, the sensitivity level for direct interferometric detection approaches the single-molecule limit.

The molecular sensitivity of interferometric imaging combined with its high spatial resolution at high magnification, makes it an attractive alternative to atomic force microscopy (AFM) for broad-area molecular imaging on surfaces. A comparison of MI2 against AFM on the same protein spot is shown in Fig. 5.14. The MI2 acquisition took approximately 30 s of frame averaging, while the AFM scan took many hours. The MI2 image is referenced against the adjacent land, while the AFM reference drifted over the large distances and long times. While features can be seen in common between MI2 and AFM, the fidelity of the MI2 image is much higher, with absolute height reference.

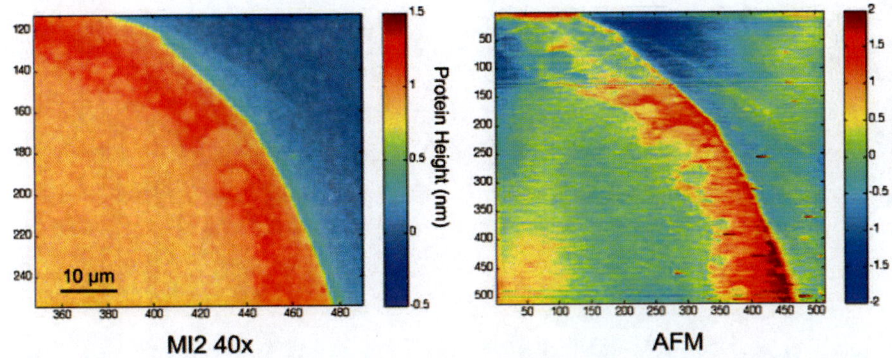

Fig. 5.14 Comparison of MI2 with AFM at approximately equal lateral resolutions. The MI2 image was acquired in 30 s compared with several hours for AFM, which loses absolute reference over wide areas

5.4.3 Biosensor Applications

The molecular sensitivity of interferometric imaging enables kinetic binding applications in which affinity capture molecules capture target analytes out of solution. The central trade-off for molecular interferometric imaging is the simplicity of the detection system traded off against the small-signal and high background. Therefore, applications that are most appropriate for MI2 are high-concentration (typically above 10–50 ng/mL) affinity measurements of relatively large molecular weight targets (typically above 20–50 kDa).

A requirement for accurate assays is the printing of highly uniform surface-immobilized capture molecules. Surfactants in the printing solution can improve print quality by slowing down the reaction between protein molecules and the surface functional groups and enable the proteins to print more uniformly. Figure 5.15 shows spots printed on butyraldehyde surface chemistry using a print solution with PBS pH 8.0 with 0.003% surfactant. Figure 5.16 shows the pixel height distribution of the protein spot and its adjacent SiO_2 surface (land) in Fig. 5.15. The protein height variation within a spot has a standard deviation of 115 pm. Because each protein spot contains more than 10,000 pixels, the standard error on the average protein spot height is approximately 1 pm.

Real-time binding experiments performed under the small-signal and high background conditions require intensity normalization. Instead of performing a spatial shift of the wafer, as described above, a time-shift can be used for normalization. The signal is therefore

$$\begin{aligned}\frac{\Delta I}{I} &= \frac{I_2(x,y;t) - I_1(x,y;0)}{I_1(x,y;0)} \\ &= \mathbb{R}_d(\lambda, n_{\text{ambient}})(h(x,y;t) - h(x,y;0)) \end{aligned} \tag{5.18}$$

and the binding height is monitored in real time.

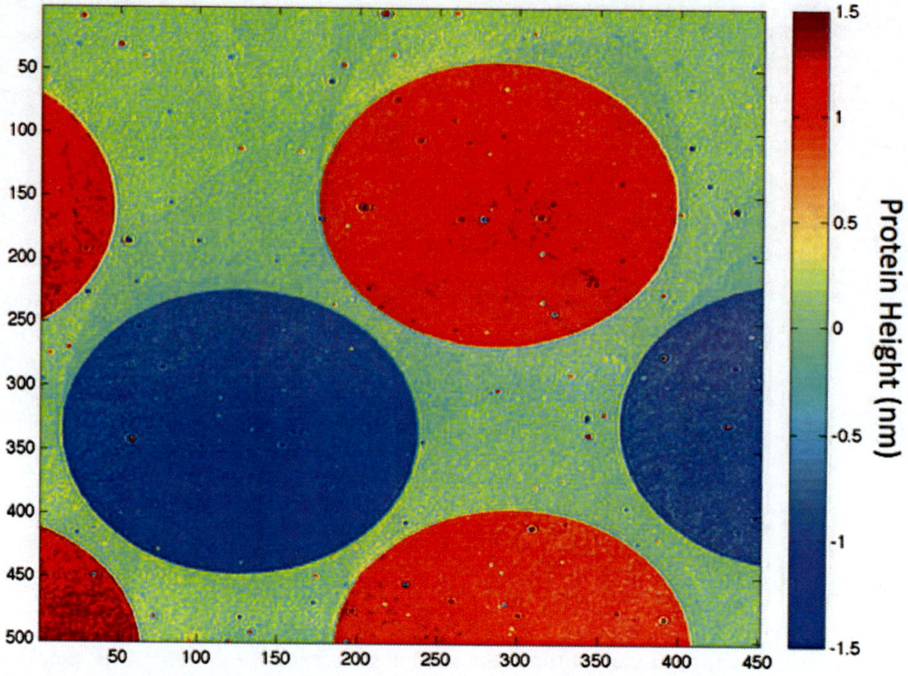

Fig. 5.15 IgG spots printed on butyraldehyde surface chemistry using a print solution with PBS at pH 8.0 with 0.003% surfactant

Real-time binding curves are shown in Fig. 5.17a for the capture of IgG antibodies by printed IgG antigen in a reverse assay for a range of concentrations from 40 to 2 μg/mL. The data are scaled by their binding time in Fig. 5.17b showing data collapse with an equilibrium rate of $3 \times 10^3\ M^{-1}\ s^{-1}$. An additional scaled binding curve in Fig. 5.17b shows Fc/AG binding with about a factor of 10 higher binding rate. Protein A/G is a fusion protein that selectively binds the Fc portion of an antibody.

An example of a real-time binding assay is shown in Fig. 5.18 for prostate-specific antigen (PSA). The printed spots in this case are protein A/G fusion protein. Because of the specificity to the base of the antibody molecule, the immobilization is highly uniform and oriented, with nearly 100% of the antibody molecules being functional (compared to 20% when the antibody is attached directly through covalent surface chemistry). In the first part of the real-time assay the antibody is introduced and binds to the protein A/G. This is followed by a short wash with buffered BSA that washes off a small portion of the bound antibodies. The chip is then exposed to PSA in buffered BSA that produces a small binding signal of 0.3 nm (compared to a loading of 1.5 nm). The molecular weight of PSA is 30 kDa, and the weight of antibody IgG is about 150 kDa. The mass ratio is 20%, and the ratio of the binding height to the initial IgG loading height is 22%, which is consistent with the capture of one antigen per antibody. After the forward assay

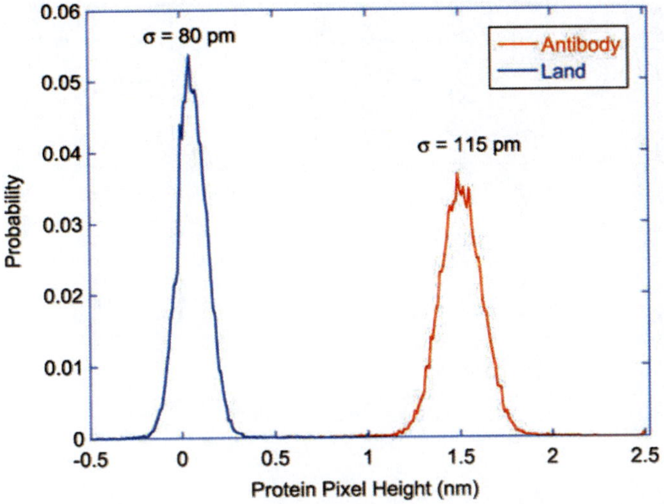

Fig. 5.16 Histogram of the spot in Fig. 5.15. The surface roughness of the land is 80 pm. The protein spot standard deviation is 115 pm

(antibody capturing a target), a sandwich assay is performed by introducing a second antibody against PSA. This binds to the capture antigen, forming an antibody-target-antibody sandwich. The mass of the second antibody is also 150 kDa, producing a large enhancement of the PSA signal.

5.5 The BioCD

The bio-optical compact disk, or BioCD, is a class of label-free interferometric optical biosensors that spin at high speed and are probed by a focused laser [11]. They operate under the condition of phase quadrature that converts surface-bound biolayers to intensity modulation that is detected by a high-speed photodetector. The high-speed scanning and the interferometric sensitivity enable thousands of molecular density measurements to be performed quickly across large-area disks.

5.5.1 Spinning Interferometric Biochips

Multichannel acquisition of protein binding, whether through imaging with pixel-array detectors (CCD cameras) or through wavelength selection (spectrometers), has a multiplex advantage on signal-to-noise, but has a throughput disadvantage. This is because it takes time and memory to acquire and store multichannel information. Label-free assays only gain an advantage over conventional labeled

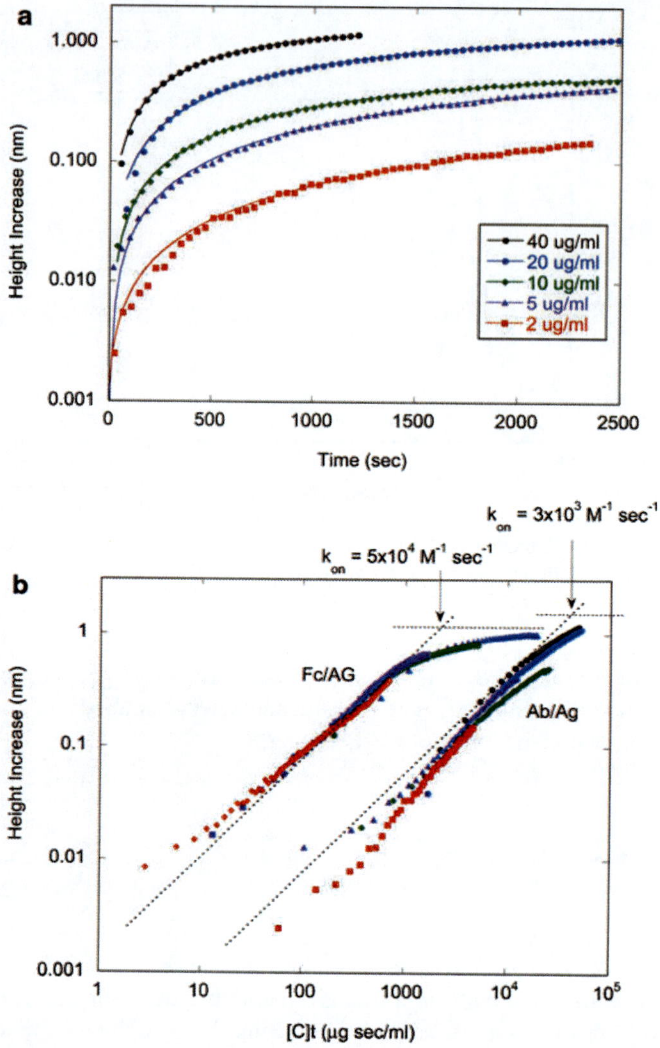

Fig. 5.17 Binding kinetics. (**a**) Height response as a function of concentration and time for antibody/antigen binding. (**b**) Height response as a function of scaled time for antibody/antigen binding and for Fc/AG binding of antibody to protein A/G

assays if they can be performed as high-throughput assays of high-abundance target analytes. The surface area of an interferometric biosensor provides the resource for thousands of optical measurements. Protein spots printed with 100 pL of antibody solution have a characteristic size of 100 μm. A disk with a diameter of 100 mm can support approximately 100,000 such antibody spots.

Spinning a BioCD has more advantages than simply acting as a fast conveyor belt. High-speed optical sampling suppresses the $1/f$ noise that is ubiquitous in all

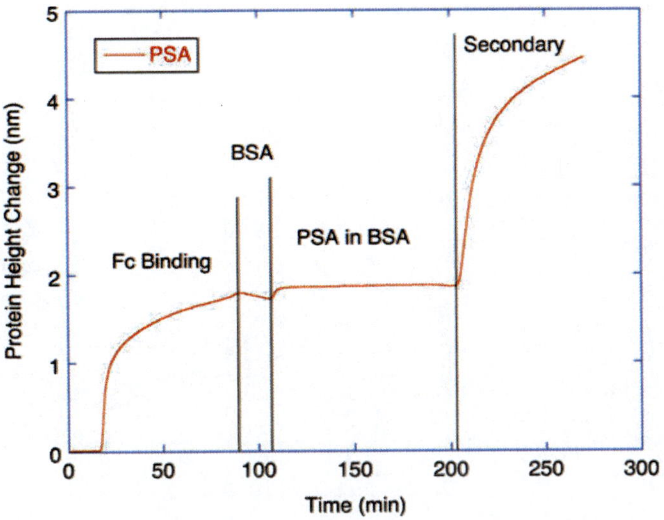

Fig. 5.18 Real-time binding curve for prostate-specific antigen (PSA). The initial loading is the anti-PSA antibody binding to protein A/G. The BSA wash shows an off rate for the antibody. The PSA concentration was high at 1 μg/mL. The second antibody provides a large mass-tag enhancement

measurement systems. The power spectrum of a BioCD signal is shown in Fig. 5.19. This BioCD carries periodic stripes of printed antibody separated by bare surface (land). The power spectrum exhibits a fundamental harmonic at 40 Hz with higher overtones superposed on top of $1/f$ noise that peaks at DC (zero frequency). The BioCD signal is compared with a bare disk. At the detection frequency, the antibody signal is 40 dB larger than the bare land. In this case, the "noise floor" is not noise at all, but is background. In the case of the bare disk it is surface roughness, and in the case of the BioCD it is inhomogeneous protein printing. This background is measurable and repeatable. On the other hand, the noise spike at DC is stochastic noise. The high-speed sampling places the detection frequency far from this $1/f$ noise of the detection system.

A key aspect of the BioCD that enables broad-area scanning is the reliance on phase quadrature. The BioCD is neither a resonant interferometer, nor relies on multiple passes of the optical field as in a thick Fabry-Perot. The BioCD has a finesse of unity (the lowest possible finesse of any interferometer), and hence has a low absolute protein responsivity. But what is gained is the ability to have a flexible operating point that can be maintained in fabrication across broad areas. For instance, in the case of in-line quadrature (as used in molecular interferometric imaging), the nominal oxide thickness for quadrature is 120 nm at a wavelength of 633 nm. However, this oxide thickness can vary by 20% across a 100 mm diameter silicon wafer and yet have no negative impact on the sensitivity of the measurements. This is in stark contrast to resonant configurations or thick oxide layers that have operating points (optimal thicknesses or optimal wavelengths) that can drift out of the operating range of the biosensors.

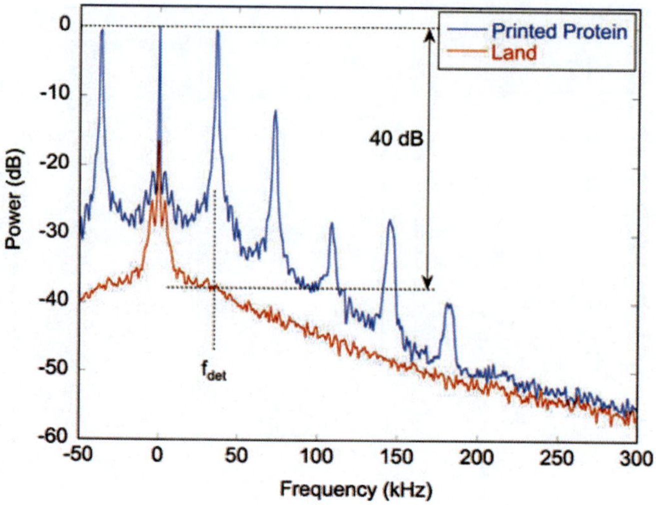

Fig. 5.19 Power spectrum of a periodic protein pattern on a BioCD compared with the noise (surface roughness) of the land. The peak of the 1/*f* noise at DC is not within scale

5.5.2 *Molecular Sensitivity, Sampling, and Scaling*

The molecular sensitivity of the in-line quadrature condition was studied in Sect. 5.4.2 for interferometric imaging. The identical disks used for MI2 can be spun as a BioCD for fast laser scanning. The noise and sensitivity conditions for rapid point scanning on a BioCD are sufficiently different than for broad field imaging and merit reexamination here. The chief question is one of trade-off of resolution against throughput. By having a larger focal area for the laser and using larger scan steps in radius, shorter scan times are permitted, but at the expense of spatial resolution of the antibody spots on the disk. An example of BioCD scanning resolution is shown in Fig. 5.20 for two different antibody spots (with different protein printing pathologies). The beam focal radius was 1, 5, and 10 μm with a constant pitch (change in radius) of 1 μm for the six images on the left, and a pitch of 10 μm on the rightmost figures.

The picometer precision of the spot height metrology is not the limiting factor in the precision of an immunological assay. Real assays involve considerable chemical protocols, all under wet chemistry conditions, or with dry-down residues, and these are the factors that limit the sensitivity of an assay. The scaling of assay sensitivity under practical chemistry limitations is shown in Fig. 5.21. Several assays are presented, including label-free PSA detection, sandwich PSA detection, and rabbit IgG control assays. Sandwich assays, which are not label-free (but are fluorescence-free), significantly improve the detection limits. The minimum

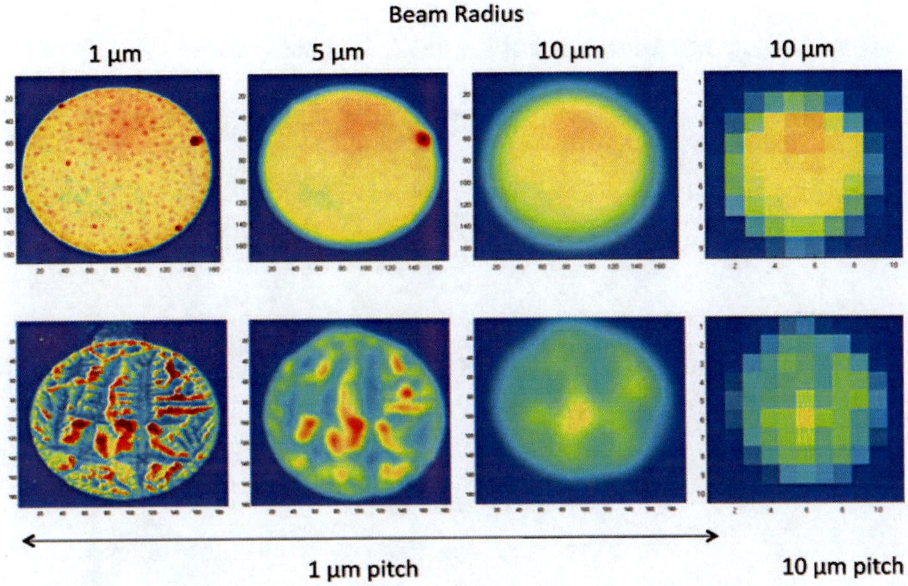

Fig. 5.20 Raw BioCD scan data of two different 100 µm-diameter and 2 nm high antibody spots as a function of beam radius and pitch. The three left columns have a pitch of 1 µm. For the rightmost column, both the beam radius and the pitch are 10 µm

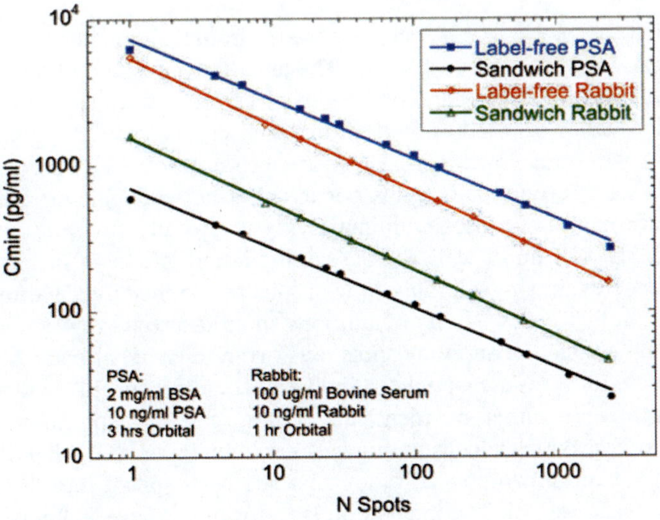

Fig. 5.21 Scaling dependence of the minimum detectable concentration (LOD) as a function of the number of antibody spots per assay. Two analytes are shown, PSA and rabbit IgG. The label-free forward assay is compared with sandwich assays for each

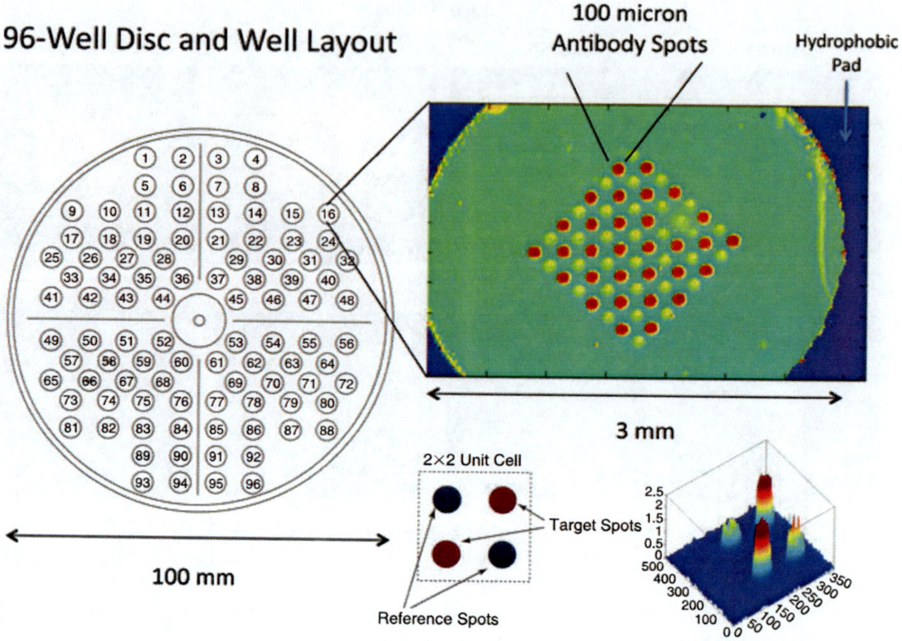

Fig. 5.22 A 96-well BioCD layout with wells defined by a hydrophobic pad. In each well are 4 × 4 arrays of 2 × 2 unit cells consisting of target and reference spots. The reference spots measure background and nonspecific binding. The well image is a raw data scan after incubation against 100 ng/mL of haptoglobin analyte

concentration for label-free PSA detection varied from 6 ng/mL at one spot per assay to 200 pg/mL for 2,500 spots per assay. The label-free detection of the larger rabbit IgG molecule was slightly better. For a sandwich PSA assay, the minimum detectable concentration varies from 600 pg/mL at one spot per assay to 25 pg/mL for 2,500 spots per assay (ten assays per disk). However, the sandwich assays are not label-free, and the use of 2,500 antibody spots per assay is not economically feasible. The nominal performance of a BioCD immunoassay is typically 1 ng/mL or higher.

An example of a multi-well BioCD configuration is shown in Fig. 5.22. The 100 mm-diameter silicon wafer has 96 wells that are defined by a hydrophobic pad. Within each well is an 8 × 8 array of antibody spots arranged in a 4 × 4 array of unit cells that have two target antibody spots plus two reference antibody spots per unit cell. The reference spots are isotype antibodies that are not specific to the target. The reference spots can either be identical, or they can be two different types of antibodies, one seeking to measure nonspecific binding and the other to measure a control analyte of known concentration that has been spiked into the sample for internal assay calibration. The incubation lasts for one to several hours, depending on the anticipated concentration of analyte, after which the disk is washed and dried. The image of one well in the figure, after incubation with haptoglobin analyte at 100 ng/mL, shows a typical raw data scan. The reference spots were direct-printed

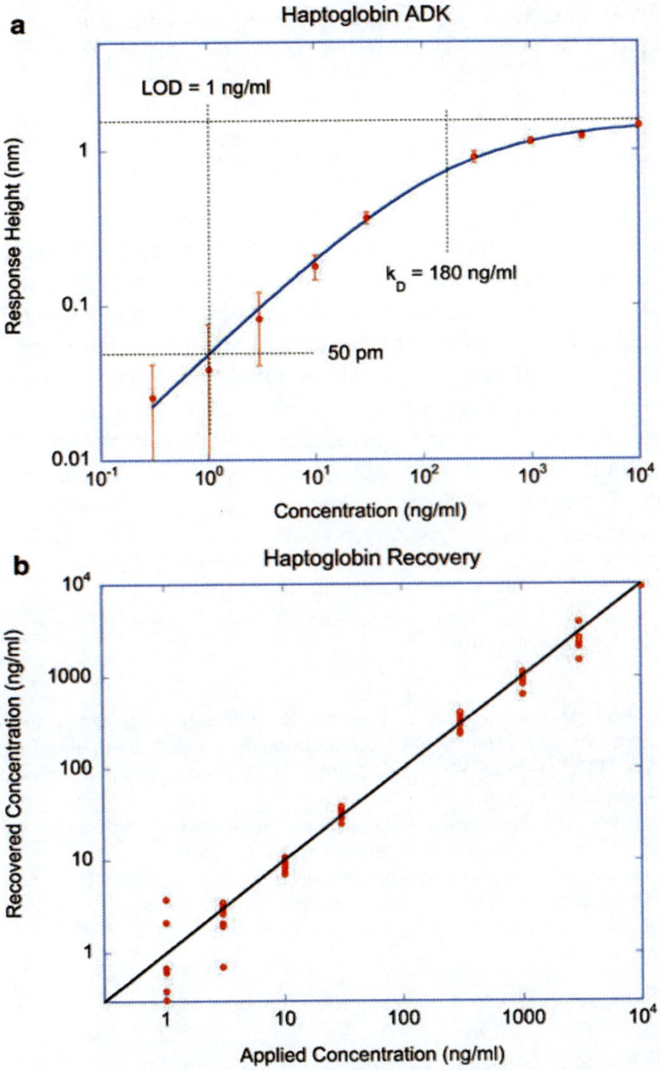

Fig. 5.23 (**a**) Standard concentration curve for a haptoglobin assay on the BioCD. (**b**) Concentration recovery defined as the measured concentration of each well compared with the applied concentration

IgG antibodies, while the target spots were direct-printed protein A/G followed by capture of anti-haptoglobin antibodies onto the A/G spots. The target spots show about 1 nm of binding in these data, while the reference spots show no significant binding.

A concentration curve for haptoglobin, called a standard curve, is shown in Fig. 5.23a. The surface height detection limit is 50 pm at approximately 1 ng/mL. The value of k_D is 180 ng/mL, but this is an effective equilibrium constant because the 1 h incubation does not drive the assay to equilibrium, especially at the lower concentrations. The curve is a "stretched" response function, defined by

$$h = h_{\mathrm{sat}} \frac{[C]^e}{[C]^e + (k_D^{\mathrm{eff}})^e} \tag{5.19}$$

where h_{sat} is the saturation height for high concentrations, and e is a stretch exponent. Although, this functional form is often observed for cooperative binding reactions, in this application it reflects the different approaches to equilibrium values for the different concentrations at the fixed incubation time. The stretch exponent in Fig. 5.23a that best fits the data is $e = 0.6$, which is typical for such static incubation experiments. Once a standard curve has been established, a concentration recovery experiment is performed in which the concentration measured in each well is plotted against the applied concentration. The recovery graph is shown in Fig. 5.23b for the standard curve in Fig. 5.23a. The recovery in this example is better than 80% across a dynamic range of three orders of magnitude in concentration.

Selected Bibliography

H. Arwin, S. Welinklintstrom, and R. Jansson, R.: Off-null ellipsometry revisited – basic considerations for measuring surface concentrations at solid liquid interfaces. J. Colloid Interface Sci. **156**, 377–382 (1993) (This journal article gives a good mathematical background to off-null ellipsometry)

Cooper, M. A. Label-free biosensors: techniques and applications. Cambridge University Press, Cambridge (2009) (A modern review of label-free biosensors)

Nolte D.D.: Review of centrifugal microfluidic and bio-optical disks," Rev. Sci. Instrum. **80**, 101101 (2009) (A recent review of the BioCD and related devices for multiplexed protein arrays)

References

1. Anderson, N.L., Anderson, N.G.: The human plasma proteome – history, character, and diagnostic prospects. Mol. Cell. Proteomics **1**(11), 845–867 (2002).
2. Arwin, H., Welinklintstrom, S., Jansson, R.: Off-null ellipsometry revisited - basic considerations for measuring surface concentrations at solid liquid interfaces. J. Colloid Interface Sci. **156**(2), 377–382 (1993).
3. Defeijter, J.A., Benjamins, J., Veer, F.A.: Ellipsometry as a tool to study adsorption behavior of synthetic and biopolymers at air-water-interface. Biopolymers **17**(7), 1759–1772 (1978).
4. Jin, G., Jansson, R., Arwin, H.: Imaging ellipsometry revisited: developments for visualization of thin transparent layers on silicon substrates. Rev. Sci Instrum. **67**(8), 2930–2936 (1996).
5. Fei, Y.Y., Landry, J.P., Sun, Y.S., Zhu, X.D., Wang, X.B., Luo, J.T., Wu, C.Y., Lam, K.S.: Screening small-molecule compound microarrays for protein ligands without fluorescence labeling with a high-throughput scanning microscope. J. Biomed Optics **15**(1) (2010).

6. Gao, T.J., Lu, J.H., Rothberg, L.J.: Biomolecular sensing using near-null single wavelength arrayed imaging reflectometry. Anal. Chem. **78**(18), 6622–6627 (2006).
7. Gao, T., Rothberg, L.J.: Label-free sensing of binding to microarrays using Brewster angle straddle interferometry. Anal. Chem. **79**(20), 7589–7595 (2007).
8. Jenison, R., Yang, S., Haeberli, A., Polisky, B.: Interference-based detection of nucleic acid targets on optically coated silicon. Nat. Biotechnol. **19**, 62–65 (2001).
9. Ozkumur, E., Needham, J.W., Bergstein, D.A., Gonzalez, R., Cabodi, M., Gershoni, J.M., Goldberg, B.B., Unlu, M.S.: Label-free and dynamic detection of biomolecular interactions for high-throughput microarray applications. Proc. Natl. Acad. Sci. U S A. **105**(23), 7988–7992 (2008).
10. Zhao, M., Wang, X., Nolte, D.D.: Molecular interferometric imaging. Optics Express **16**(10), 7102–7118 (2008).
11. Nolte, D.D.: Review of centrifugal microfluidic and bio-optical disks. Rev. Sci. Instrum. **80**(10), 101101 (2009).

Chapter 6
Diffraction-Based Interferometric Biosensors

Diffraction and interference are closely related physical phenomena, and diffraction-based biosensors constitute one class of interferometric biosensors. Diffraction arises because of interference. More specifically, diffraction arises in a wavefront-splitting interference configuration. In such a configuration, when one propagating set of wavefronts interacts with a biological sample or film, and another set does not, this affects their mutual diffraction pattern in a way that can be detected and related back to the amount of biological material in the sensor. Central issues for this type of interferometric biosensor include responsivity to immobilized or captured biomolecules (whether signals are linear or quadratic in the captured molecules) and sensitivity (limit-of-detection) depending on background levels and noise. Many diffraction-based biosensors have the advantage of being background-free (little or no background signal in the absence of captured molecules), but at the cost of having a quadratic responsivity that vanishes rapidly for decreasing molecule densities. The performance of these biosensors, and the applications they may be chosen for, depends on the trade-off between signal contrast and sensitivity. There are several distinct classes of diffraction-based biosensors. These include protein gratings on planar surfaces; proteins immobilized on microstructured surfaces, cantilevers, or capillaries; and bead-based biosensors that capture beads in distinct spatial patterns that diffract a probe beam.

6.1 Planar Diffractive Biosensors

The simplest class of diffraction-based biosensors use patterns of biomolecules or capture reagents on planar surfaces. The spatial patterns diffract a probe beam into a far-field diffraction pattern with intensities that depend on the density of immobilized or captured molecules on the surface. The principle of a planar

D.D. Nolte, *Optical Interferometry for Biology and Medicine*, Bioanalysis 1,
DOI 10.1007/978-1-4614-0890-1_6, © Springer Science+Business Media, LLC 2012

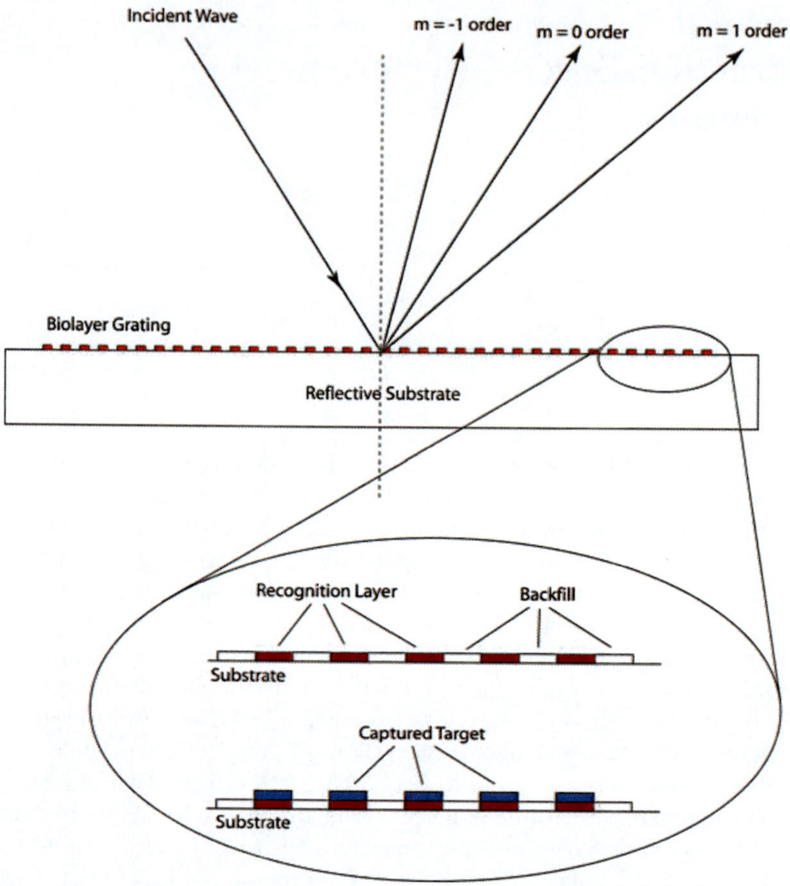

Fig. 6.1 A biolayer grating on a planar surface. The incident light is diffracted into a zeroth order (reflected wave) and two first-order beams (and possibly higher orders). The first-order diffracted beams have no background in the absence of bound target because of the planar surface. Once target molecules bind, a physical diffraction grating is formed that diffracts the probe beam

molecular grating is illustrated in Fig. 6.1. A surface is patterned with a recognition molecule and backfilled with an inert molecule of similar size and density. In the beginning state, the surface remains planar and little or no diffraction signal is produced. After target molecules are captured by the recognition layer, a physical diffraction grating is formed that diffracts the probe beam into diffraction orders that are detected and related to the amount of target molecules that are captured. Because diffraction detection is based on intensity and not field, the functional dependence of the diffracted signal is quadratic in the amount of bound target analyte, scaling as $(d/\lambda)^2$ where d is a typical molecular layer thickness and λ is the wavelength of light.

6.1.1 Diffraction Efficiency of Biolayer Gratings

Molecular gratings diffract a probe beam into diffraction orders whose intensities depend quadratically on the surface relief height (or molecular density). From (2.45) a free-standing dielectric square-wave grating of index n embedded in a medium of index n_{m} has the transmitted diffraction efficiency

$$\eta = \frac{k_0^2(n - n_{\mathrm{m}})^2 d^2}{\pi^2}, \tag{6.1}$$

where k_0 is the free-space k-vector and d is the thickness of the grating elements. This diffraction efficiency is quadratic in the thickness (or the optical phase). As an example, a 4 nm thick dry antibody biolayer in air with $(n - 1)d \approx 1$ nm yields a diffraction efficiency of $\eta = 2.5 \times 10^{-5}$. Operating a biosensor in water (or buffer) is essential for real-time binding measurements. If the antibody layer is immersed in water with $n = 1.33$, then the 4 nm thick antibody layer has $(n - n_{\mathrm{w}})d \approx 0.2$ nm, and the diffraction efficiency falls to $\eta = 1 \times 10^{-6}$, which approaches the limit of detectability for diffraction measurements.

When the grating is supported on a substrate immersed in a medium with index n_{m} and with a reflection coefficient r_0, the reflection coefficient is given in (4.50) as

$$\begin{aligned} r_{\mathrm{biolayer}} &= r_0 + \frac{(1 + r_0)^2}{n_{\mathrm{m}} \cos\,\theta_{\mathrm{m}}} \frac{\pi \mathrm{i}}{\lambda} (n_{\mathrm{m}}^2 - n_{\mathrm{p}}^2) h(x) \\ &= r_0 + \phi' h(x), \end{aligned} \tag{6.2}$$

where

$$\phi' = \frac{(1 + r_0)^2}{n_{\mathrm{m}} \cos\,\theta_{\mathrm{m}}} \frac{\pi \mathrm{i}}{\lambda} (n_{\mathrm{m}}^2 - n_{\mathrm{p}}^2) \tag{6.3}$$

is the coefficient for the biolayer height profile $h(x)$. Equation (6.2) has simply an additive term on the original surface reflection coefficient. The Fraunhofer diffraction integral for a square grating of periodicity Λ is

$$\begin{aligned} \frac{E_0}{\Lambda} \int_0^{\Lambda/2} \phi' h_0 \mathrm{e}^{-\mathrm{i}k_m x \sin\theta}\, \mathrm{d}x &= \frac{E_0 \phi' h_0}{-\mathrm{i}k_m \Lambda \sin\theta} [\mathrm{e}^{-\mathrm{i}k_m \Lambda \sin\theta/2} - 1] \\ &= \frac{E_0 \phi' h_0}{-\mathrm{i}k_m \Lambda \sin\theta} [\mathrm{e}^{-\mathrm{i}k_m \Lambda \sin\theta/4} - \mathrm{e}^{\mathrm{i}k_m \Lambda \sin\,\theta/4}] \mathrm{e}^{-\mathrm{i}k_m \Lambda \sin\,\theta/4} \\ &= \frac{E_0 \phi' h_0}{2} \mathrm{sinc}(k_m \Lambda \sin\theta/4) \mathrm{e}^{-\mathrm{i}k_m \Lambda \sin\theta/4}. \end{aligned} \tag{6.4}$$

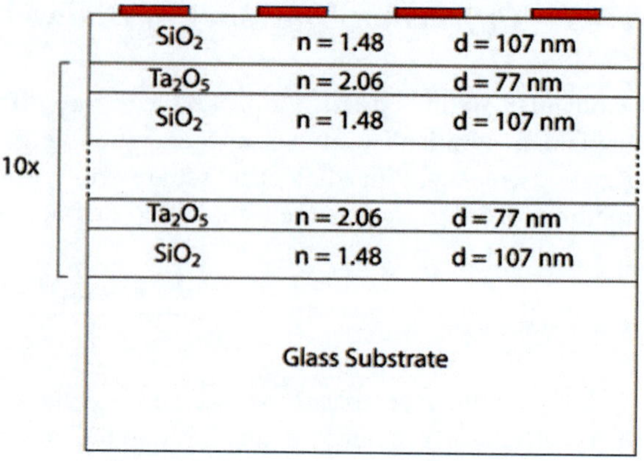

Fig. 6.2 Antinodal surface designed for $\lambda = 633$ nm composed of a multilayer stack that produces a field maximum (antinode) at the surface carrying the biolayer grating

The first diffraction order occurs at $\sin\theta = \lambda/n_\mathrm{m}\Lambda$, at which the diffracted field is

$$E_\mathrm{d} = E_0 \frac{\phi' h_0}{\pi} \mathrm{e}^{-ik_m\Lambda\sin\theta/4} \tag{6.5}$$

and the diffraction efficiency is

$$I_\mathrm{d} = \frac{|\phi'|^2 h_0^2}{\pi^2}. \tag{6.6}$$

The highest sensitivity to an incident optical field is obtained on an antinodal surface, as shown in Fig. 6.2, designed for a probe wavelength of $\lambda = 633$ nm. The antinodal surface has a high-reflectance and a maximum electric field at the surface for maximum interaction with the surface dipoles of the molecular film. The diffraction efficiency in this case is

$$\eta = \left(\frac{2}{\pi}\right)^2 k_0^2 (n^2 - n_\mathrm{m}^2)^2 d^2 \tag{6.7}$$

which is $\eta = 4.5 \times 10^{-4}$ for the 4 nm antibody layer in air, or 20 times larger for than a free-standing film. In water, the diffraction efficiency is $\eta = 1 \times 10^{-5}$, which is 30 times larger than for a free-standing film in water.

The surface patterning of a diffraction-based biosensor and the result of incubating the chip with target analyte is illustrated in Fig. 6.3. The recognition layers are printed as stripes (antibodies, aptamers, or DNA) and the surface is blocked with an inert material of similar mass, density, and hydrophilicity as the recognition layer. The similarity of the specific and nonspecific capture molecules balances the tendency of the surface to bind molecules indiscriminately (nonspecific binding). In principle, the similarity between the recognition layer and the

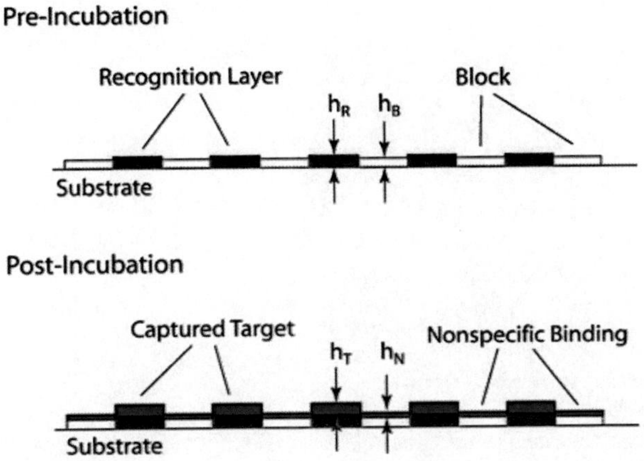

Fig. 6.3 Patterning of a diffraction-based biosensor. The recognition molecules are patterned into stripes, and the surface is blocked with an inert molecular film that has the same hydrophobicity. After incubation against the target analyte, the recognition molecules have bound target molecules. The blocked surface also can bind some molecules nonspecifically

blocking layer causes similar amounts of nonspecific binding to both types of stripes. During diffraction, this common mode of nonspecific binding is mostly subtracted away. For instance, the pre- and post-diffraction efficiencies are

$$\eta_{pre} = \frac{1}{\pi^2}k^2(n - n_m)^2(h_{R-h_B})^2,$$
$$\eta_{post} = \frac{1}{\pi^2}k^2(n - n_m)^2(h_R + h_T + h_N - (h_B + h_N))^2 \tag{6.8}$$

where h_R, h_T, h_B and h_N are the reference, target, back-fill and non-specific binding heights, respectively. The nonspecific binding is subtracted by the balance between the recognition and blocking layers. The differential change in diffracted intensity is

$$\begin{aligned}\Delta I &= \eta_{post} - \eta_{pre} \\ &= \frac{1}{\pi^2}k^2(n - n_m)^2[(h_R + h_T - h_B)^2 - (h_R - h_B)^2] \\ &= \frac{1}{\pi^2}k^2(n - n_m)^2[h_T + 2(h_R - h_B)]h_T\end{aligned} \tag{6.9}$$

which is quadratic in the height of the captured target analyte, but with a linear term that depends on the difference between the recognition layer and the blocking layer. If the recognition and blocking layer thicknesses are equal, then the background diffraction is zero, and the relative diffraction is purely quadratic. However, the $2 \times (h_R - h_B)$ bias can provide linear responsivity that has advantages over quadratic responsivities by making it more sensitive to sub-monolayer biolayers.

Diffraction-based biosensors originated with Tsay et al. [1], who used UV photoinactivation to produce periodic stripes of antigen or antibody and detected choriogonadotropin in serum. The sensitivity of this technique was in the range of 60 μg/mL. The concept was extended to two analytes by Goh et al. [2] with detection

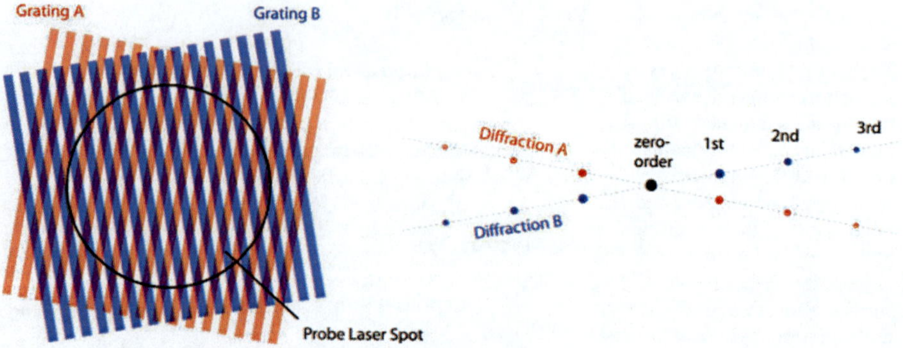

Fig. 6.4 Example of a multiplexed (duplex) biolayer grating biosensor. Two functionalized gratings are printed and capture analyte. The probe laser is diffracted into distinct Fourier components and orientations set by the grating orientations

limits approaching 1 μg/mL. Two analytes can be detected by using cross-printing of two biological species to form a two-dimensional diffraction grating with distinct far-field diffraction patterns that are associated with one or the other grating. This principle is illustrated in Fig. 6.4. Amplification of the diffraction signal is achieved by labeling the target molecules with gold particles [3] or by using horseradish peroxidase [4] to form absorption gratings to push detection limits below ng/mL, but this approach is not label-free.

Printed protein gratings also have been utilized to capture cells [5], which produce much larger scattering signals and can be imaged directly. Absorptive diffraction-based biosensors have been fabricated out of pH-sensitive gels that change absorption (color) dependent on the pH of a test sample [6].

6.1.2 Differential Phase Contrast

For sub-monolayer applications using diffraction grating-based biosensors, it is necessary to "bias" the diffraction so that the signal depends linearly on film thickness. In the extreme case, it is possible to bias the diffraction by $\pi/2$ to bring the signal into a quadrature condition with maximum linear dependence on film thickness (or surface density). One way to achieve a $\pi/2$ bias in diffraction is through a differential phase contrast approach [7] that requires no fabrication of the sample surface, other than providing a sharp edge to a biolayer, as in Fig. 6.5. When a focused Gaussian beam is centered on the edge of the biolayer, it is diffracted by the phase step into an asymmetric far-field interference intensity. The angular position of the maximum intensity change is set by the quadrature condition

$$w_0 \sin\theta_Q = \lambda/4, \tag{6.10}$$

where w_0 is the Gaussian beam radius and θ_Q is the quadrature angle.

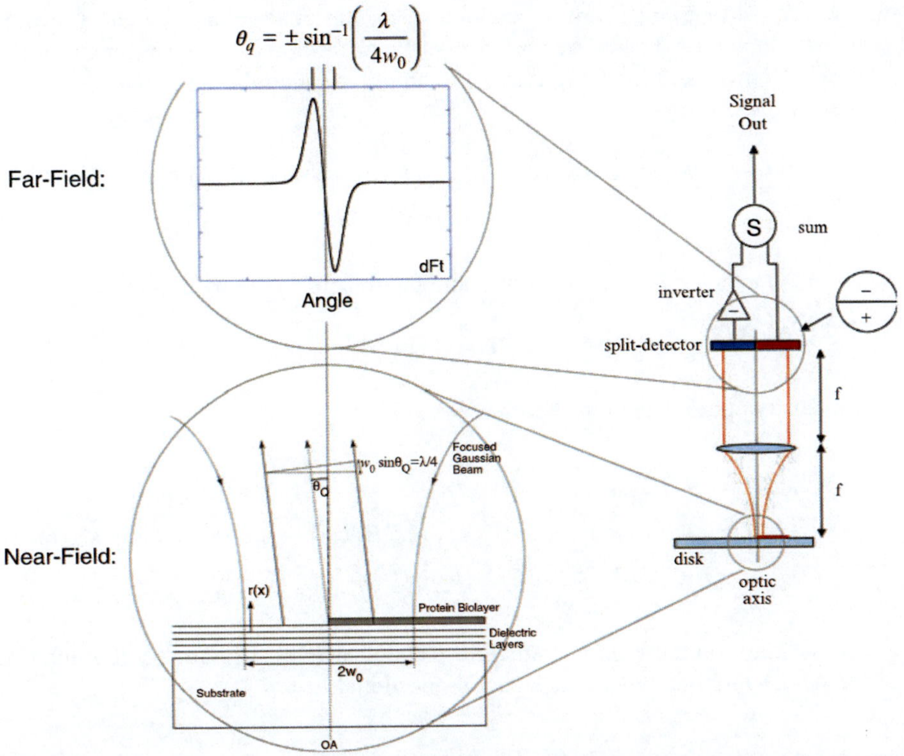

Fig. 6.5 Phase contrast detection of a sharp biolayer edge. A phase quadrature condition exists in the far-field diffraction when $w_0 \sin\,\theta_Q = \lambda/4$. The edge diffraction interferes with the Gaussian beam and creates an asymmetric far-field intensity that is detected by a split detector with a differencing output

The biolayer signal is calculated by considering the dimensionless incident electric field of a focused Gaussian beam

$$g(\rho) = g(x, y) = \frac{1}{\sqrt{\pi}w_0} \mathrm{e}^{-\rho^2/2w_0^2} \tag{6.11}$$

and its two-dimensional Fourier transform

$$G(k_x, k_y) = 2\sqrt{\pi}w_0 \mathrm{e}^{-w_0^2k^2/2}. \tag{6.12}$$

The two-dimensional diffraction problem is considered in the Fraunhofer regime. The modulated near-field, using (4.50), is

$$\begin{aligned} E(x,y) &\approx g(x,y)\left[r_0 + \mathrm{i}(1+r_0)^2 \frac{\pi}{n_\mathrm{m}\lambda}(n_\mathrm{m}^2 - n_\mathrm{p}^2)h(x-vt,y)\right] \\ &= g(x,y)[r_0 + \mathrm{i}\phi' h(x-vt,y)], \end{aligned} \tag{6.13}$$

where the surface topology is contained in the real-valued height function $h(x- vt)$, including the motion of the surface, for instance on a BioCD disk. This equation is valid for small phase and amplitude modulation.

The far field is

$$\begin{aligned} E(k_x, k_y) &= r_0 G(k_x, k_y) + \mathrm{i}\phi' \mathrm{FT}[g(x, y) h(x - vt, y)] \\ &= r_0 G(k_x, k_y) + \mathrm{i}\phi' H(k, t), \end{aligned} \tag{6.14}$$

where FT stands for Fourier transform and the transformed function is

$$H(k, t) = \mathrm{FT}[g(y, x) h(y, x - vt)]. \tag{6.15}$$

The intensity at the detection (Fourier) plane is

$$\begin{aligned} I(k_x, k_y; t) &= |r_0 G(k_x, k_y) + \mathrm{i}\phi' H(k, t)|^2 \\ &\approx R|G(k_x, k_y)|^2 + \mathrm{i}G(k_x, k_y)[r_0^* \phi' H(k, t) - r_0 \phi' * H^*(k, t)] \\ &= R|G(k_x, k_y)|^2 + 2G(k_x, k_y)\mathrm{Im}(r_0 \phi' * H^*(k, t)). \end{aligned} \tag{6.16}$$

A split detector on the Fourier transform plane detects the asymmetry through the difference channel to generate a signal proportional to

$$i_{\mathrm{d}}(t) \propto G(k_x, k_y)\mathrm{Im}(r_0 \phi' H(k, t)) \tag{6.17}$$

and more precisely is [7]

$$i_{\mathrm{d}}(t) = \frac{1}{2}|r_0 \phi'| \left[\frac{I(x)}{I(0)} \otimes \frac{\mathrm{d}h(x - vt)}{\mathrm{d}x} \right], \tag{6.18}$$

where the signal is proportional to the convolution of the incident beam profile with the spatial derivative of the height profile. The maximum differential phase contrast condition arises when maximizing

$$Re[(1 + r_0)^2 r_0^*] = Re(r_0)(1 + R) + 2R. \tag{6.19}$$

The maximum occurs at $r = +1$ (antinode surface), and has a zero response when $r = -1$ (node surface). The differential phase contrast response is shown in Fig. 6.6 as a function of reflection amplitude and phase.

An example of a photolithographically defined avidin pattern is shown in Fig. 6.7 detected on a bioCD using differential phase contrast. The disk is a dielectric mirror with an antinodal surface field condition. The spatial periodicity of the patterned protein in Fig. 6.7a is demodulated using side-band demodulation, and is shown

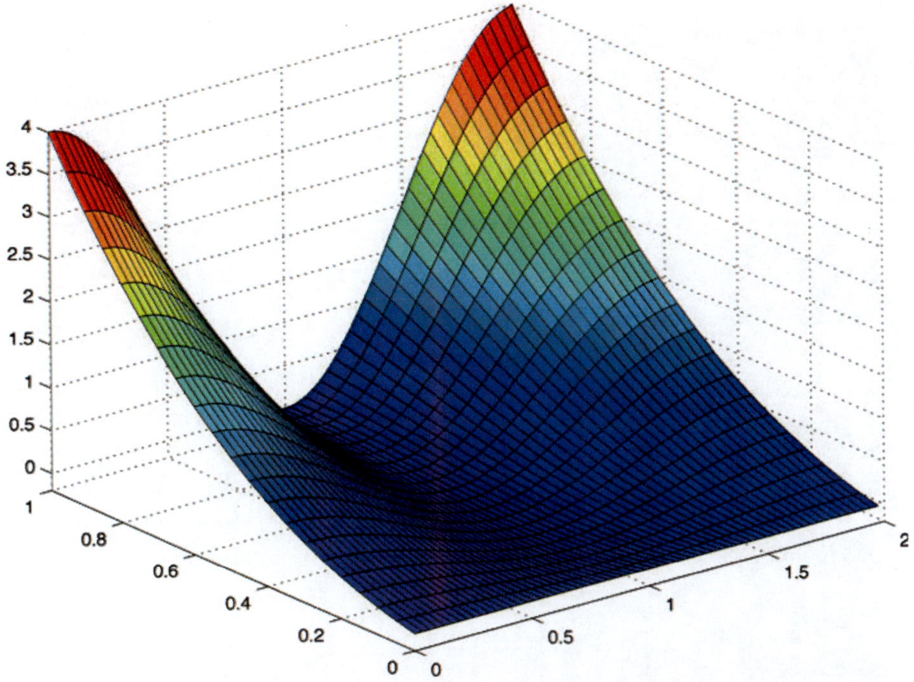

Fig. 6.6 Response of differential phase contrast as a function of reflection amplitude and phase. The maximum occurs when $r = +1$ on an antinodal surface with 100% reflectance

after demodulation in Fig. 6.7b. The surface height repeatability is histogrammed in Fig. 6.7c for successive scans comparing the demodulated with the raw surface height measurements. The height repeatability is 15 pm for the demodulated data out of 3 nm protein height, with a signal-to-background ratio of 200:1.

6.2 Microstructure Diffraction

Planar substrates can be replaced by nonplanar fabricated microstructures to yield a different class of diffraction-based biosensor. In this case, physical structures that diffract light capture biomolecules that modify their diffraction patterns and intensities.

6.2.1 *Micro-diffraction on Compact Disks*

One of the most ubiquitous microstructured objects of modern life is the compact disk (CD). The surface of a compact disk carries over a billion individual null interferometers – microstructures, called pits, that diffract light with a set phase

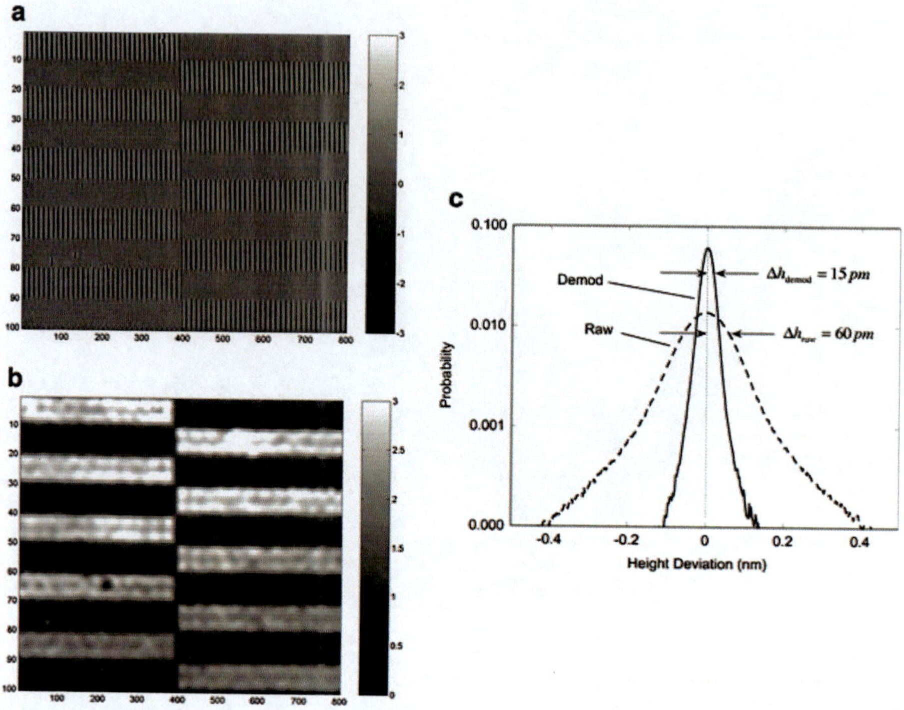

Fig. 6.7 Differential phase contrast detection of patterned avidin protein. (**a**) Raw surface scanning data. (**b**) Demodulated data. (**c**) Histograms of surface height comparing raw with demodulated data. From ref. [7]

shift of pi radians. When a focused laser beam straddles a pit, the phase shift causes destructive interference that cancels the light intensity at a detector. When the laser beam falls on the surrounding area around the pit (called the land), it is reflected with high-reflectance. As a focused laser beam scans across land and pits, the intensity flips back and forth between maxima and minima, encoding ones and zeros of binary data.

The pit structure of a conventional CD is shown in Fig. 6.8. The pits (actually ridges, because the polycarbonate is embossed by a metal master and read through the transparent carbonate) are arranged on tracks with a 1.6 μm pitch. The minimum pit length is 0.83 mm with a width of 0.5 μm. A focused laser spot moves with a linear velocity of 1.4 m/s at the rim. Two tracking satellite laser spots (with monitoring detectors) keep the probe spot focused and on track with active voice-coil actuators in a closed feedback loop.

The null interferometric condition is generated by diffraction from the pits. The pit depth is a quarter wavelength, and upon reflection this causes a half-wavelength delay. Roughly half the intensity of the beam falls on the pit and half on the land in a 50/50 wavefront-splitting interferometric configuration. On the optic axis in the far

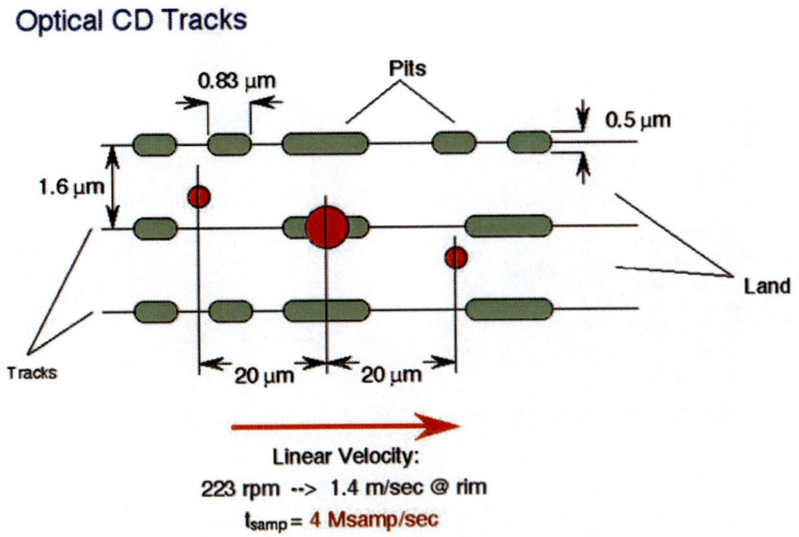

Fig. 6.8 Optical pit arrangement on a conventional compact disk (CD). The pits are arranged in tracks with a 1.5 micron pitch. The pits are 0.5 micron wide and a minimum of 0.83 micron long. A focused laser spot is maintained on a single track from two satellite spots and active feedback. A conventional CD can have up to a billion pits, each acting as an individual null interferometer

field, the two fields from the pit and the land combine with a pi phase shift that causes destructive interference – a null.

The interferometric surface of a conventional CD provides a convenient means to rapidly scan for biological molecules. Several approaches to CD-based interferometric biosensors have been pursued. The disk has been used as a high-speed scanning system for fluorescent [8] or absorption [9] readout. The first explicit use of the null interferometric pits as biosensors was by La Clair [10]. The binding of target analyte to the disk was detected as a bit error signal between digital prescan and post-incubation scans. The non-inertial forces of a rotating disk also have been employed for centrifugal microfluidics [11–13].

A drawback of conventional CD surfaces for biosensing is the null interferometric condition that is established for digital readout. As for the diffraction grating-based biosensors, this leads to a trade-off between background (contrast) and sensitivity. The null interferometric condition has the advantage of low-background, but the deviation of the intensity caused by biolayer accumulation is quadratic in the layer thickness, leading to weak sensitivity to submonolayer thicknesses.

For high sensitivity to very dilute molecular coverage, a phase quadrature condition is more favorable. In micro-diffraction BioCD biosensors [14], the principle of wavefront-splitting interferometry of the pit (ridge) was used to convert the null interferometric condition into quadrature conditions [15], illustrated in Fig. 6.9. A Gaussian beam straddles a reflecting ridge on a high-reflectance land in a 50/50 wavefront-splitting configuration. The height of the ridge sets a phase difference

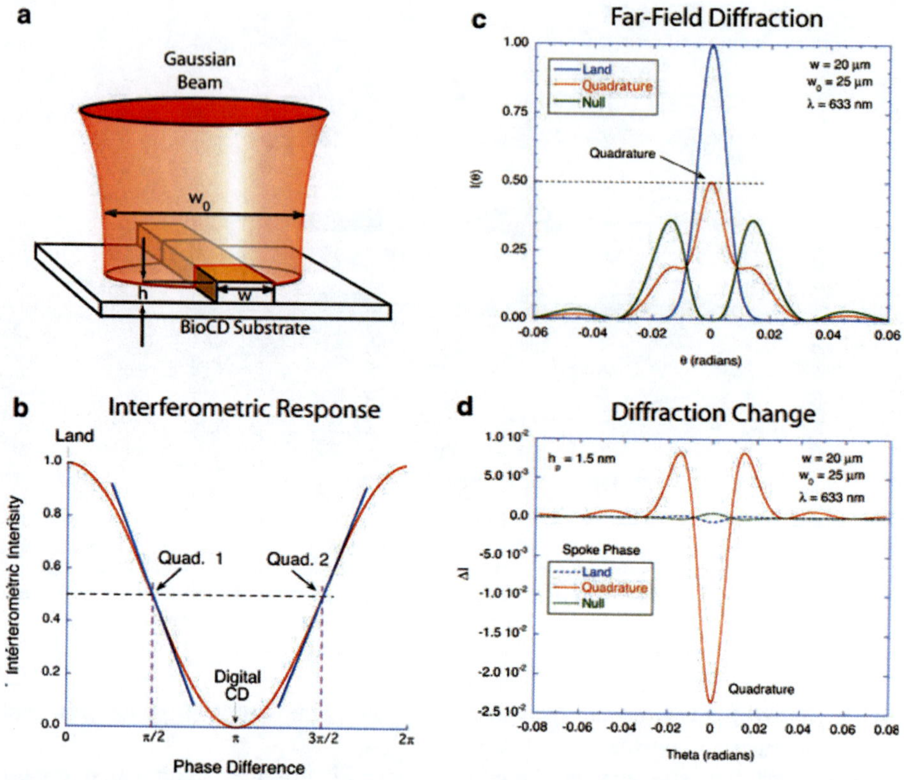

Fig. 6.9 (**a**) A focused Gaussian beam straddles a ridge of height h in a 50/50 intensity splitting configuration. (**b**) The intensity on the optic axis in the far field. The condition of phase quadrature (eighth-wave ridge height) is at the half-intensity point of the interferometric response curve. This is the condition of highest sensitivity to added material. (**c**) Diffraction patterns for ridge heights of zero, eighth-wave, and quarter-wave. (**d**) Change in the far-field diffraction with the addition of 1.5 nm of protein. From ref. [15]

between the partial waves reflected from the ridge relative to the land. At zero height, the ridge begins in a high-reflectance land condition. As the ridge height increases, it causes partial destructive interference that reduces the detected intensity. When the ridge height reaches a quarter wave, the intensity on the optic axis in the far field is zero because of perfect destructive cancelation – the null condition. The quadrature condition is located half-way between the land and the null condition at the half-intensity position on the sinusoidal interferometric response curve. At this position, any change in the relative height of the ridge and the land causes a maximum change in the detected intensity.

The far-field diffraction from the ridge is shown in Fig. 6.9c for the land, quadrature and the null conditions. When a protein layer is added to the ridge, the far-field diffraction changes, as shown in Fig. 6.9d. By spatially aperturing the far-field diffracted pattern to include only the central part of the diffraction peak, a signal linear in

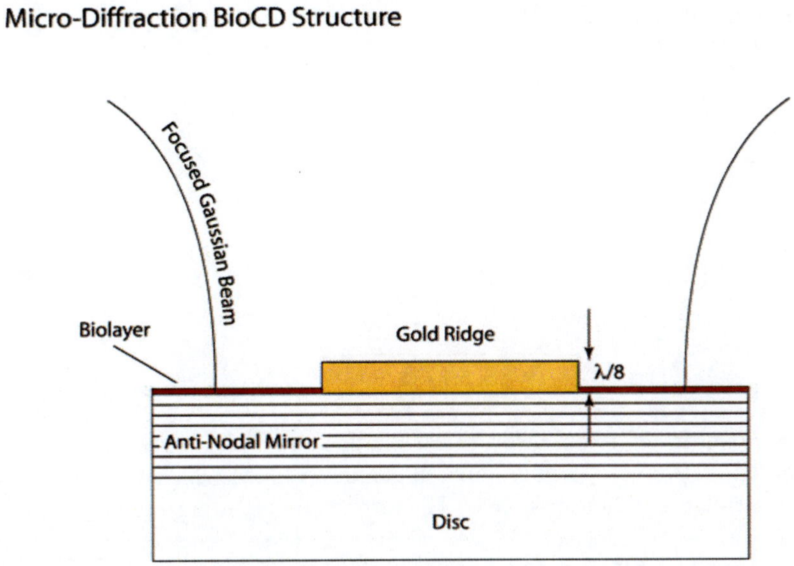

Fig. 6.10 Optimized micro-diffraction BioCD structure. The sensor surface is on the land that is an antinodal surface (maximum electric field). The gold ridge is a nodal surface and is relatively insensitive to nonspecifically bound materials, but the $\lambda/8$ ridge height establishes a quadrature condition for sensitive linear detection of the biolayer on the land

the immobilized protein is obtained. An additional quadrature condition occurs for a ridge height of 3/8-wavelength. The intensity changes upon biolayer accumulation on the ridge are equal but opposite [16] for the two opposite quadratures.

An optimized micro-diffraction BioCD structure is shown in Fig. 6.10. This design is based on an antinodal surface established by a multilayer quarter-wave mirror. The sensor surface is on the land and the antinodal surface maximizes the phase offset caused by the biolayer. The $\lambda/8$ gold ridge has a nodal surface condition (electric field is zero). This renders essentially invisible any molecules that attach to the gold and sets a positive-quadrature condition (increasing signal with increasing biolayer thickness) on the land. The micro-diffraction BioCD demonstrated opposite quadratures [16] and detection down to ng/mL [17].

6.2.2 Micro-Cantilevers

Micro-electromechanical system (MEMS) devices such as micro-cantilevers are dynamic structures that can be tuned actively to perform as diffractive optical balances with arbitrary biases.

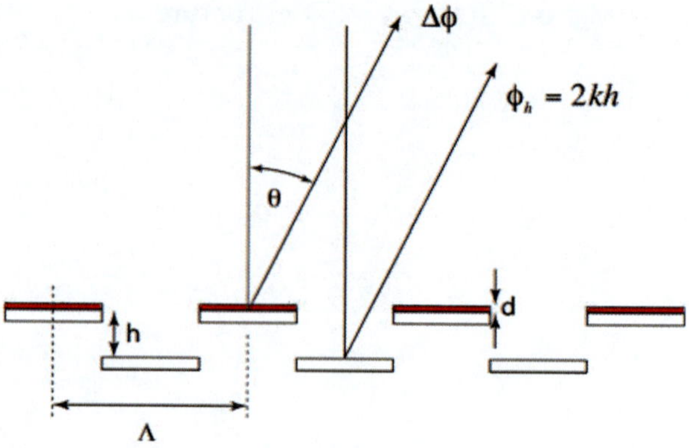

Fig. 6.11 Cantilevers with a relative displacement *h* and a biolayer thickness *d* on alternating fingers

An example of a biosensor based on this principle is shown in Fig. 6.11 composed of two sets of interdigitated cantilevers [18] with an adjustable offset distance *h* that acts as a phase bias. The diffracted field of a single pair is

$$\begin{aligned} E_{\mathrm{P}} &= \frac{-\mathrm{i}2r_0E_0}{k\Lambda\sin\theta}\left[\sin\left(\frac{k\Lambda}{2}\sin\theta+\frac{\Delta\phi_{\mathrm{b}}-2kh}{2}\right)-\sin\left(\frac{\Delta\phi_{\mathrm{b}}-2kh}{2}\right)\right]\mathrm{e}^{\mathrm{i}kh} \\ &= \frac{-\mathrm{i}2r_0E_0}{k\Lambda\sin\theta}\left[\sin\left(\frac{k\Lambda}{2}\sin\theta\right)\cos\left(\frac{\Delta\phi_{\mathrm{b}}-2kh}{2}\right)+\sin\left(\frac{\Delta\phi_{\mathrm{b}}-2kh}{2}\right)\left(\cos\left(\frac{k\Lambda}{2}\sin\theta\right)-1\right)\right]\mathrm{e}^{\mathrm{i}kh} \\ &= -\mathrm{i}r_0E_0\left[\operatorname{sinc}\left(\frac{k\Lambda}{2}\sin\theta\right)\cos\left(\frac{\Delta\phi_{\mathrm{b}}-2kh}{2}\right)-\operatorname{cosc}\left(\frac{k\Lambda}{2}\sin\theta\right)\sin\left(\frac{\Delta\phi_{\mathrm{b}}-2kh}{2}\right)\right]\mathrm{e}^{\mathrm{i}kh} \\ &= -\mathrm{i}r_0E_0F(\theta) \end{aligned} \tag{6.20}$$

using the definitions

$$\operatorname{sinc} x = \frac{\sin x}{x} \qquad \operatorname{cosc} x = \frac{1-\cos x}{x}. \tag{6.21}$$

For a Gaussian beam illumination of the periodic elements, the total field is

$$E(\theta) = -\mathrm{i}r_0E_0F(\theta)P(\theta, w_0/\Lambda), \tag{6.22}$$

where $P(\theta, w_0/\Lambda)$ is the grating function from (2.32) that is strongly peaked when

$$k\Lambda\sin\theta_{\mathrm{m}} = m2\pi \tag{6.23}$$

which acts as a spatial filter on $F(\theta)$ in (6.22). Expanding the effect of the biolayer, and using the spatial filtering, gives the field amplitudes in the odd-numbered diffraction orders

$$E_{\mathrm{m}} = -\mathrm{i}r_0E_0\mathrm{e}^{\mathrm{i}kh}\left[\left(1-\frac{1}{2}\left(\frac{\Delta\phi_{\mathrm{b}}}{2}\right)^2\right)\sin(kh)+\frac{\Delta\phi_{\mathrm{b}}}{2}\cos(kh)\right]\operatorname{cosc}(m_{\mathrm{odd}}\pi), \tag{6.24}$$

where the behavior of sinc(x) and cosc(x) are shown in Fig. 6.12.

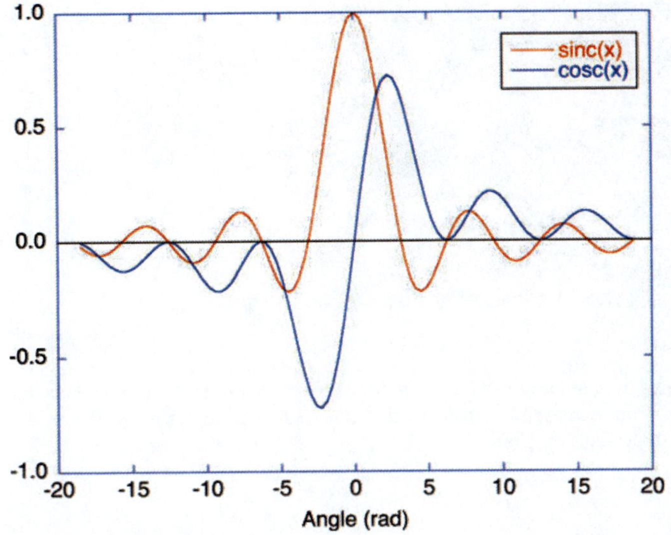

Fig. 6.12 Graphs of sinc(x) and cosc(x)

The mth-order diffracted intensity is

$$I_m = RI_0 \operatorname{cosc}^2(m_{\mathrm{odd}}\pi)\left[\sin^2(kh) + \frac{\Delta\phi_b}{2}\sin(2kh) + \left(\frac{\Delta\phi_b}{2}\right)^2\cos(2kh)\right], \quad (6.25)$$

where the phase from the biolayer from (4.53) is

$$\Delta\phi_b = \frac{(1+r_0)^2}{2r_0}\frac{k_0}{n_m\cos\theta_m}(n_m^2 - n_b^2)d. \quad (6.26)$$

As the bias height h increases from zero, there is a trade-off between the linear sensitivity to $\Delta\phi_b$ and the device contrast. When $h = \lambda/8$, the linear sensitivity is a maximum, and the system is in phase-quadrature, but has the lowest contrast. When $h = \lambda/4$, then the sensitivity is purely quadratic, but with essentially no background. The contrast of the biosensor is

$$C(h) = \frac{\Delta\phi_b\sin(2kh) + \frac{1}{2}\Delta\phi_b^2\cos(2kh)}{1-\cos(2kh)} \quad (6.27)$$

which is background-free when $2kh$ is an odd-integer of π, or when h equals an odd-integer of $\lambda/4$.

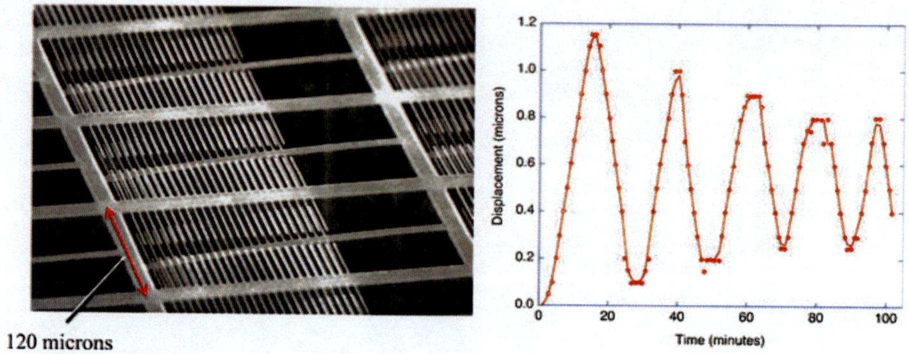

Fig. 6.13 Interdigitated micro-cantilevers for chemical vapor sensing. The data show the diffraction signal as a function of time caused by the relative deflection of the cantilevers as they absorbed mercury. With permission from ref. [19]

There have been several demonstrations of micro-cantilever sensors. A chemical vapor sensor was demonstrated in ref. [19] for mercury vapor that caused the cantilever to deflect when it absorbed mercury. The structure and the diffraction results are shown in Fig. 6.13, but the structure did not have a uniform offset. On the other hand, thermal sensing was demonstrated with a structure that could be tuned using a uniform offset to bias the sensor to maximum detection performance [18].

6.3 Bead-Based Diffraction Gratings

In the previous sections, we saw that molecular gratings produce very small diffraction efficiencies that are quadratic in the accumulating layer density. An alternative for molecular detection was provided by micro-cantilever gratings that could be tuned to quadrature for high sensitivity, or could be tuned to low-background for high contrast. Large signals can be obtained by using the deflection of one set of interdigitated cantilevers caused by molecular binding. However, the fabrication of the cantilevers is relatively complicated.

An alternative approach to diffraction biosensors that can have large signals without complex fabrication uses beads that bind specifically to target molecules, and which can then be captured onto surface patterns to diffract a probe beam. The bead size can be comparable to a wavelength, and can be in the Mie resonant scattering regime to produce large signals, while the probe and target molecules have molecular dimensions.

Consider stripes of beads immobilized on a surface with reflectance r_0. From (4.25), the far field of a surface density N of beads in a Gaussian beam is

$$E = \mathrm{i}\frac{kw_0^2\sqrt{S_0}}{f} r_0 \left[\frac{2\pi NS}{r_0 k^2} - 1\right] \exp\left(-\frac{2\pi^2 w_0^2}{(f\lambda)^2}(x^2 + y^2)\right), \tag{6.28}$$

where the scattering amplitude from the double interaction model is (4.21)

$$S_{\mathrm{DI}} = S_{\mathrm{b}}(\theta)(\mathrm{e}^{-\mathrm{i}k\delta^+} + F^2(ka)r_0^2\mathrm{e}^{\mathrm{i}2\phi}\mathrm{e}^{\mathrm{i}k\delta^+}) + S_{\mathrm{f}}(\theta)2r_0\mathrm{e}^{\mathrm{i}\phi}F(ka)(\mathrm{e}^{-\mathrm{i}k\delta^-} + \mathrm{e}^{\mathrm{i}k\delta^-}) \tag{6.29}$$

with $k = n_{\mathrm{m}}k_0$. The validity of (6.28) requires a dilute density of particles that does not significantly modify the incident beam (first Born approximation). This limit corresponds to low concentrations which are of most interest when studying detection limits. The large signal limit, with saturated bead densities, involves multiple scattering that is not captured in this analysis.

If the beads are in stripes of periodicity Λ, then the far field is given by

$$E = -\mathrm{i}\frac{kw_0^2\sqrt{S_0}}{f} r_0 \left[e^{-((2\pi w_0/\lambda)\theta)^2} - \frac{2\pi\mathcal{N}S_{DI}}{r_0 K^2}\mathrm{sinc}\left(K\frac{a}{2}\sin\theta\right)P_{\mathrm{G}}(\theta, w_0/\Lambda)\right], \tag{6.30}$$

where $P_{\mathrm{G}}(\theta, w_0/\Lambda)$ is the periodic Gaussian diffraction pattern of (2.32). Note that the reflected Gaussian has the same profile as the Gaussian functions in $P_{\mathrm{G}}(\theta, w_0/\Lambda)$. Hence there is interference in the zero-order, which represents a heterodyne condition, with possibly linear sensitivity. The higher order diffraction signals are in a homodyne condition with quadratic sensitivity. The diffracted intensity is

$$\begin{aligned} I(\theta) = I_0 \Bigg[& \left(R - \left(\frac{4\pi\mathcal{N}}{k^2}Re(r_0 S_{\mathrm{DI}})\right)\right) e^{-((2\pi w_0/\lambda)\theta)^2} \\ & + \left(\frac{2\pi N|S_{\mathrm{DI}}|}{k^2}\right)^2 \mathrm{sinc}^2\left(\mathrm{k}\tfrac{\mathrm{a}}{2}\sin\theta\right)\mathrm{P}_{\mathrm{G}}^2(\theta, \mathrm{w}_0/\Lambda)\Bigg] \end{aligned} \tag{6.31}$$

in which the first term is the zero-order interference, and the second term is for the diffraction orders that are quadratic in the bead surface density.

Bead-based diffraction gratings can assemble on prepatterned protein stripes, such as biotinylated BSA, when there are target molecules, such as avidin, on microbeads. The protein stripes are not visible on gold. But as microbeads assemble, diffraction can be detected with high sensitivity. For a probe beam that straddles approximately 25 stripes, diffraction can be detected when there is on average only a single bead per stripe that is bound. This approach has been used to detect *S*-adenosyl homocysteine on magnetic beads that were pulled to the prepatterned surface using a magnetic field. A detection limit down to 300 pM was achieved for this small 400 Da molecule [20]. Microscopic images of the accumulating beads are shown in Fig. 6.14.

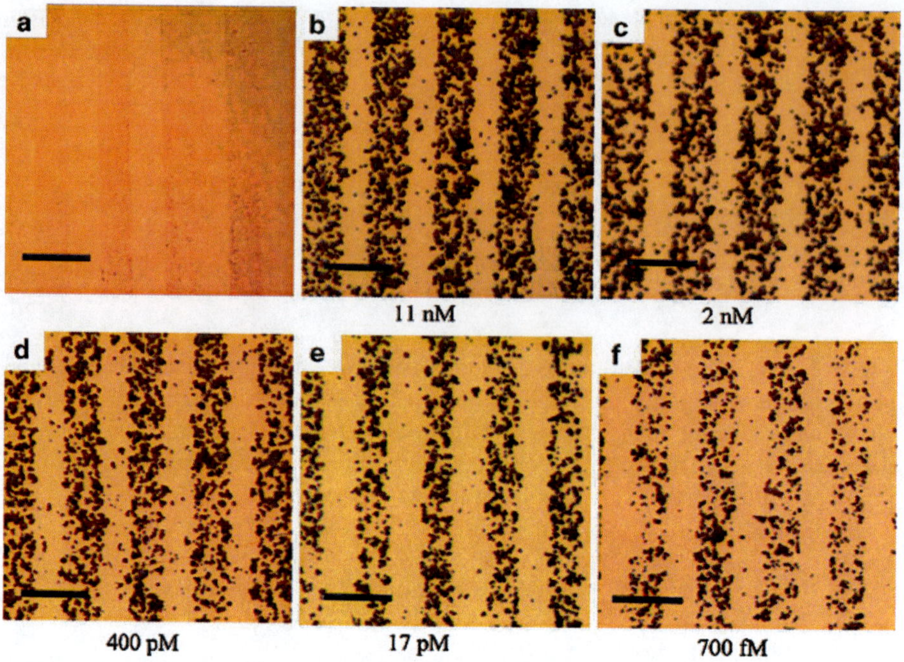

Fig. 6.14 Magnetic beads on printed BSA stripes on gold. The beads had antibodies to the target analyte SAH, which bound to the aptamer on the surface. A diffraction efficiency of 2% was observed even for the lowest concentration. The BSA stripes are seen in (**a**), and decreasing target concentrations run from (**b**) through (**f**). From ref. [20]

References

1. Tsay, Y.G., Lin, C.I., Lee, J., Gustafson, E.K., Appelqvist, R., Magginetti, P., Norton, R., Teng, N., Charlton, D.: Optical biosensor assay (Oba). Clin. Chem. **37**(9), 1502–1505 (1991)
2. Goh, J.B., Loo, R.W., McAloney, R.A., Goh, M.C.: Diffraction-based assay for detecting multiple analytes. Anal. Bioanal. Chem. **374**(1), 54–56 (2002)
3. Goh, J.B., Tam, P.L., Loo, R.W., Goh, M.C.: A quantitative diffraction-based sandwich immunoassay. Anal. Biochem. **313**(2), 262–266 (2003)
4. Loo, R.W., Tam, P.L., Goh, J.B., Goh, M.C.: An enzyme-amplified diffraction-based immunoassay. Anal. Biochem. **337**(2), 338–342 (2005)
5. St John, P.M., Davis, R., Cady, N., Czajka, J., Batt, C.A., Craighead, H.G.: Diffraction-based cell detection using a microcontact printed antibody grating. Anal. Chem. **70**(6), 1108–1111 (1998)
6. Nakajima, F., Hirakawa, Y., Kaneta, T., Imasaka, T.: Diffractive optical chemical sensor based on light absorption. Anal. Chem. **71**(13), 2262–2265 (1999)
7. Zhao, M., Cho, W., Regnier, F., Nolte, D.: Differential phase-contrast BioCD biosensor. Appl. Opt. **46**, 6196–6209 (2007)
8. Kido, H., Maquieira, A., Hammock, B.D.: Disc-based immunoassay microarrays. Anal. Chim. Acta **411**(1–2), 1–11 (2000)

9. Alexandre, I., Houbion, Y., Collet, J., Demarteau, J., Gala, J.L., Remacle, J.: Compact disc with both numeric and genomic information as DNA microarray platform. Biotechniques **33**, 435–439 (2002)
10. La Clair, J.J., Burkart, M.D.: Molecular screening on a compact disc. Org. Biomol. Chem. **1**, 3244–3249 (2003)
11. Madou, M., Zoval, J., Jia, G.Y., Kido, H., Kim, J., Kim, N.: Lab on a CD. Ann. Rev. Biomed. Eng. **8**, 601–628 (2006)
12. Haeberle, S., Zengerle, R.: Microfluidic platforms for lab-on-a-chip applications. Lab Chip. **7**(9), 1094–1110 (2007)
13. Ducree, J., Haeberle, S., Lutz, S., Pausch, S., von Stetten, F., Zengerle, R.: The centrifugal microfluidic bio-disk platform. J. Micromech. Microeng. **17**(7), S103–S115 (2007)
14. Varma, M., Nolte, D.D., Inerowicz, H.D., Regnier, F.E.: Multi-analyte array micro-diffraction interferometry. In: Bornhop, B.J., et al. (ed.) Microarrays: Design, Fabrication and Reading, vol. 4626, pp. 69–77. SPIE (2002)
15. Nolte, D.D.: Review of centrifugal microfluidic and bio-optical disks. Rev. Sci. Instrum. **80**(10), 101101 (2009)
16. Varma, M.M., Nolte, D.D., Inerowicz, H.D., Regnier, F.E.: Spinning-disk self-referencing interferometry of antigen-antibody recognition. Opt. Lett. **29**(9), 950–952 (2004)
17. Varma, M.M., Inerowicz, H.D., Regnier, F.E., Nolte, D.D.: High-speed label-free detection by spinning-disk micro-interferometry. Biosen. Bioelectron. **19**(11), 1371–1376 (2004)
18. Savran, C.A., Sparks, A.W., Sihler, J., Li, J., Wu, W.C., Berlin, D.E., Burg, T.P., Fritz, J., Schmidt, M.A., Manalis, S.R.: Fabrication and characterization of a micromechanical sensor for differential detection of nanoscale motions. J. Microelectromech. Syst. **11**(6), 703–708 (2002)
19. Thundat, T., Finot, E., Hu, Z., Ritchie, R.H., Wu, G., Majumdar, A.: Chemical sensing in Fourier space. Appl. Phys. Lett. **77**(24), 4061–4063 (2000)
20. Acharya, G., Chang, C.L., Holland, D.P., Thompson, D.H., Savran, C.A.: Rapid detection of S-adenosyl homocysteine using self-assembled optical diffraction gratings. Angew. Chem. Int. Ed. **47**(6), 1051–1053 (2008)

Chapter 7
Interferometric Waveguide Sensors

The interferometric interaction of light with a biological layer is linearly proportional to the length along which the electric fields interact with the sample. Therefore, responsivity can be improved by increasing the interaction length. One way to accomplish this is to have the light wave pass back and forth through the layer many times, as in a cavity. Another way to accomplish this is to propagate the probe light along the biolayer, which is the topic of this chapter on interferometric waveguide sensors. These types of interferometric biosensors operate on principles of waveguide confinement combined with evanescent field profiles.

7.1 Evanescent Confinement

Evanescent fields are exponentially decaying solutions to Maxwell's equations subject to boundary conditions at an interface or surface. They have characteristic decay lengths that tend to be comparable to a wavelength and hence are highly localized near the interface. These fields are preferentially sensitive to refractive index or dipole density in close proximity to the interface, for instance caused by binding of analyte to recognition molecules immobilized on the surface. Another reason why they are valuable is because the effect on the evanescent wave can be integrated over long interaction lengths, providing sensitive sampling of small changes to the surface. The simplest evanescent wave arises from total internal reflection (TIR), and the simplest configuration to exploit the evanescent wave is a waveguide structure.

D.D. Nolte, *Optical Interferometry for Biology and Medicine*, Bioanalysis 1,
DOI 10.1007/978-1-4614-0890-1_7, © Springer Science+Business Media, LLC 2012

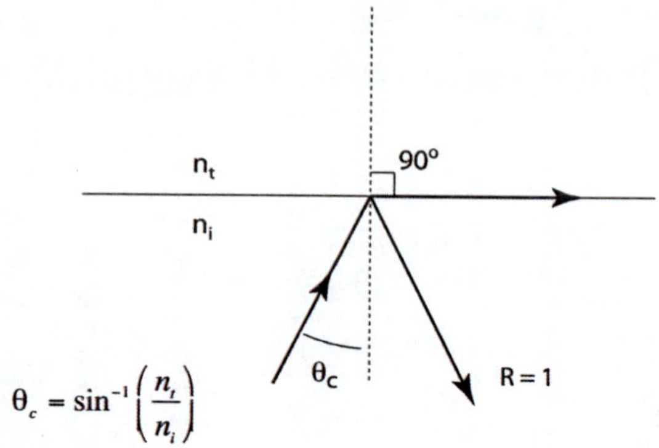

Fig. 7.1 Critical angle for total internal reflection (TIR)

7.1.1 Total Internal Reflection (TIR)

When light is incident on an interface between two dielectric media, in which the incident medium has a slower light speed than the transmitting medium, then there is a critical angle obtained from Snell's Law

$$\frac{\sin(\pi/2)}{\sin\theta_c} = \frac{n_i}{n_t} \tag{7.1}$$

at which the transmitted wave travels parallel to the interface (transmitted angle of $\pi/2$). This critical angle is

Critical angle for TIR:

$$\theta_c = \sin^{-1}\left(\frac{n_t}{n_i}\right) \tag{7.2}$$

For incident angles larger than this value, there is no real-valued transmitted angle, and all the light is reflected. This is the phenomenon of TIR (Fig. 7.1).

Under the condition of TIR of a plane wave, there is an evanescent wave confined to the interface between the two media. The energy in this wave is established during the initial transients after the wave impinges on the surface, and is sustained in steady state for continuous-wave excitation even when the steady state reflected power equals the incident power.

To find the localization of the evanescent wave, conservation of transverse momentum (or transverse k-vector) is expressed as

$$k_i \sin\theta_i = k_{t,\parallel} \tag{7.3}$$

The magnitude of the k-vector in the second medium is

$$k_{t,\parallel}^2 + k_{t,\perp}^2 = k_t^2 \tag{7.4}$$

Solving for $k_{t,\perp}$ gives

$$\begin{aligned} k_{t,\perp} &= \sqrt{k_t^2 - k_{t,\|}^2} \\ &= \sqrt{k_t^2 - k_i^2 \sin^2 \theta_i} \\ &= k_t \sqrt{1 - \frac{n_\mathrm{i}^2}{n_\mathrm{t}^2} \sin^2 \theta_i} \end{aligned} \tag{7.5}$$

For TIR, the second term is greater than unity, making the argument in the square root negative, and making the perpendicular component of k_2 imaginary.

$$\begin{aligned} k_{t,\perp} &= \mathrm{i} k_t \sqrt{\frac{n_\mathrm{i}^2}{n_\mathrm{t}^2} \sin^2 \theta_i - 1} \\ &= \mathrm{i}\beta, \end{aligned} \tag{7.6}$$

where β is the inverse decay length of the field

$$\begin{aligned} E_t &= E_t^0 \exp(\mathrm{i}\vec{k}_t \cdot \vec{r}) \\ &= E_t^0 \exp(\mathrm{i}k_i \sin\theta_i x) \exp(-\beta z) \end{aligned} \tag{7.7}$$

The decay length into the second medium is

$$\delta = \frac{1}{\beta} = \frac{\lambda}{2\pi n_\mathrm{t}} \frac{1}{\sqrt{(n_\mathrm{i}^2/n_\mathrm{t}^2)\sin^2\theta_i - 1}} \tag{7.8}$$

The shortest decay length (strongest localization at the interface) occurs as θ_i approaches $\pi/2$. The decay length in this limit is

$$\delta_{\min} \approx \frac{\lambda}{2\pi} \frac{1}{\sqrt{n_\mathrm{i}^2 - n_\mathrm{t}^2}} = \frac{\lambda}{2\pi} \frac{1}{\sqrt{2\bar{n}}} \frac{1}{\sqrt{n_\mathrm{i} - n_\mathrm{t}}} \tag{7.9}$$

For an interface between glass and water at 633 nm, this decay length is approximately 150 nm, shown in Fig. 7.2. This is the shortest decay length. For incident angles that approach the critical angle, the decay length extends further into the second medium. Therefore, for most practical TIR biosensors, the sensitivity extends several hundred nanometers into the second medium.

Although the amplitude of the reflection coefficient is clamped at unity for angles larger than the critical angle, the phase varies continuously as a function of angle. The phase is obtained from the Fresnel's equations of Chap. 4; and for TE and TM polarizations are

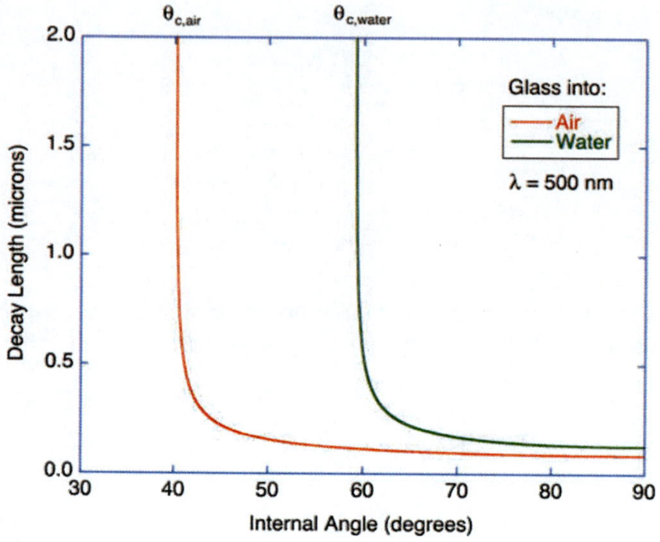

Fig. 7.2 TIR decay length vs. angle for glass into air or into water. While the decay length can be very large near the critical angle, typical values are on the order of a fraction of a wavelength

$$\tan\frac{\phi_r}{2} = \frac{\sqrt{\sin^2\theta - \sin^2\theta_c}}{\cos\,\theta} \quad \text{TE polarized} \tag{7.10}$$

$$\tan\frac{\phi_r}{2} = \frac{\sqrt{\sin^2\theta - \sin^2\theta_c}}{\cos\theta\sin^2\theta_c} \quad \text{TM polarized} \tag{7.11}$$

The reflection phases are plotted in Fig. 7.3a for the two polarizations for glass to air and glass to water.

An important aspect of this phase arises for biosensors, for instance from glass to liquid, when the refractive index of the liquid changes, either due to chemical densities or reactions, or because of accumulating molecules at the solid surface. The change in the phase per unit change in the refractive index of the second medium is shown in Fig. 7.3b. This is the refractive index responsivity of the reflected phase. The units are in radians per refractive index unit (RIU). The high phase responsivity, combined with the near-surface nature of the evanescent wave, is one reason why waveguide sensors are among the most important optical biosensors.

7.1.2 Dielectric Waveguide Modes

A planar dielectric waveguide can be as simple as a plane-parallel slab of high-refractive index embedded in lower index material. For instance, a slab of glass in air performs as a dielectric waveguide when light is launched at its end, as in Fig. 7.4.

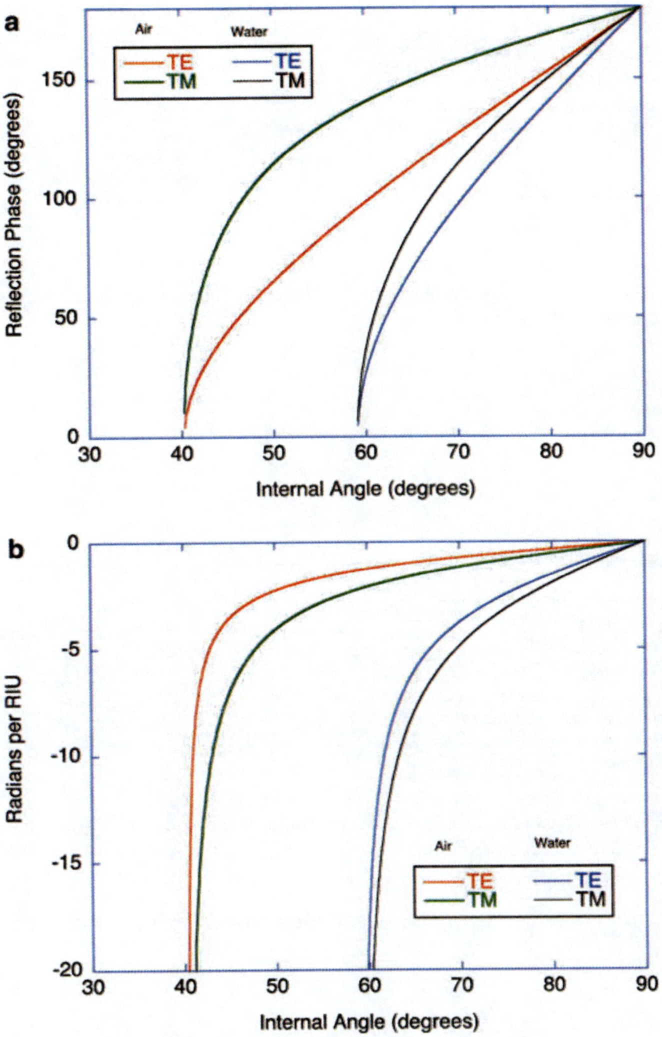

Fig. 7.3 (**a**) Phase shift on reflection for TIR for glass to air and glass to water. (**b**) Change in phase (in radians) per refractive index unit for TIR TE and TM modes

When the external angle is sufficiently small, the ray inside the slab makes an angle $\bar{\theta}$ (shown in Fig. 7.5), that is larger than the critical angle for TIR, and the rays reflect multiple times confined within the slab.

To form a propagating mode inside the waveguide, all the rays must interfere constructively. To find the angle $\bar{\theta}$ for which this constructive interference occurs, the distance AC–AB in Fig. 7.5 represents the path difference between two successive reflections that brings the reflected wave vector parallel to the original. This distance is

$$\mathrm{AC} - \mathrm{CB} = 2d \sin \bar{\theta} \tag{7.12}$$

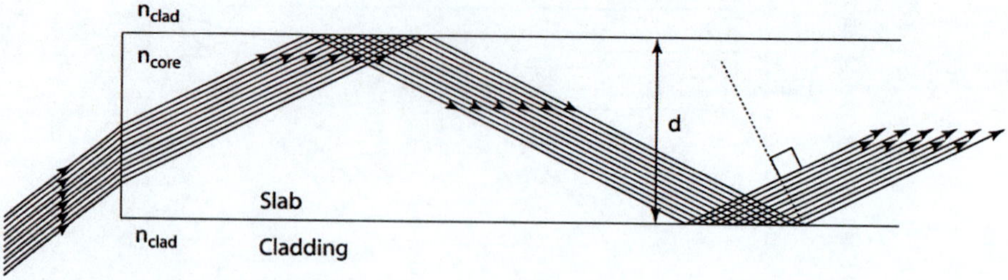

Fig. 7.4 To form a propagating mode, all the internally reflected rays of a slab waveguide must be in phase to interfere constructively

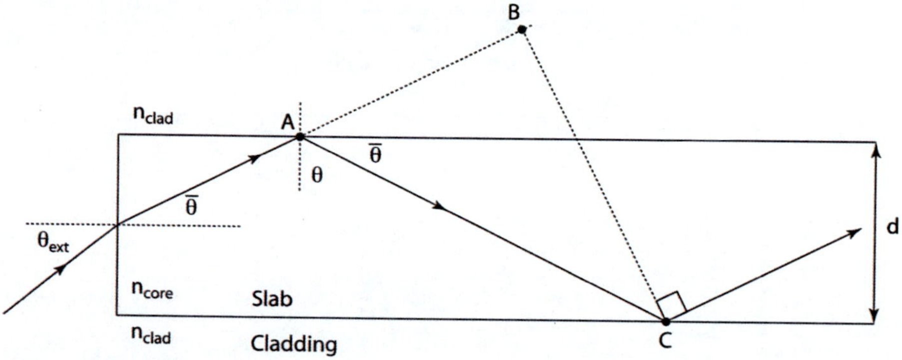

Fig. 7.5 Internally reflected rays of a slab waveguide. θ is the incident angle on the cladding, while $\bar{\theta}$ defines the *k*-vector orientation

For constructive interference, the associated optical phase must be an integer number of 2π with

$$\frac{2\pi}{\lambda} 2n_{\text{core}} d \sin \bar{\theta}_{\text{m}} - 2\phi_{\text{r}} = 2\pi m \tag{7.13}$$

The phase shift ϕ_{r} is given in (7.10) and (7.11). For instance, for TE polarization this is

$$\tan \phi_{\text{r}}/2 = \sqrt{\frac{\sin^2(\bar{\theta}_{\text{c}})}{\sin^2 \bar{\theta}_{\text{m}}} - 1} \tag{7.14}$$

Taking the tangent of (7.13) leads to the self-consistency condition

$$\tan\left(\frac{\pi}{\lambda} n_{\text{core}} d \sin \bar{\theta}_{\text{m}} - \pi m/2\right) = \tan(\phi_{\text{r}}/2) \tag{7.15}$$

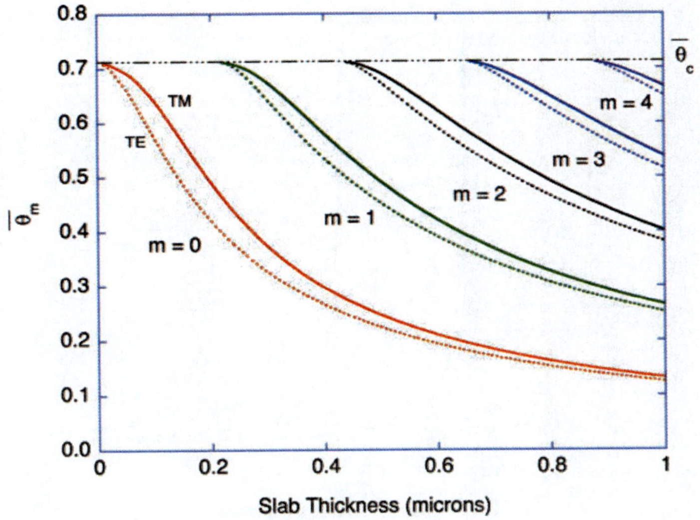

Fig. 7.6 Simulated values for θ_{m}. Glass core with water cladding

which is solved for θ_{m} for a given integer mode number m and shown in Fig. 7.6. The propagation constant $\beta_{\mathrm{m}} = k_z$ is given by

$$\beta_{\mathrm{m}} = k_z = n_{\mathrm{core}} k_0 \cos\, \bar{\theta}_{\mathrm{m}} \tag{7.16}$$

With the solutions for θ_{m}, the full field profile of the waveguide mode can be obtained. The field internal to the core is

$$u_{\mathrm{m}}^{\mathrm{core}}(y) \propto \begin{cases} \cos\left(\frac{2\pi \sin \bar{\theta}_{\mathrm{m}}}{\lambda} n_{\mathrm{core}} y\right), & m = 0, 2, 4, \ldots \\ \sin\left(\frac{2\pi \sin \bar{\theta}_{\mathrm{m}}}{\lambda} n_{\mathrm{core}} y\right), & m = 1, 3, 5, \ldots \end{cases} \tag{7.17}$$

and in the cladding is

$$u_{\mathrm{m}}^{\mathrm{clad}}(y) \propto \begin{cases} \exp(-\gamma_{\mathrm{m}} y), & y>0 \\ \exp(\gamma_{\mathrm{m}} y), & y<0 \end{cases} \tag{7.18}$$

These are continuous at the boundary, such that

$$u_{\mathrm{m}}^{\mathrm{core}}(d/2) = u_{\mathrm{m}}^{\mathrm{clad}}(d/2) \tag{7.19}$$

and the field is normalized by

$$\int_{-\infty}^{\infty} u_{\mathrm{m}}^2(y)\, \mathrm{d}y = 1 \tag{7.20}$$

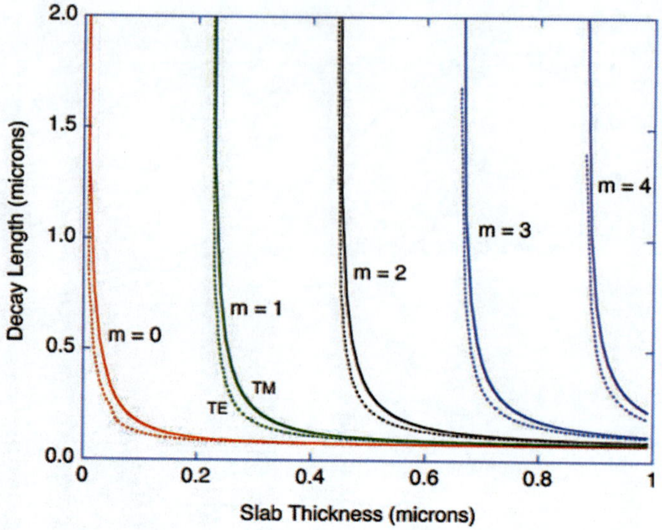

Fig. 7.7 Simulated values for $1/\gamma_{\mathrm{m}}$. Glass core with water cladding

The decay constant γ_{m} is related to the transverse propagation constant β_{m} through

$$\gamma_{\mathrm{m}}^2 = \beta_{\mathrm{m}}^2 - n_{\mathrm{core}}^2 k_0^2 \tag{7.21}$$

yielding

$$\gamma_{\mathrm{m}} = n_{\mathrm{core}} k_0 \sqrt{\cos^2 \bar{\theta}_{\mathrm{m}} - \sin^2 \theta_{\mathrm{c}}} \tag{7.22}$$

This is plotted in Fig. 7.7 for the lowest TE and TM modes of the waveguide. Of particular interest for waveguide biosensors is the field magnitude at the surface of the cladding where a biolayer would be immobilized. One measure of how much of the field extends outside the core is through the confinement factor, defined by

$$\Gamma_{\mathrm{m}} = \frac{\int_0^{d/2} u_{\mathrm{m}}^2(y)\,\mathrm{d}y}{\int_0^{\infty} u_{\mathrm{m}}^2(y)\,\mathrm{d}y} \tag{7.23}$$

where $(1-\Gamma_{\mathrm{m}})$ provides the measure of the field extending into the sensing region outside the core.

If the refractive index of the cladding is changed to $n_{\mathrm{clad}} + \Delta n$, the propagation constant is changed by

$$\Delta\beta_{\mathrm{m}} = n_{\mathrm{core}} k_0 \frac{d \cos \theta_{\mathrm{m}}}{\mathrm{d}n} \Delta n \tag{7.24}$$

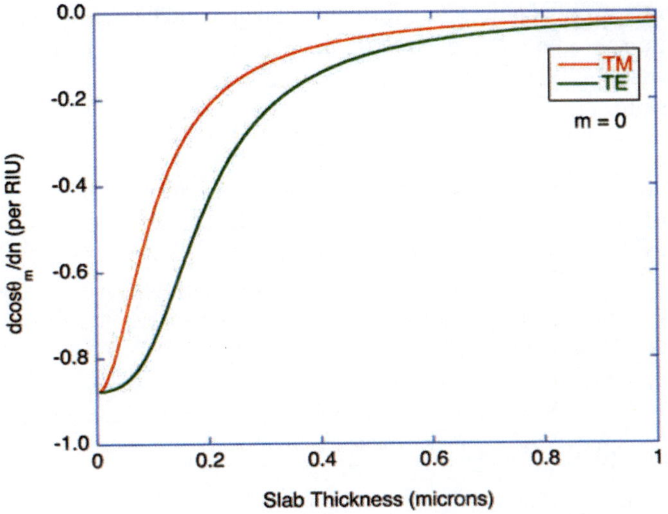

Fig. 7.8 Simulated values for $d\cos\bar{\theta}_m/dn$ for the $m = 0$ mode of a glass core with water cladding

The sensitivity of the cosine to a change in the cladding index is shown in Fig. 7.8. In the limit of the thinnest slabs, the change in the cosine is approximately equal to the change in index. For a waveguide of length L, the change in phase of the transmitted waveguide mode is

$$\Delta\phi = \Delta\beta_m L \tag{7.25}$$

Here we see the importance of the continuous interaction of the evanescent wave with the sample along the length L of the sensor. Waveguides easily can be operated over lengths of millimeters, compared with the very short interaction length of nanometers of thin film biosensors of Chap. 6. For instance, for a uniform index change of $\Delta n = 0.01$ for the cladding, the change in phase for a 1 mm long waveguide is on the order of 100 rad.

Waveguide biosensors sense the refractive index of the cladding medium, which often consists of water and uniformly distributed reagents. As reagents change in concentration, or if two or more interact to form new compounds, the refractive index of the cladding medium changes uniformly and causes a shift in the phase of the transmitted light. A common configuration for waveguide biosensors uses recognition molecules that are immobilized as a nanometer-scale layer on top of the waveguide slab or core region. As these recognition molecules bind target molecules out of solution, the layer thickness increases, causing a shift in the transmitted phase. Because the accumulating layer is not homogeneously distributed in the medium, the shift in the phase is not due to a homogeneous change in the cladding medium, but is localized. This is an advantage, because the evanescent wave is preferentially sensitive to the surface and will not be sensitive to changes in the fluid cladding far from the waveguide. The equivalent change in a bulk index that would correspond with the change caused by the film of thickness d_{film} on one side of the cladding is approximately

$$\Delta n_{\text{eff}} = 2\gamma_m d_{\text{film}}(n_{\text{film}} - n_{\text{clad}}) \tag{7.26}$$

The effective refractive index of an equivalent homogeneous cladding medium is proportional to the difference between the medium index and the film index, and is proportional to the ratio of the film thickness to the decay length of the evanescent wave. More accurate computational calibration of a waveguide structure requires multilayer simulations for the reflected phase that would be used in (7.15) to find the guided modes.

7.2 Waveguide Couplers

Having established the properties of the guided modes of a slab waveguide, the next issue is how to couple light into and out of a waveguide with high efficiency and few stray reflections (that can cause stray interference fringes). The three most common methods for launching guided modes in waveguides are illustrated in Fig. 7.9. These are: (1) prism coupling to achieve high-angle coupling into the guided mode, (2) grating coupling by diffracting a wave into the direction of

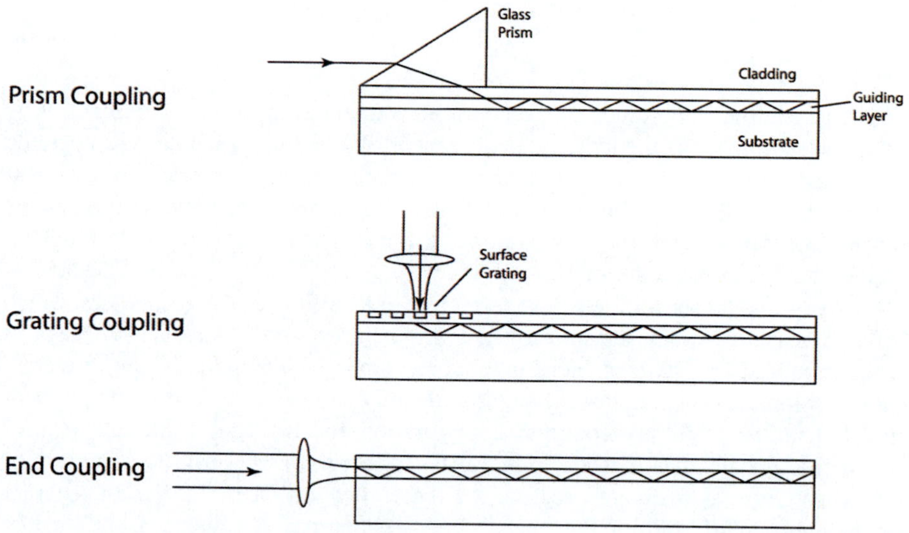

Fig. 7.9 Different methods for coupling into (and out of) planar waveguides. Prism coupling couples light in from a high index material at an angle that is above the TIR angle along the waveguide. Grating coupling uses diffraction from the grating to launch high-angle modes that are confined in the waveguide. End-coupling is conceptually straightforward, but more difficult to implement accurately

the guided mode, and (3) end-coupling by focusing a laser through a lens onto the waveguide.

End-coupling is the most straightforward approach, but is the least stable in terms of alignment and mechanical stability. For efficient coupling, the mode shape of the focused beam should be very similar to the transverse mode structure of the lowest-order guided mode. The coupling efficiency is related to the overlap integral

$$J = \left| \int_{-d/2}^{d/2} G(z) F(z) \, \mathrm{d}z \right|^2 \tag{7.27}$$

where $G(z)$ is the normalized transverse field profile of the guided mode, and $F(z)$ is the normalized profile of the focused laser. When $G(z)$ and $F(z)$ are most similar, the overlap integral is a maximum. An alternate way to consider coupling efficiency is through the numerical aperture of the coupling optics compared with the numerical aperture of the waveguide. The cut-off angle inside the guide sets a maximum allowable range of angles on the input face. The numerical aperture is defined as the sine of the maximum acceptance angle θ_a

$$\mathrm{NA} = \sin\,\theta_a = \sqrt{n_{core}^2 - n_{clad}^2} \tag{7.28}$$

This waveguide numerical aperture should match the numerical aperture of the focusing optics

$$\mathrm{NA} = \frac{d}{2f} \tag{7.29}$$

where d is the diameter of the laser beam incident on the lens with focal length f. For a beam diameter of 1 mm, a focal length to match a typical numerical aperture of $\mathrm{NA} = 0.5$ is $f = 0.5$ mm, which is usually satisfied by a high magnification objective lens. The drawback of end-coupling is the high mechanical accuracy that is needed to focus the beam on the central portion of the waveguide for maximum overlap.

A more stable approach that is easier to implement is the prism coupler. In this case, the relatively high index of the prism matches the cladding layer refractive index of the guide. There is little refraction at the prism-cladding interface, allowing high internal angles to be launched into the guiding layer. The main detail is to have the beam enter the cladding just at the apex of the prism so that, once launched, the guided mode cannot escape again through the prism.

Possibly the most controllable input coupling is through a surface grating fabricated in the cladding layer. An incident beam is diffracted by the grating. If the grating period is short enough, a diffracted wave can be launched at high angle inside the guide. For instance, for a periodicity Λ, the grating vector is $K = 2\pi/\Lambda$.

A grating can couple light into a guided mode with a propagation constant β when the propagation constant equals the grating vector

$$K = \frac{2\pi}{\Lambda} = \beta = k_0 n_{\text{core}} \cos \bar{\theta}_{\text{m}} \tag{7.30}$$

or

$$\Lambda = \frac{\lambda}{n_{\text{core}} \cos \bar{\theta}_{\text{m}}} \tag{7.31}$$

As an example, for $\bar{\theta}_{\text{m}} = 0.5$ and $\lambda = 0.8\ \mu\text{m}$, the grating periodicity should be about 0.6 μm. This requires a feature size of 0.3 μm, which is easily achieved using microfabrication techniques.

7.3 Waveguide Structures

There are several possible structures for solid-state waveguides. Some of these are illustrated in Fig. 7.10. The planar waveguide is equivalent to a free-space optical system. Waves can be launched in different directions to cross and interfere. However, the absence of lateral confinement prevents the light from being directed

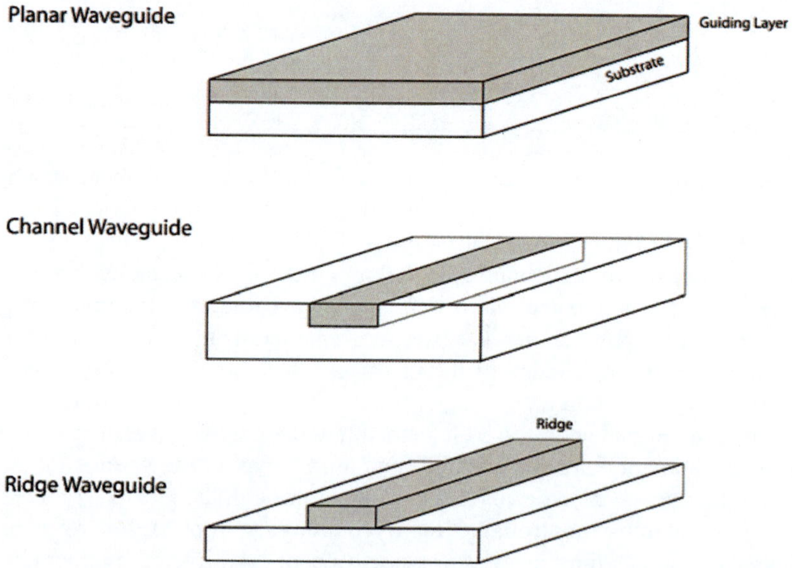

Fig. 7.10 Examples of waveguide structures for sensing. The planar waveguide is simplest, but has no lateral confinement. A channel waveguide with lateral confinement can be fabricated by ion implantation. A ridge waveguide can be fabricated by masking and etching, or by masking and deposition

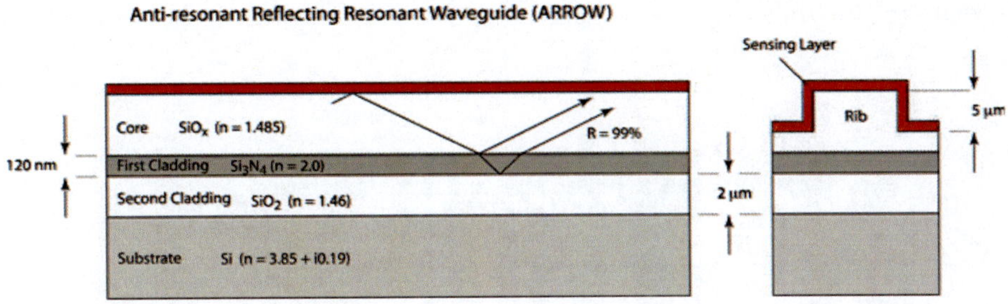

Fig. 7.11 Antiresonant reflecting optical waveguide (ARROW) structure. The evanescent field extends from the core into the outer medium around the rib structure

in the plane other than by line of flight. For lateral direction control, a channel waveguide or a ridge waveguide can be fabricated using ion implantation, etching and/or deposition. These waveguides can be designed to guide light in transverse directions on the substrate. This is the basis of integrated optics, which is the optical analog of integrated circuits, with waveguides substituting for the wires of the integrated circuit. There are waveguide structures that are designed specifically for biosensing applications. Examples of these are antiresonant waveguide (ARROW) structures and the resonant mirror (RM).

7.3.1 Antiresonant Waveguide (ARROW)

A novel waveguide structure with several advantages for biosensing applications is an antiresonant reflecting optical waveguide (ARROW). This structure can be fabricated with much larger guide dimensions of several microns instead of fractions of a micron (required for single-mode performance of slab waveguides). This larger format makes it easier (more stable) to couple into and out of the waveguide, compared with the tight fabrication and tight focusing tolerances required by conventional single-mode waveguides.

The basic operating principle of the ARROW structure is the use of a leaky waveguide structure that attenuates modes higher than the fundamental mode. This makes the structure perform like a single-mode structure in spite of its large dimensions. The structure is shown in Fig. 7.11 [1]. The core of the waveguide consists of SiO_x with $n = 1.485$ fabricated as a ridge waveguide with a rib depth of about 5 μm. The core rests on two cladding layers: a thin first layer with a high index composed of Si_3N_4 with $n = 2.0$ that is 120 nm thick, on top of a second low-index cladding that is much thicker at 2 μm and composed of SiO_2 with $n = 1.46$. The first cladding layer constitutes an antiresonant Fabry-Perot with reflectance (at the high incident angle) above 99% for the lowest-order TE mode. The reflectance is much lower for TM modes and all higher TE modes. Therefore, these modes leak into the silicon substrate and are absorbed. The disadvantage of the ARROW

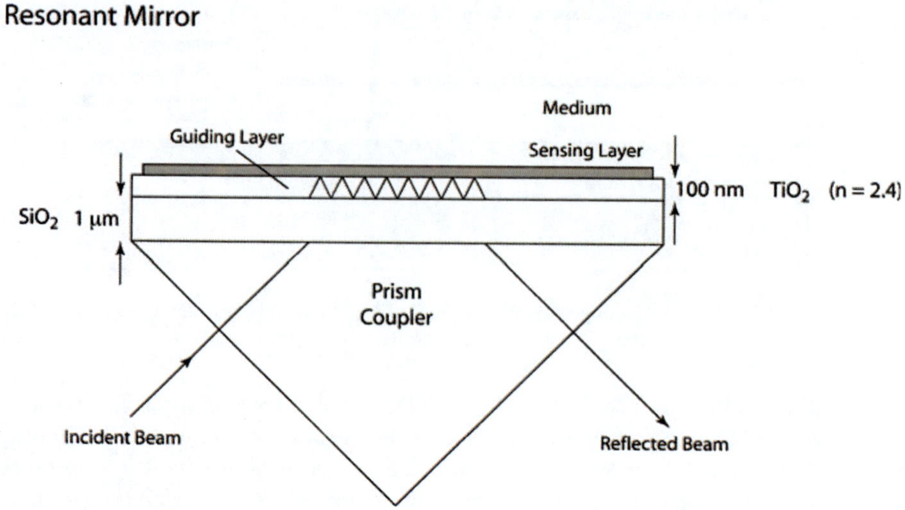

Fig. 7.12 The resonant mirror biosensor. The high-index guiding layer replaces the metal layer of an SPR sensor

structure for biosensing is a large confinement factor that confines most of the field to the core, with an effective refractive index that is significantly smaller than for conventional small-core waveguide sensors.

7.3.2 The Resonant Mirror

A biosensor structure that is based on the different phase velocities of TE and TM modes in dielectric waveguiding layers is the resonant mirror (RM). It does not strictly operate in a guided-mode configuration, but instead uses TIR like a surface-plasmon configuration. The structure uses polarization control to generate destructive interference between the field components of TE and TM modes.

The resonant mirror structure is shown in Fig. 7.12. It consists of a high-index guiding layer in contact with an overlying medium and resting on a coupling layer. The configuration and operation is very similar to prism coupling into and out of a waveguide [2]. The incident angles are beyond the critical angle for TIR, and the reflected intensities are unity. However, when the incident light is matched to a guided mode, there is a strong shift in the reflected phase of the light coupled out as the coupling angle is tuned. The phase can change by nearly 2π over only an angular change of about a degree. Because the reflected amplitude is unity, this phase shift must be detected through interference. This interference is provided by the fact that the TE and TM modes propagate at different velocities and consequently have different resonant coupling angles. For instance, as the reflected phase of the TM mode changes rapidly near one resonant angle, the TE mode reflected phase remains constant. Therefore, the resonant mirror is operated using both TE and TM excitation with a 45° polarizer, and is detected through a 45° polarizer at the output to detect the interference between the two modes. The rapid phase shift of one polarization reflection relative to the other produces destructive interference that is observed as a sharp dip in the reflectance as the angle is tuned through the resonant angle. Because the mode propagation is sensitive to bound molecules on the TIR surface, small amounts of binding on the sensor surface causes the reflectance dip to shift.

The simulated performance of a resonant mirror is shown in Fig. 7.13. The reflectance of TE and TM polarized light is plotted as a function of angle. The critical angle for both polarizations is 42°, above which the reflectance is unity. However, through a 45° analyzer, the TE and TM fields interfere, producing the reflectance in (b). The TE-mode interference is extremely sharp in the simulation. In practice, experimental losses broaden the interference. However, the resonant dip is about an order of magnitude sharper than for surface plasmon resonance (SPR) sensors that experience attenuation in the metal films, although the angle shift caused by binding is about three time smaller than SPR because less of the resonant mirror guided mode is in the evanescent wave in the overlying medium. The overall performance in terms of shift per resonance width is about an order of magnitude larger for the resonant mirror relative to SPR sensors. In addition, the guiding materials can be more robust to aqueous environments than metals. Despite these advantages, the resonant mirror has not made inroads into the biosensor marketplace that continues to be dominated by SPR-related biosensors.

7.4 Mach–Zehnder Interferometric Waveguide Sensors

One of the simplest interferometers to construct using a planar waveguide is the Mach–Zehnder interferometer discussed in Chap. 1. For example, a planar waveguide Mach–Zehnder can use a broad grating coupler to "illuminate" two parallel sensing regions using a single input beam. One region is the sensor with immobilized recognition molecules, while the other is a standard reference. The sensor region of this chip is immersed in a sample containing target analytes. As the analytes bind specifically to the recognition strip, balanced by nonspecific binding on the reference strip, the relative phase of the two guided modes shifts according to (7.25). The two beams are coupled out of the sensor chip by an output coupling grating. The waves interfere in the two output ports to produce intensity shifts that directly track the amount of binding on the sensor strip. The sensor length in this example [3] may be typically 1 mm, leading to phase shifts of approximately 10 times 2π for several nm of bound protein.

An integrated optics approach to a Mach–Zehnder biosensor is shown in Fig. 7.14 in which a ridge waveguide structure is fabricated with two splitters that separate and recombine the waves on the optical chip. By bringing the splitters onto

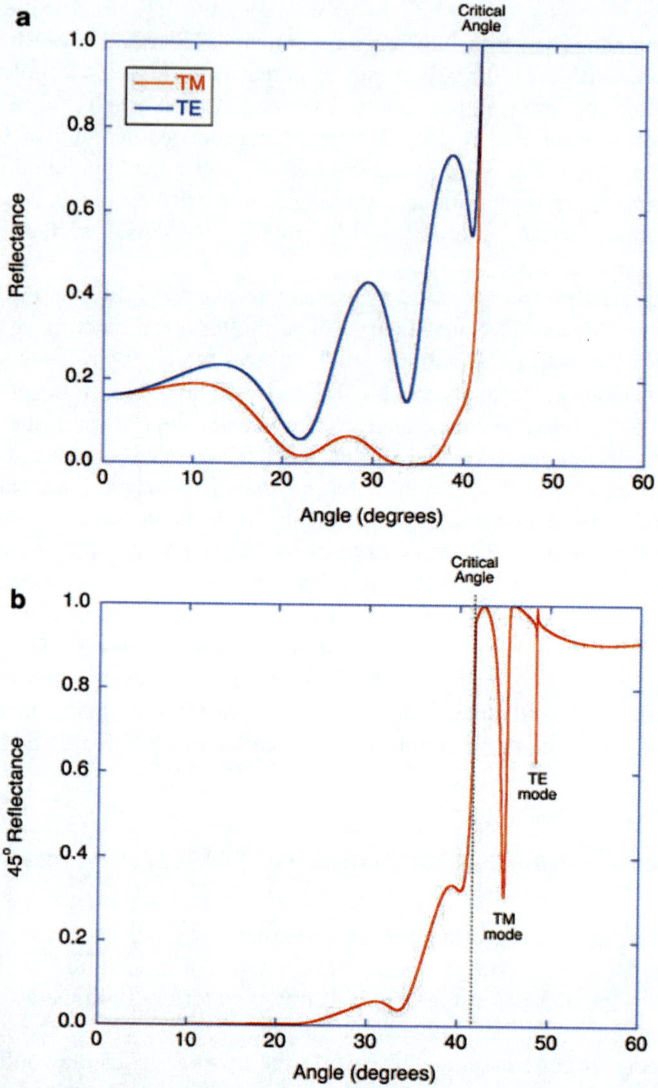

Fig. 7.13 Resonant mirror performance. The TM and TE reflectances are shown in (**a**). They both share the same critical angle of 42°. The reflectance through a 45° polarizer is shown in (**b**). The strong phase shift that occurs as the angle is tuned through the resonant coupling angle for each guided mode causes an interference dip at each resonant angle. Binding from the overlying medium shifts the resonance

the chip, the entire device becomes much more compact. The sensor arm has the cladding removed to bring the sample to be sensed into contact with the guiding core. The reference arm can be fixed (with cladding), or can be an additional sensing area with nonspecific recognition molecules to cancel out nonspecific

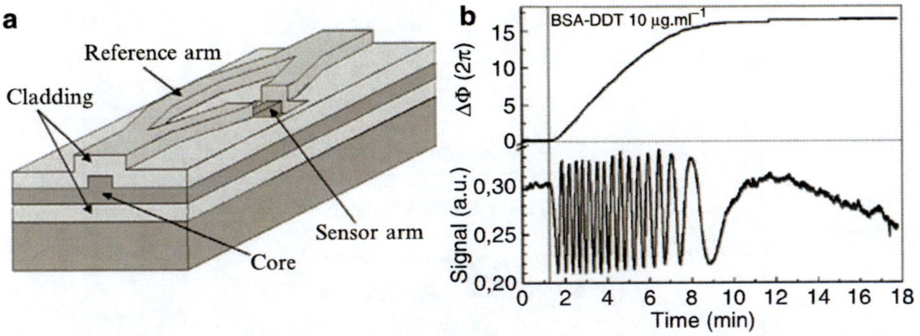

Fig. 7.14 Integrated Mach–Zehnder ridge-waveguide interferometer on a chip. The cladding in (**a**) has been removed from the sensor region to bring the sample into contact with the waveguide core. The phase change and signal are shown in (**b**). Reprinted with permission from ref. [4]

binding from the interference signal. This device exhibits 10–15 times 2π phase rotation upon saturated binding of target molecules.

7.5 Young's-Type Fringe-Shifting Interferometers

Mach–Zehnder interferometers have only a single detection channel and hence require large modulation for accurate measurement of phase shifts caused by accumulating molecules on the sensor region. A closely related waveguide approach with a large number of detection channels is the Young's double slit configuration. This uses two waveguide channels, just like the Mach–Zehnder. But instead of recombining the channels prior to detection, the light is launched into free space from the two separated channels to generate interference fringes on a pixel-array detector. An example is shown in Fig. 7.15 that is similar to the first half of a Mach–Zehnder, but launches the modes into free space. The interference fringes shift as the effective index over the sensing region is modified by a biochemical reaction. Because the fringes are sampled by many pixels, the fringe position can be located with high precision, and hence very small changes in the effective index can be detected.

A similar two-channel Young's-type interferometer uses two waveguide slabs situated vertically, as in Fig. 7.16. The sensing slab is in communication with an ambient medium, while the lower slab is the reference slab. The waves from the slabs launch into free space and are detected on a pixel-array detector [5]. This configuration is particularly valuable when both polarizations TE and TM are used in the slabs [6], because the effective index n_{eff} is different for the two polarizations, and it is possible to separate the $(n - n_m)$ term and the thickness d from the $(n - n_m)d$ product when the biolayer is not too thin. This configuration is called dual-polarization interferometry (DPI) and has been used to quantify protein refractive index values.

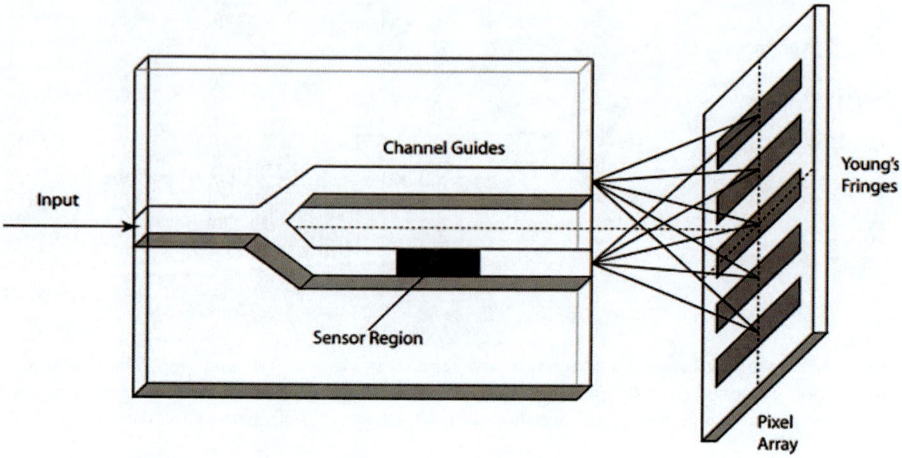

Fig. 7.15 Young's-type interferometer with channel waveguides. Accumulation of a biolayer on the sensor region shifts the fringe positions

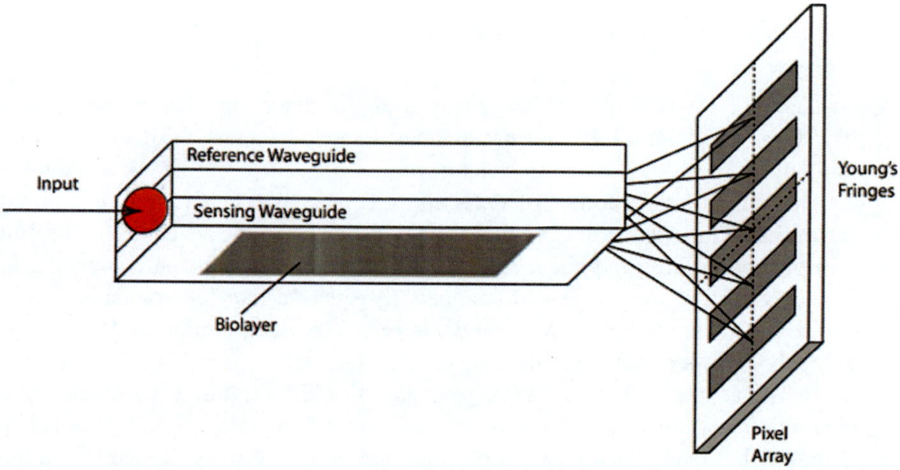

Fig. 7.16 Dual-polarization interferometer (DPI) using over-and-under slab waveguides with free-space detection of Young's fringes that shift as a biolayer accumulates on the sensing waveguide

7.6 Guided-Mode Resonance (GMR) Sensors

SPR sensors based on gold films are the most common label-free sensors. These sensors have sensitivities down to 1 pg/mm^2 and rely on the sharp reflectance dip when the external light couples into the surface plasmons. The plasmons are evanescent waves that sense refractive index changes close to the surface, which

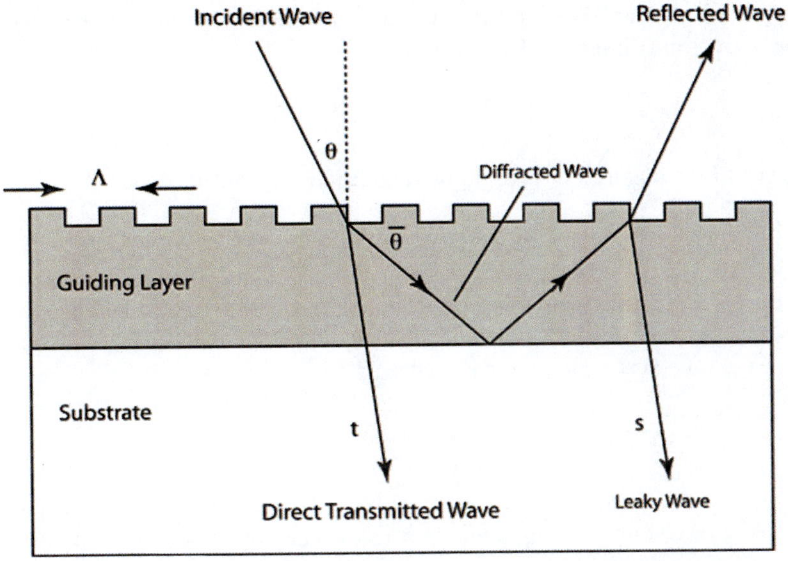

Fig. 7.17 Guided-mode resonant (GMR) structure. At resonance, the transmitted fields t and s are π out of phase and interfere destructively

causes the resonant feature in the reflectance to shift, and is detected either by a shift in angle or a shift in wavelength. However, the width of the resonance, which determines its responsivity, is dictated by the properties of the gold film and is hence not a "designable" feature of the sensor.

Alternative interference structures that have many of the attributes of SPR sensors, but which use dielectric gratings that can control the width of the resonance, are guided-mode resonance (GMR) sensors. These are waveguide structures that have subwavelength gratings that serve several simultaneous functions: grating coupling into the waveguide, Bragg reflection of the guided modes, and output coupling of the Bragg-reflected modes. The basic GMR structure is shown in Fig. 7.17 consisting of a low-index leaky-mode waveguide layer and a grating layer [7]. An incident plane wave diffracts into a first diffraction order that is coupled to the leaky mode of the waveguide. This guided mode is diffracted by the grating into reflected and transmitted directions. Under resonant conditions, the original transmitted wave and the transmitted leaky waves are π radians out of phase, which causes destructive interference in the transmitted direction. This produces a narrow but deep transmittance dip under resonance conditions.

The condition for resonant coupling into the waveguide is given by the diffraction equation

$$n_m k_0 \sin\theta + mK = n_g k_0 \cos\bar{\theta} = \beta \tag{7.32}$$

where n_m is the effective index of the overlying medium, k_0 is the free-space k-vector, θ is the incident angle, m is the diffraction order, $K = 2\pi/\Lambda$, n_g is the index of the guiding layer and $\bar{\theta}$ is the internal angle. The total phase difference between the transmitted fields t and s is

$$\phi_{st} = 2n_g k_0 d + 2\phi_1 + 2\phi_2 \tag{7.33}$$

which can be compared directly to the condition for guided modes

$$2n_g k_0 d + 2\phi_1 + 2\phi_2 = m2\pi \tag{7.34}$$

Therefore, when the incident light couples resonantly into the guided mode, $\phi_{st} = \pi$, and the transmission is quenched by destructive interference. The transmittance can be approximately parametrized as

$$\frac{T}{T_0} = \frac{\Delta^2}{s^2 + \Delta^2} \tag{7.35}$$

where Δ is the detuning parameter from resonance, and s is set by the diffraction efficiency of the grating. When s is made very small, the transmittance shows a very sharp dip at the resonance wavelength (or angle).

The resonance condition at normal incidence and for first-order diffraction is simply

$$K = \beta$$

$$\Lambda = \lambda / n_{\mathrm{eff}} \tag{7.36}$$

where the dependence on the effective mode index arises explicitly. Therefore, a change in the effective mode index is observed as a shift of the resonance condition, which can be measured at normal incidence by measuring the transmittance (or reflectance) spectrum.

This dependence on n_{eff} is the basis of resonant waveguide structures that perform in a way analogous to SPR sensors, but which have a larger design space to engineer tight resonance conditions across broad areas. A biosensor structure based on GMR is shown in Fig. 7.18 [8]. This device is operated in reflection with the light incident from the substrate and diffracted at the grating layer that also supports a biolayer that accumulates out of solution. At normal incidence and at resonance, the grating diffracts the first-order into the guided mode. However, the same grating causes this guided mode to Bragg-reflect back (second-order Bragg reflection [9]), where it is coupled out and counter-propagates back along the incident wave. Small changes in the biolayer thickness cause a shift in the resonant reflectance spectrum that is detected using a spectrometer. The width of the reflectance peak can be only a few nanometers, which represents a much sharper resonance than achieved by SPR. In addition, the ability to operate at normal incidence (instead of the high angles used in SPR) considerably simplifies the detection optics and also enables direct imaging.

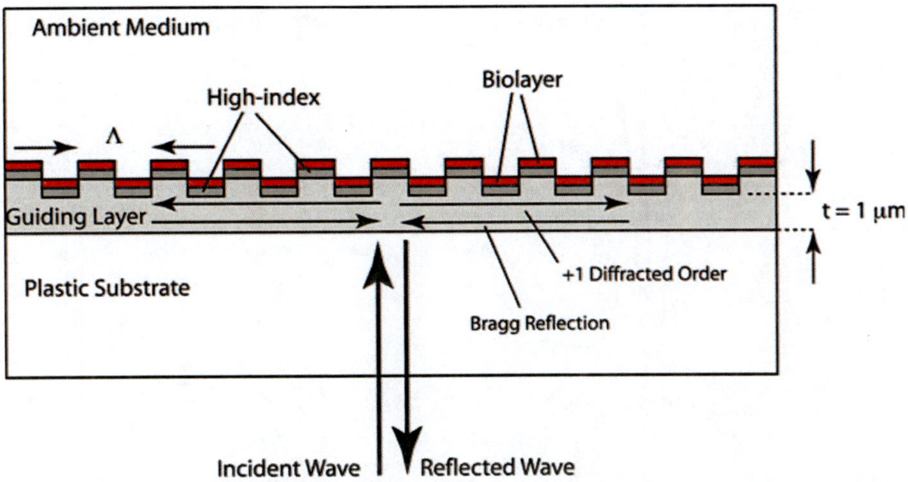

Fig. 7.18 A resonant grating biosensor operating at normal incidence in reflectance. The grating supports the biolayer and diffracts the incident wave into the guided mode, where it is Bragg-reflected back and coupled out back along the direction of the incident wave. At resonance, the reflectance approaches unity, but with a wavelength bandwidth of only a few nanometers

There are numerous variants on the resonant grating waveguide biosensors because the design space is large, and different configurations may have different advantages for fabrication or detection [10, 11]. One variant is called optical waveguide light mode spectroscopy (OWLS) that operates at oblique angle and can excite more than one mode [12]. Accurate measurements of protein refractive index can be made using this approach [13]. These grating waveguide systems could also be combined with an SPR system by placing 40 nm of Gold on the grating [14].

The sharpness of resonance features provides a clear advantage for biosensor responsivity by creating a response function with a very high slope dependence on a local refractive index. However, it must also be kept in mind that high responsivity does not always translate to low detection limits because chemical recognition chemistry and residues may be limiting, and higher responsivity also can be accompanied by higher system noise. Nevertheless, the strong responsivity of resonant biosensors has the advantage of strong signals from small numbers of analyte which can make them very sensitive under appropriate conditions.

7.7 Optofluidic Biosensors

Capillary tubes have the advantage of co-confinement of light with the liquid fluid containing analyte. The liquid cores of capillaries or hollow fibers have small volumes that need only very small amounts of sample solution, and light can be

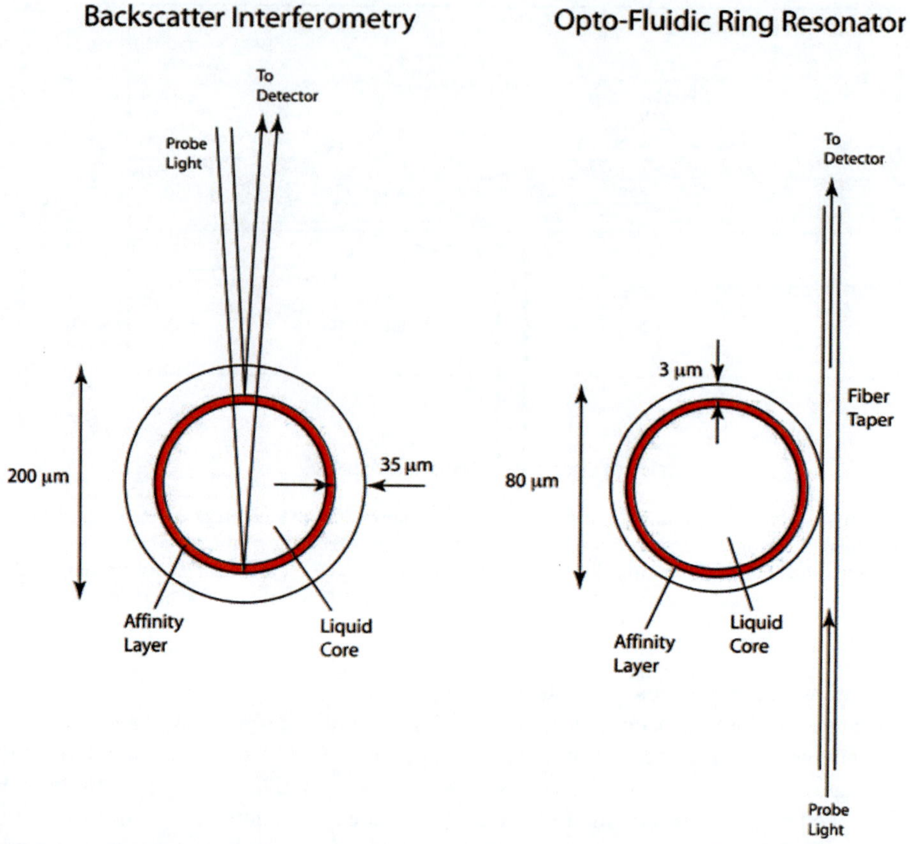

Fig. 7.19 Optofluidic biosensor configurations using liquid-core optical capillaries with internal-surface affinity capture layers. Backscatter interferometry (BSI) detects reflected light from the capillary with a pixel array that captures interference fringes in the far field. The optofluidic ring resonator (OFRR) couples light from a fiber taper into a waveguide mode that senses molecules binding on the inner surface

concentrated on the thin biosensing layer. Examples of liquid core biosensor configurations are shown in Fig. 7.19.

Backscatter interferometry (BSI) uses a capillary tube with an outer diameter of about 200 μm and a wall thickness of 35 μm. The affinity capture layer is deposited on the inner surface of the tube in contact with the liquid core containing analyte molecules. Light is incident externally, which is refracted by the cylindrical geometry, with partial waves generated by the interfaces that interfere in the far field and produce interference fringes at a pixel-array detector [15]. As analyte binds to the affinity capture layer, the interference fringes shift proportional to the amount captured. BSI also has been used to detect biochemical reactions in free-solution [16].

A fiber-coupled resonant configuration is the optofluidic ring resonator (OFRR), shown in Fig. 7.19. A hollow fiber is melted and stretched to a small outer diameter of about 80 μm and the wall is etched to several microns. A fiber taper couples probe light into the resonant waveguide modes of the glass ring. The affinity capture layer is on the inside in contact with the liquid core that contains the analyte molecules. The ring waveguide electric field modes have evanescent tails that probe the internal interface where molecules accumulate on the affinity layer. The cavity resonance can have very large Q factors in the millions, with correspondingly small resonant linewidths of picometers. Small amounts of captured analyte can cause the resonance wavelength to shift by a significant fraction of the linewidth, leading to high responsivity of tens of nanometers per RIU and a detection limit around 1 pg/mm^2 for these resonant biosensors [17–20].

7.8 Ring and Microdisk Resonators

The OFRR is one example of a resonant biosensor with high Q and narrow resonance linewidths. Many resonant biosensors exist in many configurations, such as whispering gallery modes in microspheres [21, 22], silicon-on-insulator micro-ring resonators [23] and silicon micro-toroids [24]. These all benefit from evanescent fields of high-Q resonant modes probing thin layers of immobilized or captured molecules. Examples of a silicon-on-insulator micro-ring resonator and a micro-toroid are shown in Fig. 7.20. The small dimensions and small resonant linewidths translate to small amounts of detected analyte on the order of a femtogram [23].

The extremely small dimensions of the micro-resonators, combined with extremely high Q-values, raise the possibility to perform label-free interferometric detection of single molecules. The micro-toroid sensor has a particular advantage

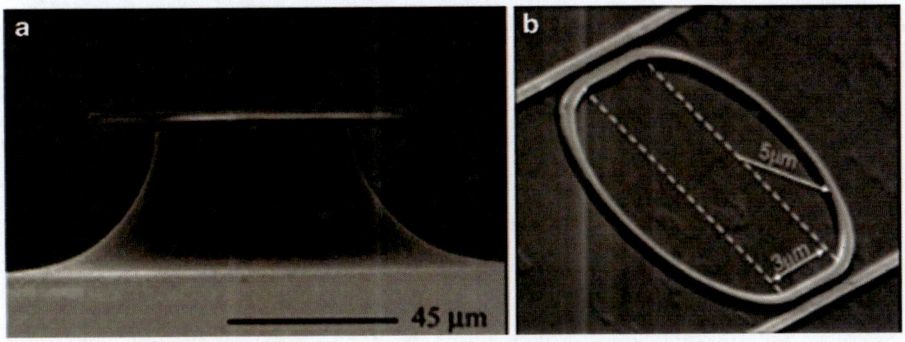

Fig. 7.20 (**a**) A micro-toroid resonator. (**b**) An optical ring resonator between two coupling waveguides. From refs. [23, 24]

for single-molecule detection because of its record high Q-values from the near-perfect silicon microfabrication [24]. The presence of a single molecule adjacent to the resonator causes a small shift of the resonant wavelength. This effect has a thermo-optic origin, in which light absorbed by the molecule from the resonant evanescent field causes a slight heating and hence shift in the cavity wavelength. The shift of the resonant wavelength caused by a single molecule is

$$\frac{\delta\lambda}{\lambda} = \frac{\sigma\lambda}{8\pi^2 n^2 \kappa V}\left(\frac{\mathrm{d}n}{\mathrm{d}T}\right) Q P I_{\mathrm{u}} \tag{7.37}$$

where σ is the molecule absorption cross-section, λ is the wavelength, n is the effective refractive index of the toroid, κ is the thermal conductivity, V is the optical mode volume, $\mathrm{d}n/\mathrm{d}T$ is the thermo-optic effect, Q is the Q-value, P is the coupled optical power, and I_{u} is an integral that depends on the mode profile. This expression can be inverted to extract the limit-of-detection on the optical cross-section that allows a single molecule to be detected. The minimum detectable optical cross-section is

$$\sigma_{\min} = \frac{V}{PQ^2}\frac{8\pi^2 n^2 \kappa}{\lambda(\mathrm{d}n/\mathrm{d}T)} \tag{7.38}$$

For example, for $Q = 2.5 \times 10^8$, $P = 1$ mW, $\lambda = 680$ nm and a toroid diameter of 80 μm, the smallest absorption cross-section is about $\sigma_{\min} = 1 \times 10^{-17}$ cm^2. Many biological molecules have absorption cross-sections larger than this value, and single-molecule detection was achieved in liquid for the detection of single IL-2 molecules in serum.

7.9 Photonic-Bandgap Biosensors

Fabry-Perot resonators have cavities and reflectors that are functionally separated. An alternative resonant structure is a distributed reflector structure, like a distributed Bragg reflector. In this case there is no physical separation between the reflection elements and the cavity elements. As biosensors, these devices can capture analyte molecules distributed across their structure. Like Fabry-Perot sensors, these distributed reflector structures can have narrow linewidths and hence high responsivity to captured analyte. But they are more compact and hence can have very small mass detection limits around a femtogram. Distributed reflectors are often referred to as photonic crystals because their characteristic spacings are fractions of a wavelength, and they can have bandgaps in their dispersion just like electronic bandgaps of ordinary crystals [25].

An example of a 1D photonic crystal sensor is shown in Fig. 7.21. The 1D structure consists of a coupling waveguide and a cavity with a distributed reflector

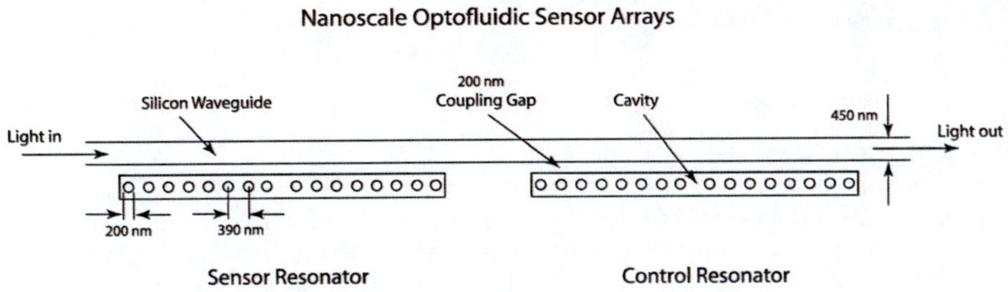

Fig. 7.21 1D photonic crystal biosensors fabricated from silicon-on-insulator operating near 1,550 nm wavelength

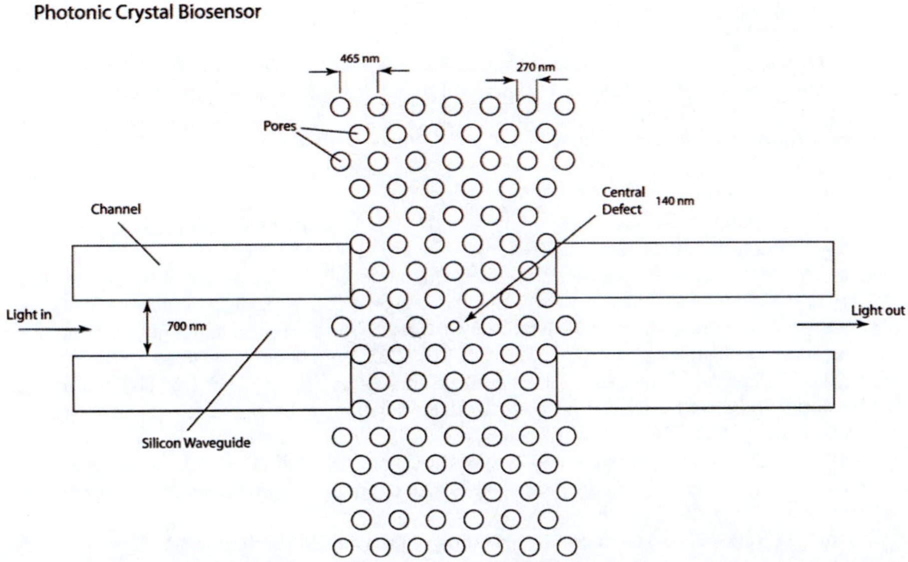

Fig. 7.22 A 2D photonic crystal sensor fabricated on 400 nm thick silicon-on-silica. The silicon waveguides are defined by etched channels, and the photonic crystal elements are etched pores. The central defect has a cavity resonance near 1,580 nm. The cylindrical pores are functionalized to capture target molecules

fabricated in silicon-on-insulator [26]. The Q of the structure was around 3,000, with relative index detection limit around 10^{-4} RIU, and a responsivity of 130 nm/RIU. The small dimensions of the device support a small mass detection limit of 35 ag. A 2D photonic crystal biosensor is shown in Fig. 7.22. The 2D photonic crystal has a lattice constant of 465 nm and a pore diameter of 270 nm. The responsivity for analyte binding inside the pores was approximately 1 nm/nm of protein captured on the inside surface of the pore with a mass detection limit of 2.5 fg [27].

References

1. Prieto, F., Sepulveda, B., Calle, A., Llobera, A., Dominguez, C., Lechuga, L.M.: Integrated Mach-Zehnder interferometer based on ARROW structures for biosensor applications. Sens. Actuat. B Chem. **92**(1–2), 151–158 (2003)
2. Cush, R., Cronin, J.M., Stewart, W.J., Maule, C.H., Molloy, J., Goddard, N.J.: The resonant mirror – a novel optical biosensor for direct sensing of biomolecular interactions.1. Principle of operation and associated instrumentation. Biosens. Bioelectron. **8**(7–8), 347–353 (1993)
3. Heideman, R.G., Kooyman, R.P.H., Greve, J.: Performance of a highly sensitive optical waveguide Mach-Zehnder interferometer immunosensor. Sens. Actuat. B Chem. **10**(3), 209–217 (1993)
4. Prieto, F., Sepulveda, B., Calle, A., Llobera, A., Dominguez, C., Abad, A., Montoya, A., Lechuga, L.M.: An integrated optical interferometric nanodevice based on silicon technology for biosensor applications. Nanotechnology **14**(8), 907–912 (2003)
5. Cross, G.H., Reeves, A.A., Brand, S., Popplewell, J.F., Peel, L.L., Swann, M.J., Freeman, N.J.: A new quantitative optical biosensor for protein characterisation. Biosens. Bioelectron. **19**(4), 383–390 (2003)
6. Swann, M.J., Peel, L.L., Carrington, S., Freeman, N.J.: Dual-polarization interferometry: an analytical technique to measure changes in protein structure in real time, to determine the stoichiometry of binding events, and to differentiate between specific and nonspecific interactions. Anal. Biochem. **329**(2), 190–198 (2004)
7. Rosenblatt, D., Sharon, A., Friesem, A.A.: Resonant grating waveguide structures. IEEE J. Quantum Electron. **33**(11), 2038–2059 (1997)
8. Block, I.D., Ganesh, N., Lu, M., Cunningham, B.T.: Sensitivity model for predicting photonic crystal biosensor performance. IEEE Sens. J. **8**(3–4), 274–280 (2008)
9. Magnusson, R., Ding, Y., Lee, K.J., Shin, D., Priambodo, P.S., Young, P.P., Maldonado, T.A.: Photonic devices enabled by waveguide-mode resonance effects in periodically modulated films. In: Nano-and Micro-Optics for Information Systems. Proceedings of the Society of Photo-Optical Instrumentation Engineers (SPIE) 5225, pp. 20–34. San Diego (2003)
10. Choi, C.J., Block, I.D., Bole, B., Dralle, D., Cunningham, B.T.: Label-free photonic crystal biosensor integrated microfluidic chip for determination of kinetic reaction rate constants. IEEE Sens. J. **9**(12), 1697–1704 (2009)
11. Yih, J.N., Chu, Y.M., Mao, Y.C., Wang, W.H., Chien, F.C., Lin, C.Y., Lee, K.L., Wei, P.K., Chen, S.J.: Optical waveguide biosensors constructed with subwavelength gratings. Appl. Opt. **45**(9), 1938–1942 (2006)
12. Voros, J., Ramsden, J.J., Csucs, G., Szendro, I., De Paul, S.M., Textor, M., Spencer, N.D.: Optical grating coupler biosensors. Biomaterials **23**(17), 3699–3710 (2002)
13. Voros, J.: The density and refractive index of adsorbing protein layers. Biophys. J. **87**(1), 553–561 (2004)
14. Chien, F.C., Lin, C.Y., Yih, J.N., Lee, K.L., Chang, C.W., Wei, P.K., Sun, C.C., Chen, S.J.: Coupled waveguide-surface plasmon resonance biosensor with subwavelength grating. Biosens. Bioelectron. **22**(11), 2737–2742 (2007)
15. Sorensen, H.S., Larsen, N.B., Latham, J.C., Bornhop, D.J., Andersen, P.E.: Highly sensitive biosensing based on interference from light scattering in capillary tubes. Appl. Phys. Lett. **89** (15) (2006)
16. Bornhop, D.J., Latham, J.C., Kussrow, A., Markov, D.A., Jones, R.D., Sorensen, H.S.: Free-solution, label-free molecular interactions studied by back-scattering interferometry. Science **317**(5845), 1732–1736 (2007)
17. Fan, X.D., White, I.M., Shopoua, S.I., Zhu, H.Y., Suter, J.D., Sun, Y.Z.: Sensitive optical biosensors for unlabeled targets: a review. Anal. Chim Acta **620**(1–2), 8–26 (2008)
18. Zhu, H.Y., White, I.M., Suter, J.D., Dale, P.S., Fan, X.D.: Analysis of biomolecule detection with optofluidic ring resonator sensors. Opt. Express **15**(15), 9139–9146 (2007)

19. White, I.M., Zhu, H.Y., Suter, J.D., Hanumegowda, N.M., Oveys, H., Zourob, M., Fan, X.D.: Refractometric sensors for lab-on-a-chip based on optical. ring resonators. IEEE Sens. J. **7** (1–2), 28–35 (2007)
20. White, I.M., Oveys, H., Fan, X.D.: Liquid-core optical ring-resonator sensors. Opt. Lett. **31**(9), 1319–1321 (2006)
21. Vollmer, F., Braun, D., Libchaber, A., Khoshsima, M., Teraoka, I., Arnold, S.: Protein detection by optical shift of a resonant microcavity. Appl. Phys. Lett. **80**(21), 4057–4059 (2002)
22. Arnold, S., Khoshsima, M., Teraoka, I., Holler, S., Vollmer, F.: Shift of whispering-gallery modes in microspheres by protein adsorption. Opt. Lett. **28**(4), 272–274 (2003)
23. De Vos, K., Bartolozzi, I., Schacht, E., Bienstman, P., Baets, R.: Silicon-on-Insulator microring resonator for sensitive and label-free biosensing. Opt. Express **15**(12), 7610–7615 (2007)
24. Armani, A.M., Kulkarni, A.P., Fraser, S.E., Flagen, R.C., Vahala, K.J.: Label-free, single-molecule detection with optical microcavities. Science **317**, 783–787 (2007)
25. Joannopoulos, J.D., Johnson, S.G., Winn, J.N., Meade, R.D.: Photonic crystals: molding the flow of light. Princeton University Press, Princeton (2008)
26. Mandal, S., Erickson, D.: Nanoscale optofluidic sensor arrays. Opt. Express **16**(3), 1623–1631 (2008)
27. Lee, M., Fauchet, P.M.: Two-dimensional silicon photonic crystal based biosensing platform for protein detection. Opt. Express **15**(8), 4530–4535 (2007)

Part III
Cellular Interferometry

Chapter 8
Cell Structure and Dynamics

The cell is the smallest autonomous unit of living tissue. A eukaryotic cell is characterized by many internal structures organized within membranes. These organelles include the nucleus, mitochondria, endoplasmic reticulum and the Golgi apparatus. There are, in addition, many internal vesicles that transport metabolic reagents and enzymes. The complex internal structures have differing protein and lipid densities and differing refractive indices that scatter light. The cellular structures are in constant motion, leading to highly dynamic light scattering (DLS).

This chapter introduces the organization of the cell, its many constituents, and how they move. The motions of the internal components of a cell, including motions of the cell membrane, produce rich dynamic behavior in observed laser speckle and in the mutual interferences among the many scattered waves arising from light interacting with the cell.

8.1 Organization of the Cell

Eukaryotic cells are highly heterogeneous objects, with numerous internal compartments that have many different functions. Within several of the compartments are additional structures, such as the nucleolus inside the nucleus or cristae and granules inside mitochondria. All of these structures present heterogeneous topologies to propagating optical wavefronts, causing refraction and diffraction of light waves. The structural and functional organization of cells influence the interaction of light with cells and living tissues.

The wavelength of light sets a natural length scale against which to compare characteristic sizes of the constituents of tissue and cells. The ratio nL/λ, where nL is an optical thickness and λ is the wavelength of the probe, can serve as an

D.D. Nolte, *Optical Interferometry for Biology and Medicine*, Bioanalysis 1,
DOI 10.1007/978-1-4614-0890-1_8, © Springer Science+Business Media, LLC 2012

expansion parameter for models of optical scattering and diffraction. When $nL/\lambda \gg 1$, ray optics dominates, and the random topologies produce caustics as well as speckle. When $nL/\lambda \ll 1$, then scattering dominates. However, when $nL/\lambda \approx 1$, there is a complicated interplay between geometric optics and diffraction.

The largest scale of the cell is the size of the cell itself, at length scales of 5–20 μm. Light interaction at this scale is well-described by geometric and ray optics. Cells act as tiny lenses with a refractive index that is different than the surrounding matrix. Collections of cells generate a "showerglass" effect, caused by refractive aberration, which is a fundamental limitation on high-resolution imaging through three-dimensional tissues. At the cell scale, physical properties include the cell volume and shape and the contact of the cell with the extracellular matrix (ECM) and with other cells in the tissue. The ECM in animal cells is composed of fibrous proteins like collagen and proteoglycan molecules with high molecular weight that contain polysaccharide chains. These molecules are synthesized by cells to organize them into tissues. The ECM is not static, but is continually remodeled by the surrounding cells.

Cell-to-cell contacts are a particularly important feature of organized tissue that is mostly lacking in conventional cell cultures. Epithelial cells have several different types of junctions. A tight junction is composed of protein ridges that isolate the extracellular space and retard the movement of molecules between cells. Structural junctions provide mechanical connections between cells. These include adherens junctions and desmosomes. Adherens junctions link the actin microfilaments of one cell to an adjacent cell, while desmosomes anchor and link intermediate filaments between cells. The desmosome is structurally very strong and these are found in large numbers in tissues that must withstand mechanical stresses such as the epidermis and cardiac tissue. The opposite of a tight junction is a gap junction that is bridged by molecules that act as pores to enable the transport of ions between the cytoplasm of neighboring cells.

Within the cell, the cytoplasm is separated into compartments at the scale of several microns. The compartments are defined by cytoskeletal structures, such as microtubules, actin microfilaments, and intermediate filaments. Within these compartments are the cellular organelles. The largest organelle is the nucleus, typically 3–5 μm in diameter, which is large relative to the wavelength of the probe light. Light scattering from the nucleus has some geometric optics behavior, but also exhibits Mie scattering signatures that can be used to measure the size of the nucleus in cells [1, 2].

The mitochondria in cells are a particularly important source of scattering. Mitochondria range in size from 1 to 5 μm in length, but sometimes can be as small as 200 nm in diameter. These sizes are comparable to probe wavelengths, leading to strong scattering. Mitochondria number densities can be very high in the cytoplasm, providing large numbers of scatterers. Because of their size and numbers, mitochondria can dominate large-angle scattering of light in tissues [3, 4]. Other organelles include endosomes and lysosomes. These organelles are not just static structures, but are constantly in motion, trafficking materials inside the cell and dynamically scattering light.

8.2 Optical Properties of Cellular Components

There are no single values for the refractive index even for all the same type of organelles in the same cell. All components of a cell are heterogeneous within the cell, and vary from one cell type to another. Ranges of values for different constituents are shown in Table 8.1, but the ranges are wide, and must be used only as a rough guide [5]. Note that the relative index between the nucleus or mitochondria and the cytoplasm is $m = 1.01$–1.04, and between the cytoplasm and the interstitial fluid is $m = 1.00$–1.01. Scattering calculations show that $m = 1.04$ can provide good fits to some data. The largest refractive indices are attributed to the cell membrane and collagen, but these constitute only a small fraction of the tissue volume. Detailed refractive index values among many types of tissues can be found in Ref. [6].

An example of Mie scattering from a particle of radius 0.5 μm in water with a relative refractive index difference of $m = 1.04$ is shown in Fig. 8.1 for two polarizations. The differential scattering cross section exhibits characteristic dips that are related to the size of the scatterer.

Table 8.1 Refractive indices of tissue components

Interstitial fluid	$n = 1.35$–1.36
Cytoplasm	$n = 1.36$–1.375
Nucleus	$n = 1.38$–1.41
Mitochondria and organelles	$n = 1.38$–1.41
Membrane	$n = 1.46(1.43$–$1.49)$
Collagen	$n = 1.46$–1.53

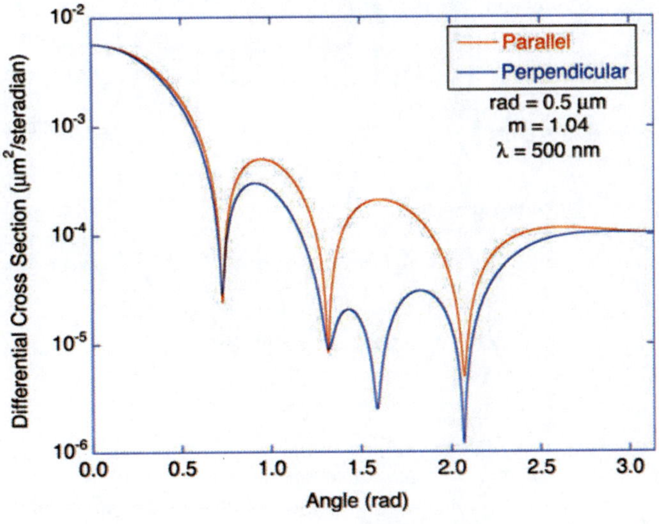

Fig. 8.1 Differential scattering cross sections for parallel and perpendicular polarizations on a 0.5-μm radius particle with $n_m = 1.33$ and relative index $m = 1.04$

8.3 The Cytoskeleton

The cytoskeleton is an interconnected network of polymer filaments that spans the cytoplasm and determines the shapes and motions of cells. It is composed of three types of filaments: microtubules, actin filaments, and intermediate filaments. They occupy the inside of the cytosol, but many interact directly with the cell membrane. The internal components and structures of the cytoskeleton in a typical mammalian cell are shown in Fig. 8.2. Microtubules radiate from the centrosome, an important organelle for mitosis that acts as a microtubule organizing center (MTOC). The microtubules terminate at the membrane at focal adhesions to the ECM and at adherens junctions between cells. They also connect to components of the nucleus during mitosis. Actin filaments form a lattice called the cortex on the inside of the cell membrane, providing strength and rigidity to the membrane. The actin cortex participates in many active membrane processes, such as cell crawling using

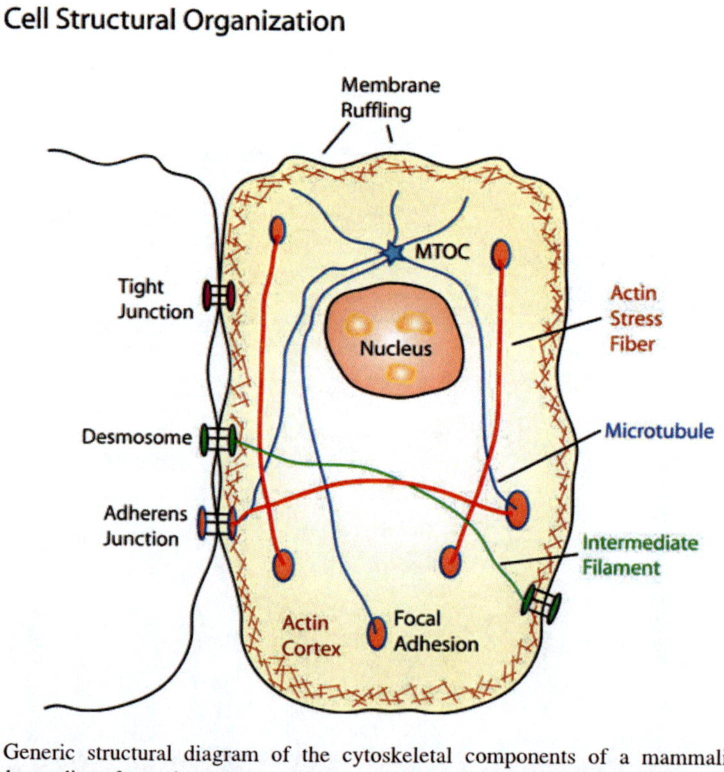

Fig. 8.2 Generic structural diagram of the cytoskeletal components of a mammalian cell. Microtubules radiate from the centrosome that acts as a microtubule organizing center. The microtubules connect to the cell membrane at focal adhesions to the extracellular matrix and at adherens junctions between cells. Actin filaments reside in a lattice network called the cell cortex around the inside of the cell membrane, and stress fibers attach to focal adhesions on the membrane

lamellepodia and filopodia, and in endo- and exocytosis or phagocytosis. Actin filaments also terminate at focal adhesions and form long-range stress fibers inside the cell.

The compressive and tensile loads on the cytoskeleton give it tensegrity: a term coined by Buckminster Fuller to describe a cohesive structure with high mechanical strength but also high flexibility. The tensegrity structure has no rigid connections among its load-bearing components, but the compressive components are linked by tensile cables in a way that maintains overall shape. This description of the cytoskeleton [7] provides a conceptual framework to understand its mechanical properties that are the basis of its dynamics and the origin of DLS.

The most active use of the cytoskeletal machinery occurs during mitosis when the entire cellular structure is reorganized prior to and during division. During mitosis, the microtubules form the mitotic spindle which is an organized mechanical structure that helps divide the intracellular contents prior to cell division. Actin plays an important role in cytokinesis at the end of mitosis when the cell membrane pinches off, and the cell physically divides. For these reasons, drugs that inhibit processes of the cytoskeleton are common anticancer agents that arrest the cell cycle by arresting motion [8, 9].

8.4 Cellular Mechanics

Cells are micromechanical systems of considerable sophistication and complexity, with both passive and active feedback that allow them to respond to their changing environment. There are also active processes that drive the cellular mechanical-chemical machinery. The principle consequence of the interaction of light with internal displacements is DLS with multiple-wave interference, intensity fluctuations, and dynamic speckle fields. How cells behave and respond, and how these processes are detected with light scattering and interferometry, are largely dictated by the structural mechanics of the cytoskeleton – the compressive/tensile network of filaments that provides strength and rigidity to the cell as well as compliance and adaptability, and provide the pathways for organelle transport.

The cytoplasm of the cell is a semiflexible polymer solution that is crowded with proteins and filaments. The total protein concentration of the cytoplasm is between 200 and 300 mg/mL, with the remainder consisting of water, both free and bound to molecules. Actin and tubulin are the most abundant proteins, with total concentrations each around 5 mg/mL in which half is free and half is bound. The cross-linked cytoskeletal networks give the cytosol elasticity, but the largely fluid nature of the cytosol gives it viscous properties. These two properties are typically experienced at different frequencies. High frequencies tend to probe the viscous properties of the cell, while low frequencies tend to probe the elastic properties of cells. The cytosol has complex rheology that responds to forces in ways that can be probed with light scattering and interferometry.

8.4.1 Brownian Motion

Small particles, such as molecules or small vesicles in cells, are buffeted by thermal fluctuations of the surrounding medium. The small momentum transfers caused by collisions with the molecules of the medium cause the particles to execute a random walk called Brownian motion. For a protein molecule, the diffusion coefficient is

$$D = \frac{k_B T}{6\pi\eta R_h}, \tag{8.1}$$

where η is the fluid viscosity and R_h is the hydrodynamic radius of a protein. The density of globular proteins is roughly constant for all types with $\rho \approx 1$ g/mL. The mass of a globular protein is roughly

$$m = \frac{4}{3}\pi {R_g}^3 \rho, \tag{8.2}$$

giving the radius of gyration

$$R_g = \sqrt[3]{\frac{3m}{4\pi\rho}}, \tag{8.3}$$

which is approximately equal to the hydrodynamic radius. Measured diffusion coefficients for globular proteins are a little smaller than this value because hydrated waters participate, and the proteins are not perfect spheres, which causes extra friction. Typical measured values are $D = 1.3 \times 10^{-6}$ cm^2/s for $m_w = 10$ kDa, and $D = 6 \times 10^{-7}$ cm^2/s for $m_w = 100$ kDa.

The viscosity of the cytoplasm for translational diffusion through a cell is not an intrinsic or single-valued property of the cell. It depends on many factors, the most important being the size of the diffusing particle, but it is also dependent on the osmolarity (and cell volume) and the structure of the cytoskeletal components in the cell. Even small molecules in the cytoplasm experience macromolecular crowding by the high density of protein molecules [10]. For small molecules, the cytoplasm viscosity is approximately 2–3 times larger than for saline solution ($\eta = 1$ mPa-sec). But as the molecules or particles increase in size, the viscosity rises and can be over 1,000 times larger than for saline when the particle size approaches 100 nm, which is the mesh size of the intermediate filament networks. The results of diffusion data from assorted living cells is summarized in Fig. 8.3 as a function of particle size. There is a strong decrease in diffusion coefficient (relative to water) with increasing radius, and this confinement effect goes beyond simple Fickian diffusion of (8.1). The light scattering knee frequency for backscattering is

$$f_{\text{back}} = \frac{8\pi n^2}{\lambda^2} D, \tag{8.4}$$

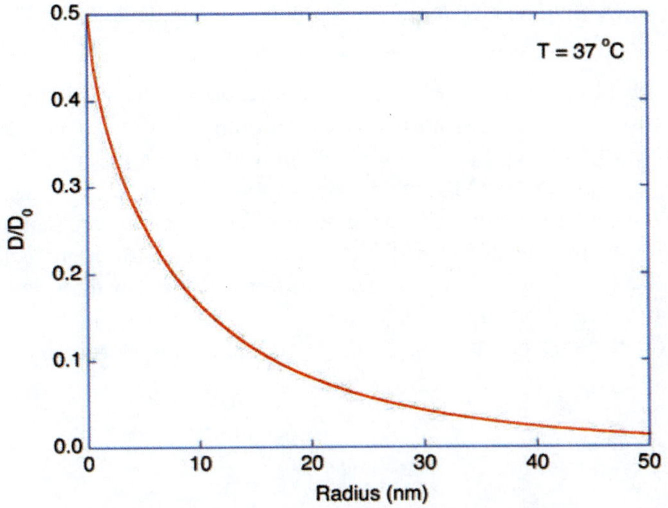

Fig. 8.3 Reduced diffusion coefficient relative to saline for particles in living fibroblasts, neurons, and myotubes as a function of particle radius

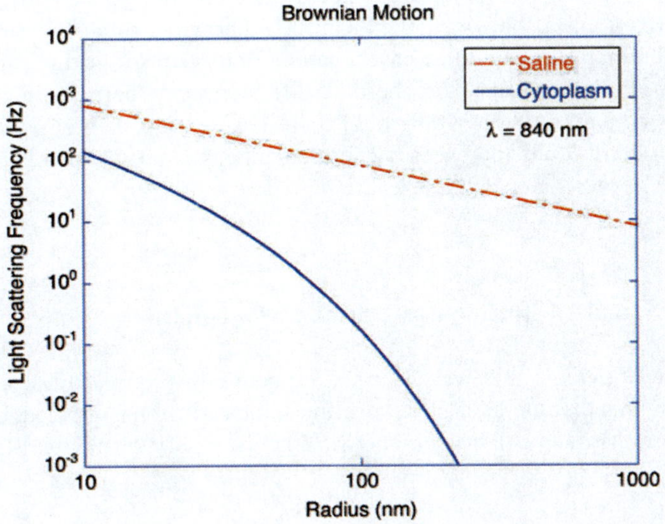

Fig. 8.4 Dynamic light scattering (backscattering geometry) knee frequency as a function of particle radius for saline solution and cytoplasm

with $\lambda = 840$ nm, and is shown in Fig. 8.4 for saline solution compared with living cytoplasm. For instance, a vesicle with a 200-nm diameter has a knee frequency around 0.1 Hz.

8.4.2 *Anomalous Diffusion*

One of the key physical attributes of intracellular motion is the deviation from simple Fickian diffusion. All components of a cell are extremely crowded, and the mesh size of the actin cortex or the filament networks prevents free motion of large particles. These effects cause significantly constrained motion that cannot be described as free diffusion. Viscosity is a complex frequency-dependent quantity that partially explains this deviation from free diffusion inside the cytoplasm. However, there are other simple mechanisms, such as compartmentalization or constrained displacements, that also cause deviation from simple diffusion.

Deviation from free diffusion is described phenomenologically through the relation

$$\Delta r^2 = D^* \left(\frac{t}{t_0}\right)^{\beta}, \tag{8.5}$$

where $D*$ is a time-independent coefficient, and t_0 is a characteristic time. The exponent for Fickian diffusion is $\beta = 1$. If $\beta < 1$, the diffusion is called subdiffusive, while if $\beta > 1$, the diffusion is called superdiffusive. Superdiffusive behavior occurs if there is persistent and correlated motion of the particle, as is the case of active transport of organelles. On the other hand, subdiffusive behavior is the most commonly encountered case when motion is constrained in some way. To give an example, consider the thermal diffusion of a particle in a harmonic potential. The harmonic potential could represent the elastic energy of a stretched membrane, or it could represent a confinement potential of an organelle in the porous meshwork of the actin cortex.

8.4.2.1 Example: Diffusion in a Harmonic Potential

Diffusion of a particle in a weak harmonic potential is equivalent to a mass on a weak spring in a thermal bath. For short times, the diffusion is Fickian, but for long times, the mean-squared displacement (MSD) must satisfy the equipartition relation

$$\frac{1}{2}k\langle x^2\rangle = \frac{1}{2}k_B T. \tag{8.6}$$

The starting point for such problems is the Langevin equation of motion under a stochastic force F'

$$\begin{aligned} m\frac{d\dot{x}}{dt} &= -kx - \alpha\dot{x} + F' \\ mx\frac{d\dot{x}}{dt} &= -kx^2 - \alpha x\dot{x} + xF', \end{aligned} \tag{8.7}$$

where, for a spherical particle of radius a, the viscous drag factor is

$$\alpha = 6\pi\eta a \tag{8.8}$$

and η is the viscosity. The term on the left of (8.7) can be rewritten to give

$$m\left[\frac{d}{\mathrm{d}t}(x\dot{x}) - \dot{x}^2\right] = -kx^2 - \alpha x\dot{x} + xF', \tag{8.9}$$

because

$$m\left\langle\frac{d}{\mathrm{d}t}(x\dot{x})\right\rangle = m\frac{d}{\mathrm{d}t}\langle(x\dot{x})\rangle. \tag{8.10}$$

From equipartition on the velocity

$$m\langle\dot{x}^2\rangle = k_\mathrm{B}T, \tag{8.11}$$

this gives

$$m\frac{d}{\mathrm{d}t}\langle(x\dot{x})\rangle = k_\mathrm{B}T - k\langle x^2\rangle - \alpha\langle x\dot{x}\rangle, \tag{8.12}$$

hence

$$\frac{1}{2}m\frac{d}{\mathrm{d}t}\left(\frac{d}{\mathrm{d}t}\langle x^2\rangle\right) = k_\mathrm{B}T - k\langle x^2\rangle - \frac{1}{2}\alpha\frac{d}{\mathrm{d}t}\langle x^2\rangle. \tag{8.13}$$

But this is just the simple differential equation for a damped harmonic oscillator with a constant force for $y = \langle x^2\rangle$

$$\frac{1}{2}m\ddot{y} + \frac{1}{2}\alpha\dot{y} + ky = k_\mathrm{B}T. \tag{8.14}$$

For small objects in cytoplasm, the inertial terms are negligible relative to the other terms (Life at Small Reynolds Number [11]), so the dynamic equation is

$$\alpha\dot{y} + 2ky = 2k_\mathrm{B}T, \tag{8.15}$$

with the general solution

$$\langle x^2\rangle = \frac{k_\mathrm{B}T}{k}\left(1 - \exp\left(-\frac{2k}{\alpha}t\right)\right), \tag{8.16}$$

where $D = k_\mathrm{B}T/\alpha$ is the diffusion coefficient. This solution at short times describes a diffusing particle with a diffusion coefficient D that asymptotes at long times to an equipartition value of $k_\mathrm{B}T/k$. The time constant to approach equilibrium is

$$\tau = \frac{3\pi\eta a}{k}. \tag{8.17}$$

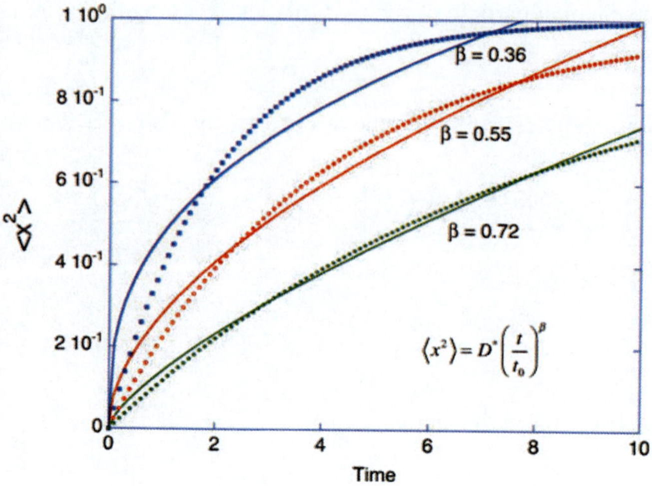

Fig. 8.5 Fit of mean-squared displacement (MSD) for constrained diffusion to the anomalous diffusion equation. The saturated MSD mimics the functional form for anomalous diffusion

For a very weak spring constant, Fickian diffusive behavior is maintained at short times, but eventually the extension of the spring constrains the displacement, and the MSD saturates.

Another simple form of constrained motion consists of a compartment of fixed linear size L. In this case

$$\langle x^2 \rangle = L^2\left(1 - \exp\left(-\frac{2k_B T}{6\pi\eta a L^2}t\right)\right), \tag{8.18}$$

and the time to approach equilibrium is

$$\tau = \frac{6\pi\eta a L^2}{2k_B T} = L^2/D. \tag{8.19}$$

Constrained motion shows clear saturation to the size set by the physical constraints (equipartition for an oscillator or compartment size for a freely diffusing particle [12]). However, if the experimental data do not clearly extend into the saturation time regime, then the fit to anomalous diffusion can lead to exponents β that do not equal unity. This is illustrated in Fig. 8.5 with asymptotic MSD compared with the anomalous diffusion equation with the exponent β. Care therefore should be exercised in the interpretation of the exponents obtained from anomalous diffusion experiments. In particular, all constrained motion leads to subdiffusive interpretations if measured at short times.

8.4.3 Cell Rheology

Rheology is the study of the mechanical properties of viscoelastic materials. Unlike purely elastic materials that have time-independent relationships between stress and strain, a viscoelastic material relaxes in response to applied stresses or strains. To study the mechanical properties of viscoelastic materials, it is necessary to apply specific strain rates and then measure the transient stress. The viscoelastic response function $G(t)$ is defined by

$$\sigma_{xy} = \int_{-\infty}^{t} G(t-t')\frac{\mathrm{d}u_{xy}}{\mathrm{d}t'}\mathrm{d}t', \tag{8.20}$$

where σ_{xy} is a shear stress, and u_{xy} is a shear strain. The integral defines the transient stress response to the strain rate. It is common to study the response in the frequency domain by applying periodic shear strains. In this case, the frequency dependence is

$$\sigma_{xy} = G'(\omega)u_{xy}(\omega t) + \frac{1}{\omega}G''(\omega)\frac{\mathrm{d}u_{xy}(\omega t)}{\mathrm{d}t}, \tag{8.21}$$

where $G'(\omega)$ is the in-phase response called the shear storage modulus, and $G''(\omega)$ is the quadrature-phase response called the shear loss modulus. The storage modulus is a measure of the elastic properties, and the loss modulus is a measure of the viscous properties. A pure fluid, like water, has $G'(\omega) = 0$ and $G''(\omega) \sim \omega$. Indeed, for a pure fluid, (8.21) can be identified with Newtonian fluid behavior that allows the identification

$$\eta(\omega) = \frac{G''(\omega)}{\omega}, \tag{8.22}$$

as the fluid viscosity. Conversely, for a purely elastic material $G''(\omega) = 0$ and $G'(\omega) = \text{const.}$

Several studies of the viscoelastic properties of the cytoplasm show common behavior for the modulus in the form

$$\begin{aligned} G'(\omega) &= A\cos(\pi\alpha/2)\omega^{\alpha} + B\cos(3\pi/8)\omega^{3/4} \\ G''(\omega) &= A\sin(\pi\alpha/2)\omega^{\alpha} + B\sin(3\pi/8)\omega^{3/4} \\ |G(\omega)|^2 &= |G'(\omega)|^2 + |G''(\omega)|^2, \end{aligned} \tag{8.23}$$

where α is typically between 0.1 and 0.3 [13, 14]. The cross-over between the two exponent behaviors occurs around a frequency of 30 Hz. An example of the frequency-dependent magnitude of the (normalized) shear modulus is shown in Fig. 8.6. The smaller frequency dependence at low frequency is a consequence of the elastic properties, while the stronger frequency dependence at high frequency

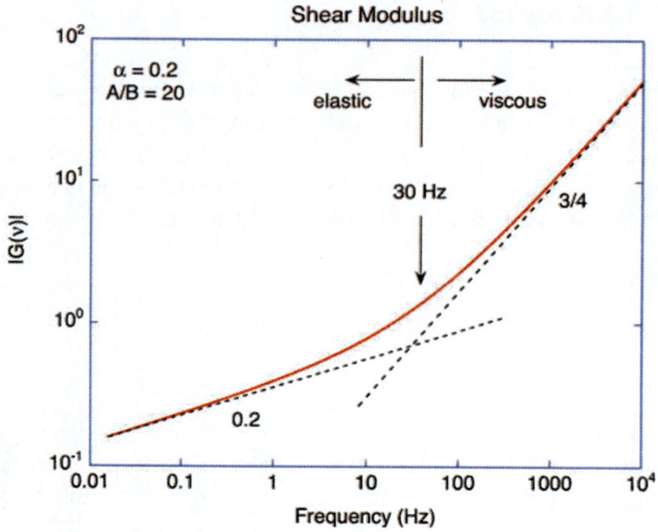

Fig. 8.6 Magnitude of the shear modulus of the cytosol as a function of frequency ($\nu = \omega/2\pi$)

reflects the viscous properties. The cross-over frequency around 30 Hz is where the elastic and viscous forces balance.

There are two main experimental approaches to measuring the complex shear modulus of the cytoplasm. One is an active (and direct) approach that applies frequency-dependent strains and measures the stress. The other is a passive (and less direct) approach that measures thermal fluctuations of the system and relates these to the modulus using thermal equilibrium arguments (i.e., the fluctuation-dissipation theorem, which does not necessarily apply for living active systems).

The passive approach often uses small beads attached to the cytoskeleton, and the mean-squared displacement (MSD) of the beads are measured either directly using particle-tracking microscopy, or by DLS, in which the MSD is extracted from the time autocorrelation function of the intensity fluctuations. In either case, it is necessary to make a connection between the MSD and the modulus. This is established through the generalized Stokes-Einstein relation (GSER), which is related to the fluctuation-dissipation theorem.

8.4.4 Generalized Stokes-Einstein Relation

The Stokes-Einstein relation, which applies to Newtonian fluids, relates the diffusion coefficient of a small probe to the fluid viscosity. For a spherical particle, this relation is

$$\langle \Delta r^2 \rangle = 2NDt = \frac{2Nk_B T}{6\pi\eta R_h} t, \tag{8.24}$$

where $\langle \Delta r^2 \rangle$ is the MSD, N is the number of dimensions (e.g., $N = 3$ for 3D), η is the fluid viscosity, and R_h is the particle hydrodynamic radius.

However, the cytosol of a cell is non-Newtonian, being viscoelastic, and a generalized form of the Stokes-Einstein is needed to establish a generalized relationship between thermal fluctuations and deterministic mechanical properties of the fluid. The GSER consists of two components [15]: an Einstein component that relates thermal fluctuations to mechanical mobility, and a Stokes component that relates the mobility to the rheological properties of the material, such as the complex shear modulus $G(\omega)$.

The Einstein component uses a general form of the fluctuation dissipation theorem [16] that connects stochastic properties, such as the fluctuating position of a probe particle, with deterministic properties, such as the displacement of the particle subject to a given force. This relationship is quite astounding because it takes the natural fluctuations that are present in an *equilibrium* system, and relates it to the rate at which a *nonequilibrium* system relaxes back to equilibrium. As such, it represents one of the most important results of nonequilibrium thermodynamics. It also has great utility, because it connects passive rheological experiments (in which particles diffuse randomly) with active rheological experiments (in which forces are directly applied to particles).

The most familiar form of the fluctuation dissipation theorem is expressed as

$$\begin{gathered}\textit{Einstein relation}:\\ \langle \Delta r^2(\omega) \rangle = \frac{2}{\omega^2} N k_{\mathrm{B}} T \,\mathrm{Re}\{M(\omega)\},\end{gathered} \tag{8.25}$$

where the restriction has been placed on the MSD spectrum to be real-valued and symmetric. This expression, which captures the Einstein component of the GSER, relates the Fourier frequency spectrum of the MSD to the equilibrium temperature and to the frequency-dependent particle mobility. The mobility captures the macroscopic rheological properties of the system. The mobility is related to the shear modulus

$$G(\omega) = i\omega\eta(\omega) = G'(\omega) + iG''(\omega), \tag{8.26}$$

through the Stokes component of the GSER, which for a spherical particle is

$$\begin{gathered}\textit{Stokes relation}:\\ M(\omega) = \frac{1}{6\pi\eta(\omega)a} = \frac{i\omega}{6\pi G(\omega)a},\end{gathered} \tag{8.27}$$

where $M(\omega)$ is the Fourier transform of the time-dependent mobility, and $\eta(\omega)$ is the frequency-dependent viscosity. Putting the complex Stokes component of (8.27) into (8.25) gives the GSER for equilibrium conditions

$$\begin{gathered}\textit{GSER}:\\ \langle \Delta r^2(\omega) \rangle = \frac{2Nk_{\mathrm{B}}T}{\omega} \frac{G''(\omega)}{6\pi |G(\omega)|^2 a}.\end{gathered} \tag{8.28}$$

Equation (8.28) can be related back to the measured shear modulus of cells given in (8.23) to predict a frequency dependence for the spectral power density of the MSD. This is found to be $\omega^{1+\alpha}$, where $\alpha = 0.1$–0.3 for frequencies below 30 Hz, and where $\alpha = 0.75$ for frequencies above 30 Hz. The behavior for the spectral power density might then expected to be between $\omega^{1.2}$ and $\omega^{1.75}$, assuming that the MSD inside living cells shows thermal-like motion. However, motion inside living cells is largely controlled by active processes that are far from thermal equilibrium [17]. Even then, it may be possible to describe the active motion by an effective temperature T_{eff} (that is larger than the actual temperature and is frequency dependent), and many of the processes described in this section remain valid phenomenologically, although with different frequency power-law dependences.

8.5 Active Intracellular Motion

The internal motions of a living cell are not thermal. Cells are active biomechanical and biochemical machines that are highly energetic, driven by ATP hydrolysis, and characterized best by internal forces. Therefore, thermal estimates of the dynamics of living cells are not usually relevant. It is more important to characterize the forces and the rheological response to the forces. Fortunately, force fluctuations are linked to velocity fluctuations, which are linked in turn to MSDs. This provides an avenue for analyzing the active processes that occur within cells.

8.5.1 Microrheology Far from Equilibrium

In many microrheology experiments, the measured quantities are the MSD $\Delta x^2(t)$ as a function of time (passive measurements of fluctuations) and the creep function *J(t)* that is defined by

$$x(t) = \int_{-\infty}^{\infty} J(t - t') \frac{\mathrm{d}F}{\mathrm{d}t'} \mathrm{d}t' \tag{8.29}$$

in which *J(t)* is the displacement caused by a step in the force (active measurements of displacements). The velocity fluctuation spectrum is related to mobility by [18]

$$\bar{S}_{\mathrm{v}}(s) = M(s)^2 \bar{S}_{\mathrm{F}}(s) = s^4 \bar{J}(s)^2 \bar{S}_{\mathrm{F}}(s), \tag{8.30}$$

which is expressed in terms of the Laplace transforms with frequency s. This expression continues to hold, even out of equilibrium. What is needed is to convert

(8.30) into one that relates the measured quantities $\Delta x^2(t)$ and $J(t)$. First, note that the MSD is related to the velocity transform as

$$s^2 \Delta \bar{x}^2(s) = 2\bar{S}_v(s). \tag{8.31}$$

The force fluctuation spectrum is then

$$\bar{S}_F(s) = \frac{\Delta \bar{x}^2(s)}{2s^2 \bar{J}(s)^2}, \tag{8.32}$$

which is the relation that is needed to find the force fluctuation spectrum from the MSD and the creep function.

The time dependence on each of these is defined as

$$\begin{aligned} \Delta x^2(t) \propto t^{\beta} \quad & \Delta \bar{x}^2(s) \propto s^{-(1+\beta)} \\ J(t) \propto t^{\alpha} \quad & \bar{J}(s) \propto s^{-(1+\alpha)} \end{aligned} \tag{8.33}$$

where the first is the conventional definition of anomalous diffusion, and the second is for the creep function (the Laplace transform of the frequency-dependent modulus). Therefore,

$$\begin{aligned} \langle F(t')F(t'+t)\rangle \propto t^{\beta-2\alpha} \\ \bar{S}_F(s) \propto s^{2\alpha-\beta-1} = s^{-\lambda}. \end{aligned} \tag{8.34}$$

However, it is also known [19] that in equilibrium, the following relations hold

$$\begin{aligned} \langle F(t')F(t'+t)\rangle_{Eq} \propto t^{-\alpha} \\ \bar{S}_F(s)_{Eq} \propto s^{\alpha-1}, \end{aligned} \tag{8.35}$$

from which, $\beta = \alpha$ can be concluded for equilibrium conditions.

Experiments on beads bound to the actin cortex through the cell membrane [18] obtained the results in Table 8.2. From these values it is clear that the equilibrium scaling does not hold at long times (low frequencies).

Table 8.2 Power-law dependence for anomalous diffusion

Super diffusion (long times >3 s)	Sub diffusion (short times <1 s)
$\bar{S}_F(s) \sim s^{-1.7}$	$\bar{S}_F(s) \sim s^{-0.8}$
$\bar{J}(s) \sim s^{-1.2}$	$\bar{J}(s) \sim s^{-1.2}$
$\alpha \approx 0.2$	$\alpha \approx 0.2$
$\beta \approx 1.1$	$\beta \approx 0.2$

Similarly, Wilhelm [17] found inside the cytosol (uncoupled to the actin cortex) at mid frequencies (0.1–10 Hz) the scaling relations

$$\begin{aligned} \lambda &= \beta + 1 - 2\alpha \\ &\approx 1.2 \\ \beta &\approx 1.4 \\ \alpha &\approx 0.6, \end{aligned} \tag{8.36}$$

instead of the equilibrium conditions

$$\begin{aligned} \lambda &= 1 - \alpha \\ \beta &= \alpha. \end{aligned} \tag{8.37}$$

These results clearly show nonequilibrium behavior of the microrheology of cells. In the case of the actin cortex, the deviations from equilibrium become larger at lower frequencies (below 0.5 Hz), while at higher frequencies, the behavior is more like equilibrium behavior. This suggests that the typical times for active motions of the actin cortex (and by association for active motion of the membrane) are longer than 2 seconds.

One way to characterize the nonequilibrium behavior of active materials is through an effective temperature. Even out of equilibrium, the ratio between the velocity and the force spectra from (8.30) yields the quantity $s^2 J(s)$. If the system was in thermodynamic equilibrium then equipartition gives

$$\bar{S}_F(s) = \frac{k_B T}{s^2 \bar{J}(s)}. \tag{8.38}$$

For a system that is out of equilibrium, one can define an effective temperature by analogy as

$$k_B T_{\text{eff}} = s^2 \bar{J}(s) \bar{S}_F(s). \tag{8.39}$$

The effective temperature of a normal cell, and one that was ATP depleted, is shown in Fig. 8.7. The effective temperature of the normal cell above 1 Hz is relatively constant around $10T_{\text{eq}}$, which is 10 times larger than the physiological temperature. Below 1 Hz, the temperature increases as high as $100T_{\text{eq}}$ around 0.1 Hz. The ATP-depleted cells showed much smaller effective temperatures (but still larger than physiological). In these experiments, beads were bound to the actin cortex through membrane receptors.

In active actin-myosin gels [20], the active component of the transport caused by molecular motors appears at low frequencies. The separation frequency is at 1 Hz, above which equilibrium and active behaviors level off. In the theoretical model of [21] the origin of the $1/\omega^2$ dependence of the displacement spectral power density is

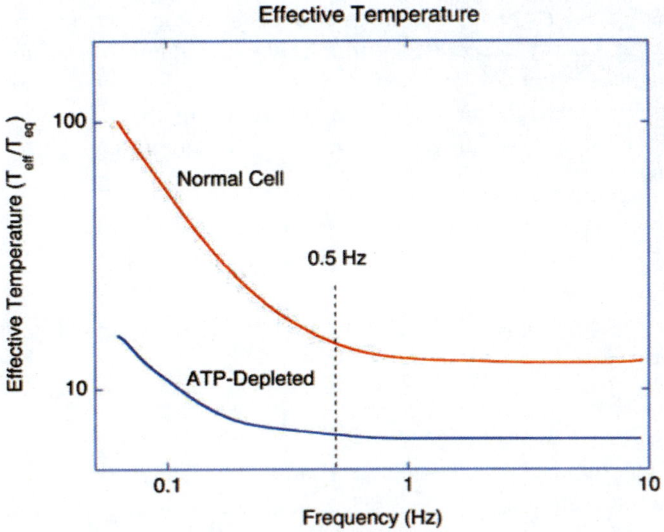

Fig. 8.7 Effective temperature defined through the excess force fluctuations as a function of frequency ($\nu = s/2\pi$). Large nonequilibrium fluctuations occur at low relaxation rates. From [18]

the sudden changes in active forces. This sharpness of the force onset and release leads to a sinc^2 dependence from the Fourier transform which decays as $1/\omega^2$ above the persistence frequency (inverse persistence time for molecular forces). From [20], in an active gel, the persistence time is about 2 s with a persistence frequency of 0.1 Hz. The active gel regime, caused by molecular motors and cytoskeletal filaments, which violate the fluctuation-dissipation theorem, is then in the range between 0.1 and 10 Hz.

8.6 Membrane Mechanics

Just as the cytoplasm is a complex viscoelastic medium, so too the cell membrane is a viscoelastic mechanical structure that has elastic bending rigidity, measured by a bending modulus κ_b, that maintains its nominal shape. The bending rigidity arises both from the elastic properties of the membrane as well as from interactions with the cytoskeleton. Viscous damping likewise has contributions both from internal friction among the lipid constituents of the membrane and from interaction with the viscoelastic properties of the cytosol and extracellular interstitial fluids. The elastic properties of membranes are described phenomenologically (without resort to molecular mechanisms), and these are combined with the viscous properties to determine thermal membrane motions.

For an oriented elastic membrane, a nearly flat membrane surface can be described in three dimensions using the coordinates $[x, y, h(x, y)]$, where the membrane is projected onto the x-y plane, and the height of the membrane is described by $h(x,y)$ [22]. Because the membrane is elastic, any deviation from a planar surface has an elastic energy associated with it, like the energy associated with the stretching and compression of a spring. For surfaces, displacements contribute to surface energies through curvatures. The curvature of the surface at a point has two principal values (the largest and smallest curvatures) C_1 and C_2. The mean curvature is

$$C_M = \frac{1}{2}(C_1 + C_2) = \frac{1}{2}\left(\frac{\partial^2 h}{\partial x^2} + \frac{\partial^2 h}{\partial y^2}\right) \tag{8.40}$$

The bending energy density of an elastic membrane is expressed in terms of this curvature as

$$U = \kappa_b {C_M}^2, \tag{8.41}$$

where the coefficient is the bending rigidity κ_b. These rigidities control the magnitudes of membrane fluctuations much like a spring constant limits the range of displacements of a spring (at a given temperature). The bending rigidity of a cell membrane is approximately $\kappa_b \approx 10k_BT \approx 4 \times 10^{-20}$ J, but it varies depending on the stiffness of the actin cortex.

For a membrane under a tension γ_0, the elastic energy of an area element dA of the membrane is given by

$$\begin{aligned} dE &= UdA \\ &= \gamma_0 dA + \kappa_b C_M^2 dA. \end{aligned} \tag{8.42}$$

The area element dA is not the cross-sectional area of the membrane (which is just dxdy), but is its actual surface area including all deviations from the plane. From differential geometry, the area element on a surface is related to the projection of the area element on the x-y plane through the expression

$$dA = \sqrt{g}\,dxdy, \tag{8.43}$$

where g plays the role of the metric tensor. For local small surface variations, the metric tensor is given by

$$g = 1 + \left(\frac{\partial h}{\partial x}\right)^2 + \left(\frac{\partial h}{\partial y}\right)^2, \tag{8.44}$$

and the area element is

$$\begin{aligned} dA &= \sqrt{1 + \left(\frac{\partial h}{\partial x}\right)^2 + \left(\frac{\partial h}{\partial y}\right)^2}\,dxdy \\ &\approx \left(1 + \frac{1}{2}\left(\frac{\partial h}{\partial x}\right)^2 + \frac{1}{2}\left(\frac{\partial h}{\partial y}\right)^2\right)dxdy = \left(1 + \frac{1}{2}(h_x^2 + h_y^2)\right)dxdy. \end{aligned} \tag{8.45}$$

For a membrane under a tension γ_0, the energy differential is now

$$\begin{aligned} \mathrm{d}E &= (\gamma_0 + \kappa_\mathrm{b} C_\mathrm{M}^2)\left(1 + \frac{1}{2}(h_x^2 + h_y^2)\right)\mathrm{d}x\mathrm{d}y \\ &= \gamma_0\left(1 + \frac{1}{2}(h_x^2 + h_y^2)\right)\mathrm{d}x\mathrm{d}y + \kappa_\mathrm{b} C_\mathrm{M}^2 \mathrm{d}x\mathrm{d}y \\ &= \left(\gamma_0 + \gamma_0\frac{1}{2}(h_x^2 + h_y^2) + \kappa_\mathrm{b} C_\mathrm{M}^2\right)\mathrm{d}x\mathrm{d}y, \end{aligned} \tag{8.46}$$

to lowest order. The first term is simply an additive constant for a fixed tension and is neglected in subsequent dynamics analysis. The second term is the surface tension effect, and the third term is the bending rigidity contribution. As the membrane undulates under active forces and thermal motions, these undulations contribute to the elastic energy of the membrane.

Motions of the cell membrane are major contributors to the overall dynamical behavior of cells and tissues. Although membranes can experience thermal undulations akin to Brownian motion, most of the membrane dynamics are active and involve interactions with the cytoskeleton. For this reason, the membrane dynamics are strongly dependent on the dynamics and rheology of the cytoskeleton and on the metabolism and ATP production of the cell.

Membrane motions span broad length and time scales, as well as a broad amplitude range. For instance, at the largest length and time scales (size of the cell and minutes to hours), membranes can form blebs and ruffles associated with endo and exo-cytosis that have large amplitudes. At intermediate length and time scales (fractions of a cell size and seconds), the membrane undulates with characteristic spatial frequencies that scatter light with characteristic damping times that can be measured in DLS intensity fluctuation autocorrelation functions. At the shortest length and time scales (molecular distances and milliseconds), the membrane is buffeted by thermal motions.

To obtain the amplitude of membrane motions, membrane undulations are decomposed into their Fourier components as

$$h(x,y) = \sum h(k)\mathrm{e}^{\mathrm{i}k\cdot\mathbf{r}}. \tag{8.47}$$

The curvature of interest in the elastic energy of (8.46) is the mean curvature of (8.40) as well as the partial derivatives in the tension term

$$h_x^2 = \left(\frac{\partial h}{\partial x}\right)^2 = -k^2h^2 \qquad h_{xx}^2 = \left(\frac{\partial^2 h}{\partial x^2}\right)^2 = -k^4h^2. \tag{8.48}$$

The total energy is then

$$\begin{aligned} E &= \frac{\gamma_0}{2}\int\left(h_x^2+h_y^2\right)\mathrm{d}x\mathrm{d}y + \frac{\kappa_b}{2}\int\left(h_{xx}+h_{yy}\right)^2\mathrm{d}x\mathrm{d}y \\ &= \gamma_0\frac{A^2}{8\pi^2}\int k^2 h(\mathbf{k})h^*(\mathbf{k})\mathrm{d}\mathbf{k} + \kappa_\mathrm{b}\frac{A^2}{8\pi^2}\int k^4 h(\mathbf{k})h^*(\mathbf{k})\mathrm{d}\mathbf{k} \\ &= \frac{A^2}{8\pi^2}\int(\gamma_0 k^2+\kappa_\mathrm{b}k^4)h(\mathbf{k})h^*(\mathbf{k})\mathrm{d}\mathbf{k}. \end{aligned} \tag{8.49}$$

When the total energy has this form, the principle of equipartition of thermal energy gives the MSD as

$$\langle h(k)h^*(k)\rangle = \frac{k_\mathrm{B}T}{A}\left(\frac{1}{\gamma_0 k^2+\kappa_\mathrm{b}k^4}\right). \tag{8.50}$$

This result has a simple physical form with a mean-squared height proportional to temperature, but inversely proportional to the area of the membrane. In the absence of membrane tension, the bending stiffness appears directly with a strong dependence on the spatial frequency. The cross-over from stiffness-dominated to tension-dominated fluctuations occurs when

$$\gamma_0 > \kappa_\mathrm{b}k^2. \tag{8.51}$$

The fluctuations are largest for the smallest spatial frequencies, and the spatial frequency k has a long-wavelength cutoff given by

$$k_\mathrm{min} = \frac{\pi}{d}, \tag{8.52}$$

where d is the diameter of the cell. Therefore, tension begins to dominate when

$$\gamma_0 \approx \kappa_\mathrm{b}\left(\frac{\pi}{d}\right)^2 \approx 4\ \mathrm{nN/m}, \tag{8.53}$$

which is seven orders of magnitude smaller than the surface tension of water.

The relaxation rate for a membrane undulation with spatial frequency k is given by [23]

$$\frac{1}{\tau_k} = \frac{\gamma_0 k + \kappa_\mathrm{b}k^3}{4\eta}, \tag{8.54}$$

where the viscosity η relates to the viscosity for fluid redistribution within the cell, which is approximately $3\eta_\mathrm{water}$ [24]. The relaxation time contributes to the amplitude fluctuation time autocorrelation function as

$$\langle h_k(0)h_k(t)\rangle = \Delta h^2{}_k \exp(-t/\tau_k). \tag{8.55}$$

The autocorrelation times for height fluctuations are not the same as measured with DLS, which is a different process. For instance, the correlation time for DLS is

$$\begin{aligned}\frac{t}{\tau_{\mathrm{DLS}}} &= q^2 \Delta h^2{}_k(t) \\ &= q^2 \frac{k_{\mathrm{B}}T}{A(\gamma_0 k^2 + \kappa_b k^4)}(1 - \exp(-t/\tau_k)) \\ &\approx q^2 \frac{k_{\mathrm{B}}T}{A(\gamma_0 k^2 + \kappa_b k^4)} \frac{t}{\tau_k},\end{aligned} \tag{8.56}$$

so that

$$\frac{1}{\tau_{\mathrm{DLS}}} \approx \left(q^2 \frac{k_{\mathrm{B}}T}{A(\gamma_0 k^3 + \kappa_{\mathrm{b}} k^4)}\right) \frac{1}{\tau_k}. \tag{8.57}$$

In DLS backscattering, the largest spatial frequencies dominate with $k_{\mathrm{min}} = \pi/d$. For example, for a 10 μm diameter cell and backscattering at $\lambda = 840$ nm, this is a characteristic DLS knee frequency of $f_{\mathrm{knee}} = 0.5$ Hz for thermal undulations of the membrane. Note that the numerical factor in the bracket of (8.57) is equal to about 30. Hence the measured DLS decorrelation rate is about 30 times faster than the membrane fluctuation decay rate for backscattering. For small-angle forward quasi-elastic light scattering (QELS), the DLS scattering rate can be slower than the membrane fluctuation decay rate because of the q^2 dependence. On the other hand, multiple scattering is dominated by backscattering, and hence thick tissues exhibit fast fluctuations.

The DLS correlation rate can be reexpressed as

$$\frac{1}{\tau_{\mathrm{DLS}}} = \frac{q^2 d}{A} \frac{k_{\mathrm{B}}T}{4\pi\eta} \approx q^2 \frac{k_{\mathrm{B}}T}{6\pi\eta a}, \tag{8.58}$$

where the cell diameter $d = 2a$, which has the interesting physical consequence that the mechanics of the cell membrane drop out and do not affect the light scattering knee frequency. The correlation time is simply diffusive (thermal) with an effective diffusion coefficient determined by the cell radius. This result reflects the passive ensemble spectroscopy nature of DLS. However, this interpretation is for short times in the autocorrelation function. For longer times, the membrane height fluctuations saturate at the value set by (8.50), which provides a means to extract mechanical parameters for long delays from light fluctuation autocorrelation functions.

Selected Bibliography

Boal, D.: Mechanics of the Cell.Cambridge (2002) (An excellent up-to-date viewpoint of the mechanical properties of cells)

Bray, D.: Cell Movements: From Molecules to Motility. Garland (2001) (A great reference text on all types of cell movements)

Evans, E.A., Skalak, R.: Mechanics and Thermodynamics of Biomembranes. CRC Press, Boca Raton (1980) (This is the classic original work on mechanics of membranes)

Reif, F.: Fundamentals of Statistical and Thermal Physics. McGraw-Hill, New York (1965) (A good source for the fluctuation-dissipation theorem)

Squires, T.M., Mason, T.G.: Fluid mechanics of microrheology. Annu. Rev. Fluid Mech. **42**, 413–438 (2010) (A good recent review on the generalized Stokes Einstein relation)

References

1. Wax, A., Yang, C.H., Backman, V., Badizadegan, K., Boone, C.W., Dasari, R.R., Feld, M.S.: Cellular organization and substructure measured using angle-resolved low-coherence interferometry. Biophys. J. **82**(4), 2256–2264 (2002)
2. Liu, Y., Li, X., Kim, Y.L., Backman, V.: Elastic backscattering spectroscopic microscopy. Opt. Lett. **30**(18), 2445–2447 (2005)
3. Mourant, J.R., Freyer, J.P., Hielscher, A.H., Eick, A.A., Shen, D., Johnson, T.M.: Mechanisms of light scattering from biological cells relevant to noninvasive optical-tissue diagnostics. Appl. Opt. **37**(16), 3586–3593 (1998)
4. Wilson, J.D., Foster, T.H.: Mie theory interpretations of light scattering from intact cells. Opt. Lett. **30**(18), 2442–2444 (2005)
5. Drezek, R., Dunn, A., Richards-Kortum, R.: Light scattering from cells: finite-difference time-domain simulations and goniometric measurements. Appl. Opt. **38**(16), 3651–3661 (1999)
6. Tuchin, V.: Tissue Optics: Light Scattering Methods and Instruments for Medical Diagnosis. SPIE, Bellingham (2000).
7. Ingber, D.E.: Tensegrity I. Cell structure and hierarchical systems biology. J. Cell Sci. **116**(7), 1157–1173 (2003)
8. Peterson, J.R., Mitchison, T.J.: Small molecules, big impact: a history of chemical inhibitors and the cytoskeleton. Chem. Biol. **9**(12), 1275–1285 (2002)
9. Jordan, M.A., Wilson, L.: Microtubules and actin filaments: dynamic targets for cancer chemotherapy. Curr. Opin. Cell Biol. **10**(1), 123–130 (1998)
10. Ellis, R.J.: Macromolecular crowding: an important but neglected aspect of the intracellular environment. Curr. Opin. Struct. Biol. **11**(1), 114–119 (2001)
11. Purcell, E.M.: Life at low Reynolds-number. Am. J. Phys. **45**(1), 3–11 (1977)
12. Ritchie, K., Shan, X.Y., Kondo, J., Iwasawa, K., Fujiwara, T., Kusumi, A.: Detection of non-Brownian diffusion in the cell membrane in single molecule tracking. Biophys. J. **88**(3), 2266–2277 (2005)
13. Hoffman, B.D., Crocker, J.C.: Cell mechanics: dissecting the physical responses of cells to force. Annu. Rev. Biomed. Eng. **11**, 259–288 (2009)
14. Massiera, G., Van Citters, K.M., Biancaniello, P.L., Crocker, J.C.: Mechanics of single cells: rheology, time depndence, and fluctuations. Biophys. J. **93**(10), 3703–3713 (2007)
15. Squires, T.M., Mason, T.G.: Fluid mechanics of microrheology. Annu. Rev. Fluid Mech. **42**, 413–438 (2010)
16. Reif, F.: Fundamentals of Statistical and Thermal Physics. McGraw-Hill, New York (1965).
17. Wilhelm, C.: Out-of-equilibrium microrheology inside living cells. Phys. Rev. Lett. **101**(2), 028101 (2008)

18. Gallet, F., Arcizet, D., Bohec, P., Richert, A.: Power spectrum of out-of-equilibrium forces in living cells: amplitude and frequency dependence. Soft Matter **5**(15), 2947–2953 (2009)
19. Lau, A.W.C., Hoffman, B.D., Davies, A., Crocker, J.C., Lubensky, T.C.: Microrheology, stress fluctuations, and active behavior of living cells. Phys. Rev. Lett. **91**(19), 198101 (2003).
20. Mizuno, D., Tardin, C., Schmidt, C.F., MacKintosh, F.C.: Nonequilibrium mechanics of active cytoskeletal networks. Science **315**(5810), 370–373 (2007)
21. MacKintosh, F.C., Levine, A.J.: Nonequilibrium mechanics and dynamics of motor-activated gels. Phys. Rev. Lett. **100**(1), 018104 (2008).
22. Boal, D.: Mechanics of the Cell. Cambridge University Press, Cambridge, (2002).
23. Shkulipa, S.A., den Otter, W.K., Briels, W.J.: Simulations of the dynamics of thermal undulations in lipid bilayers in the tensionless state and under stress. J. Chem. Phys. **125** (23), 234905 (2006)
24. Luby-Phelps, K.: Cytoarchitecture and physical properties of cytoplasm: volume, viscosity, diffusion, intracellular surface area. In: International Review of Cytology – a Survey of Cell Biology, vol 192, pp. 189–221 (2000)

Chapter 9
Interference Microscopy

Many biological targets of interest are highly translucent, such as cells and tissues, with light transport lengths of many tens of microns up to hundreds of microns. Therefore, in microscopy of surface-supported cells and thin tissue culture, the optical contrast tends to be very low. On the other hand, cells and tissues have spatially nonuniform refractive indices and spatially varying geometries that lead to optical path length differences with phase that wraps around 2π several times. These large phase deviations provide the basis for phase contrast approaches to microscopic imaging. This may be accomplished in many different ways, including interference microscopy, differential interference contrast (DIC), and holography.

9.1 Phase-Contrast Microscopy

When a transparent object is illuminated by a plane wave, light is diffracted into many partial waves with a spectrum of k-vectors. Of these partial waves, the zero-order wave represents the average transmission and phase of the target, and is structureless. Therefore, this wave can serve as a reference wave against which the other partial waves can be compared, i.e., interfere. In fact, this is the basis for the Abbe theory of image formation [1]. If nothing, other than a lens, is encountered between the transparent object and the image plane, then the relative phases of the partial waves will tend to be in quadrature, and little intensity contrast will be observed. However, if some of the diffraction orders are blocked, or shifted in phase, then intensity contrast emerges.

The principle of the Abbe theory for imaging a transparent phase grating is shown in Fig. 9.1. A transparent small-amplitude phase grating is illuminated by a plane wave and is diffracted into three principal orders: the zeroth order and two first orders. If all three waves recombine at the image plane, then the field is

$$E = 1 + \delta e^{iKx + i\pi/2} - \delta e^{-iKx - i\pi/2} = 1 + i2\delta\ \cos(Kx), \qquad (9.1)$$

D.D. Nolte, *Optical Interferometry for Biology and Medicine*, Bioanalysis 1,
DOI 10.1007/978-1-4614-0890-1_9, © Springer Science+Business Media, LLC 2012

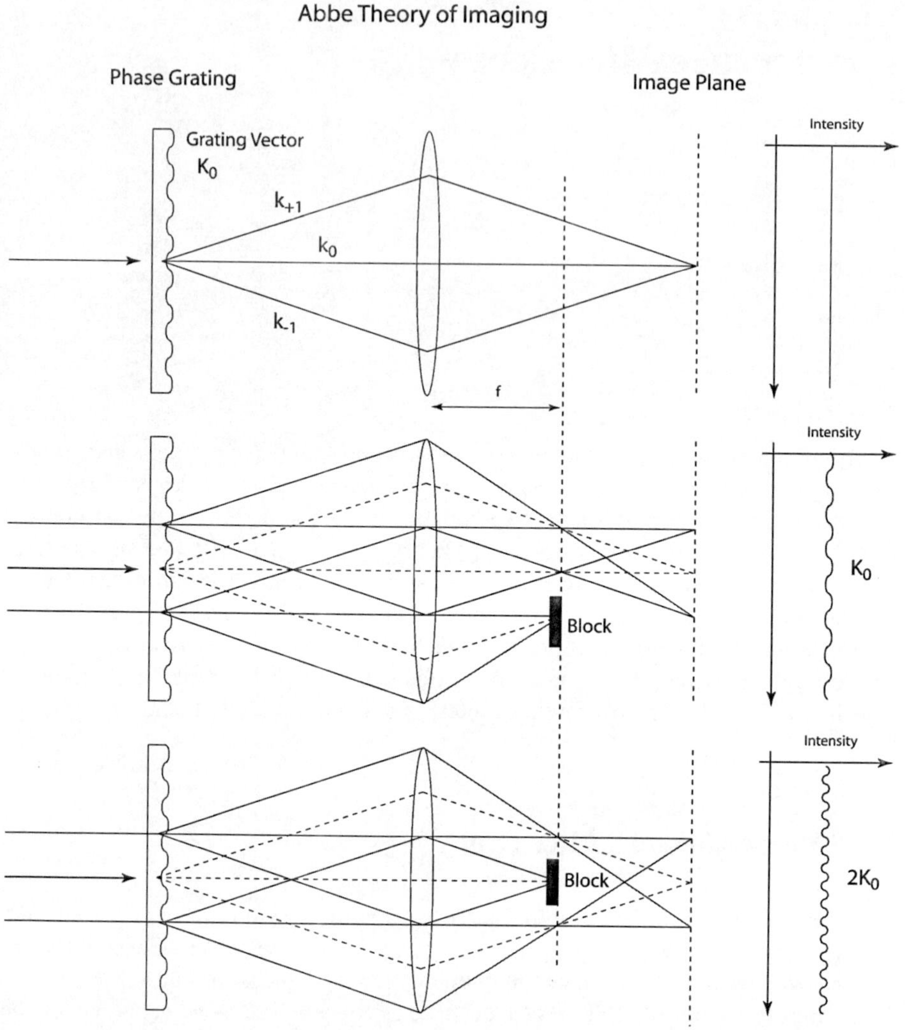

Fig. 9.1 Abbe theory of imaging for a transparent target. A phase grating is illuminated by a plane wave and is diffracted into three orders: 0 and ±1. When all three orders are recombined, the intensity is homogeneous. If one of the first orders is blocked, then the phase grating is uncovered as a fringe pattern. If the zero-order is blocked, then a second-order intensity fringe pattern emerges

where $\delta = (n-1)d$ is the optical path length modulation of the phase grating, and the field modulation is in quadrature with the zero-order. The intensity is then

$$I = |1 + i2\delta \cos(Kx)|^2 = 1 + O(\delta^2) \tag{9.2}$$

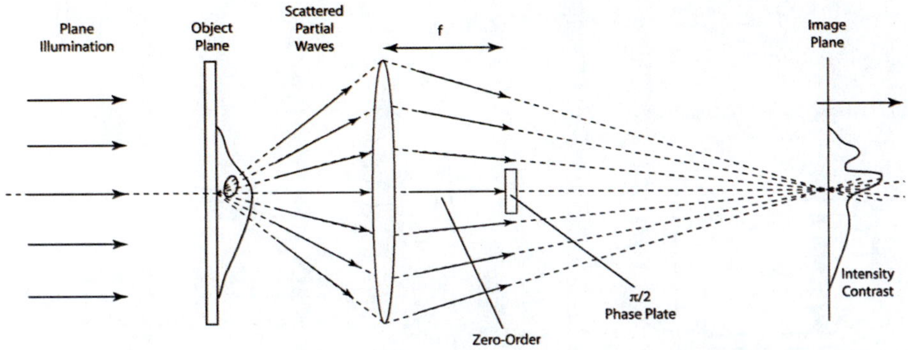

Fig. 9.2 Shifting the phase of the zero-order by $\pi/2$ brings it into phase with the scattered waves and produces intensity variations on the image plane that are proportional to the phase variations of the object

which is uniform to first order. On the other hand, if one of the first diffraction orders is blocked, then the field and intensity become

$$\begin{aligned} E &= 1 + \delta e^{iKx + i\pi/2}, \\ I &= |1 + \delta e^{iKx + i\pi/2}|^2 = 1 - 2\delta \sin(Kx) \end{aligned} \tag{9.3}$$

and the phase grating is recovered as an intensity fringe pattern with the same periodicity as the original grating. This is the principle of the "knife-edge" approach of phase-contrast imaging. In this way, an image is viewed as the result of interference of partial waves that were scattered by the object. Because any object can be described in terms of multiple spatial frequencies, the image is constructed of many interferences.

The "knife-edge" phase-contrast approach blocks half the light and lowers the image brightness. Brightness can be improved by using the phase relationship between the zero-order and the first diffraction orders of an object. The conceptual diagram of this approach is shown in Fig. 9.2. The first diffraction orders acquire phases of $\pm\pi/2$, while the zero-order does not. If the zero-order is phase shifted by $\pi/2$, then it interferes (in phase) with the scattered waves to produce intensity variations that are proportional to the phase of the object.

The phase plate on the zero-order can lead to higher contrast and lower stray reflectances if the zero-order has an annular form. This is accomplished with an annular illumination of the object and an annular phase plate in the Zernicke phase contrast microscope, shown in Fig. 9.3. The annular illumination produces the "surround" light that passes through an annular phase plate at the back focal plane of the objective lens. This surround light is also attenuated because it is much brighter than the scattered intensities from a nearly transparent object. By balancing the relative intensities, phase contrast images of high contrast can be formed.

Phase plates do not need to be static optical elements, but can be electrically addressed spatial light modulators, as in liquid crystal screens. These have

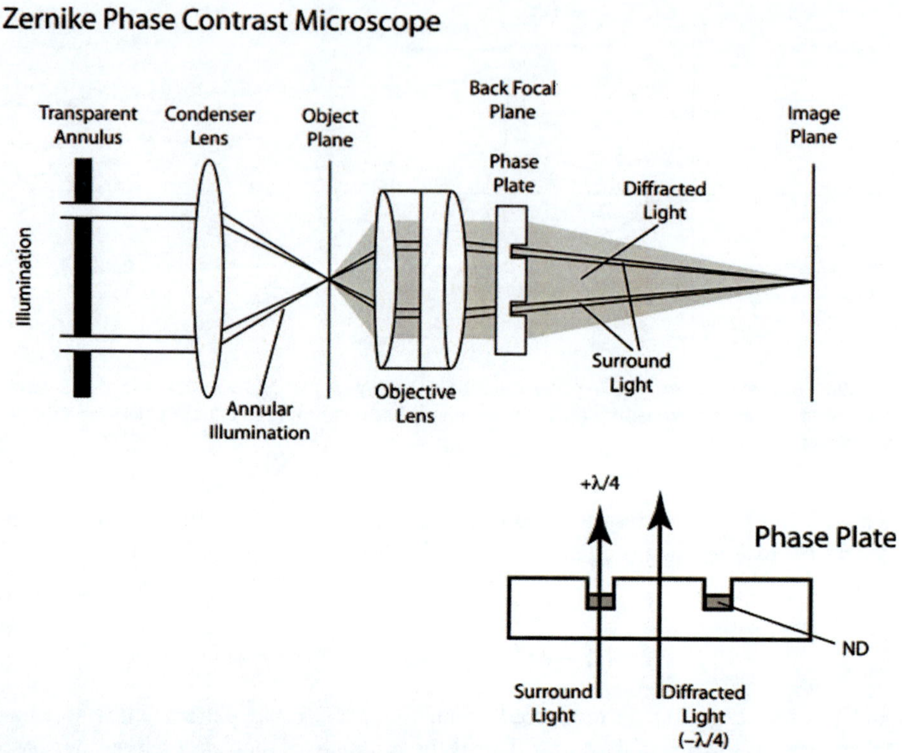

Fig. 9.3 Zernike phase-contrast microscope. The annular illumination of the object is collected by the objective lens and shifted in phase by $+\lambda/4$ by the phase plate at the back focal plane of the objective lens. The light scattered (diffracted) by the object is shifted by $-\lambda/4$ by the scattering process. When these partial waves are combined at the image plane, phase variations in the object produce intensity variations through interference of the diffracted with the surround light

the advantage that the phase shift can be tuned, providing quantitative phase microscopy, and the spatial arrangement of the phase plate can be reconfigured by computer control. An example of a programmable phase contrast microscope configuration is shown in Fig. 9.4 [2]. A correcting lens is placed at the image plane of a conventional microscope to collimate the zero-order, which is Fourier transformed to the center of a programmable phase modulator. The image field is also Fourier transformed by the same lens, and both are transformed back to the charge-coupled device (CCD). The relative phase of the interfering fields at the CCD plane is stepped by integer factors of $\pi/2$ using the phase modulator, with an image acquired at each step. The intensity at the CCD is then

$$I(x, y) = |E_0|^2 + |E_1(x, y)|^2 + 2|E_0||E_1(x, y)|\cos[\Delta\phi(x, y) + n\pi/2], \qquad (9.4)$$

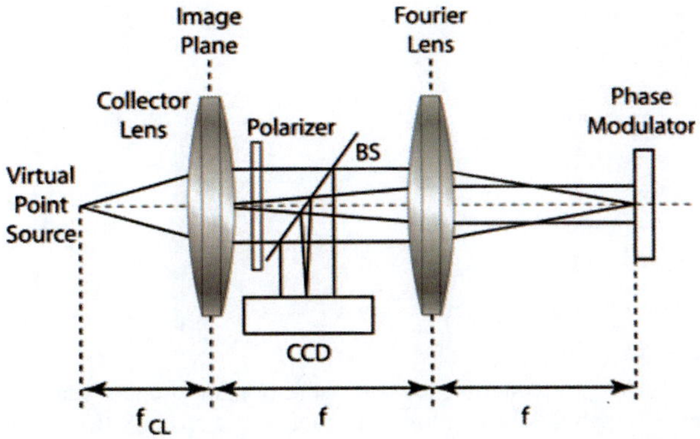

Fig. 9.4 Example of a programmable phase contrast microscopy system. The correcting lens is placed at the image plane of a microscope to collimate the zero-order. The image is Fourier transformed to the programmable phase modulator where the zero-order is phase shifted in increments of $\pi/2$. The field is then Fourier transformed to the CCD camera

where $\Delta\phi$ is the relative phase between the zero-order and the scattered fields. The relative optical phase is retrieved as

$$\Delta\phi(x, y) = \tan^{-1}\left[\frac{I(x, y; 3) - I(x, y; 1)}{I(x, y; 0) - I(x, y; 2)}\right], \tag{9.5}$$

where the four intensities are for $n = 0, 1, 2$, and 3. The absolute phase of the sample is retrieved as

$$\phi(x, y) = \tan^{-1}\left\{\frac{\beta(x, y)\sin\Delta\phi\ (x, y)}{1 + \beta(x, y)\cos\Delta\phi\ (x, y)}\right\}, \tag{9.6}$$

where the spatial contrast is

$$\beta(x, y) = \frac{I(x, y; 0) - I(x, y; 2) + I(x, y; 3) - I(x, y; 1)}{\sin\Delta\phi\ (x, y) + \cos\Delta\phi\ (x, y)}. \tag{9.7}$$

The quantitative control of the phase modulator enables a quantitative extraction of phase and amplitude from the sample, including the ability to capture dynamic effects such as cell shape changes or membrane undulations.

9.2 Differential Interference Contrast

Where the Zernicke phase contrast microscope produces an intensity contrast that is proportional to the phase of the target, the Nomarski DIC microscope, shown in Fig. 9.5, produces intensity contrast that is proportional to the gradient of the phase of the sample. This microscope produces high-contrast images with minimal halo

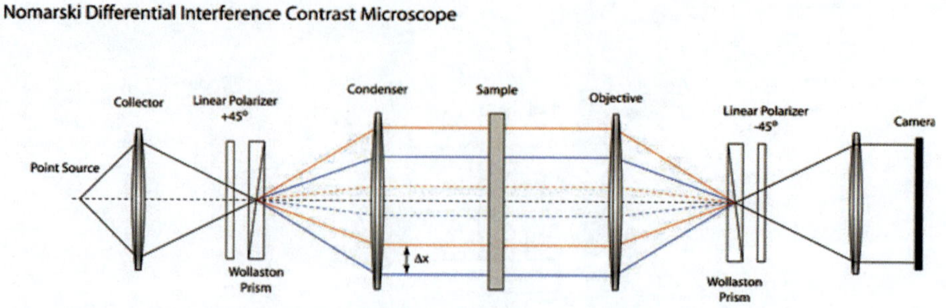

Fig. 9.5 Nomarski differential interference contrast (DIC) microscope. The first Wollaston prism separates the beams, which are then collimated, leading to a spatial offset Δx between the paths that pass through the sample. The second Wollaston prism recombines the beams. Differences between the two optical paths produce phase offsets that interfere on the camera plane

effects and thus is often chosen for less quantitative but more impressive visual depiction of cellular structure and processes. The microscope relies on double refraction by anisotropic optical crystals to split two polarizations into two different paths. Because the paths are spatially shifted, they experience different phase delays. When they are recombined and interfere, the phase delay causes an intensity change. The beam splitting is achieved using a Wollaston prism that combines two anisotropic materials in a configuration with two prisms having orthogonal optic axes. This causes each polarization to refract into angles $\pm\Delta\psi$. A condenser lens recollimates the beams with a spatial offset $\Delta x = f_c \Delta\psi$, where Δx is usually set to the diffraction-limited resolution of the microscope, and f_c is a function of the Wollaston prism design. These two orthogonally polarized beams pass through the sample and are recombined after the objective lens by a second Wollaston prism and linear polarizer. The combined beams have a phase offset given by

$$\Delta\phi(x,y) = \frac{1}{2}\Delta\vec{r}\cdot\vec{\nabla}\phi, \tag{9.8}$$

where the direction of $\Delta\vec{r}$ is defined by the orientation of the Wollaston prism. The linear polarizers introduce a phase bias that converts the phase difference into an intensity contrast as

$$I(x,y) = I_0(\sin^2\Delta\theta + \Delta\vec{r}\cdot\vec{\nabla}\phi\ \sin\Delta\theta\ \cos\Delta\theta). \tag{9.9}$$

For example, with a phase bias of $\Delta\theta = 45°$, this is

$$I(x,y) = \frac{I_0}{2}(1 + \Delta\vec{r}\cdot\vec{\nabla}\phi) \tag{9.10}$$

which has maximum linear sensitivity to small phases on a uniform background generated by the point-source illumination. The directional dependence selected by the direction of Δr produces a "shadowing" effect that gives DIC images their characteristic "3D" appearances, as in Fig. 9.6.

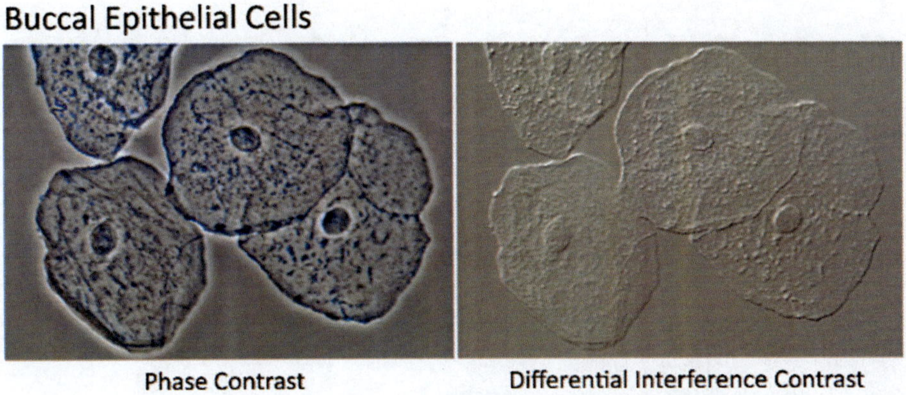

Fig. 9.6 Comparison of Zernike phase contrast and DIC of buccal epithelial cells (from http://www.microscopyu.com/galleries/dicphasecontrast/index.html)

9.3 Particle Tracking Interferometry

Phase contrast microscopy lends itself to an important application in particle tracking interferometry for bead-based imaging of beads or nanoparticles attached to cells and biomolecules. The scattered light from the bead in a focused Gaussian beam interferes in the far field with the transmitted beam, changing the diffraction pattern in a manner proportional to the displacement of the bead. By detecting the interference shift of the beam on a position-sensitive detector, sensitivity to particle displacements below a nanometer is possible.

9.3.1 Back Focal-Plane Interferometry

The simplest configuration for the detection of particle displacement is similar to the case of a molecule in a Gaussian beam described in Sect. 2.4.1. The scattering geometry is shown in Fig. 9.7. A small bead is displaced off the optic axis and scatters light that interferes with the Gaussian beam in the far field.

The scattered field collected by a Fourier transform lens of focal length f is

$$E_{\text{scat}} = \frac{k^2}{f}\left(\frac{\varepsilon - 1}{\varepsilon + 2}\right)a^3\sqrt{S_0}\exp(-\Delta x^2/2w_0^2)\exp(\mathrm{i}k\Delta xx'/f), \tag{9.11}$$

where Δx is the bead displacement from the optic axis, a is the bead radius, w_0 is the radius of the Gaussian beam at its focal waist and ε is the dielectric constant

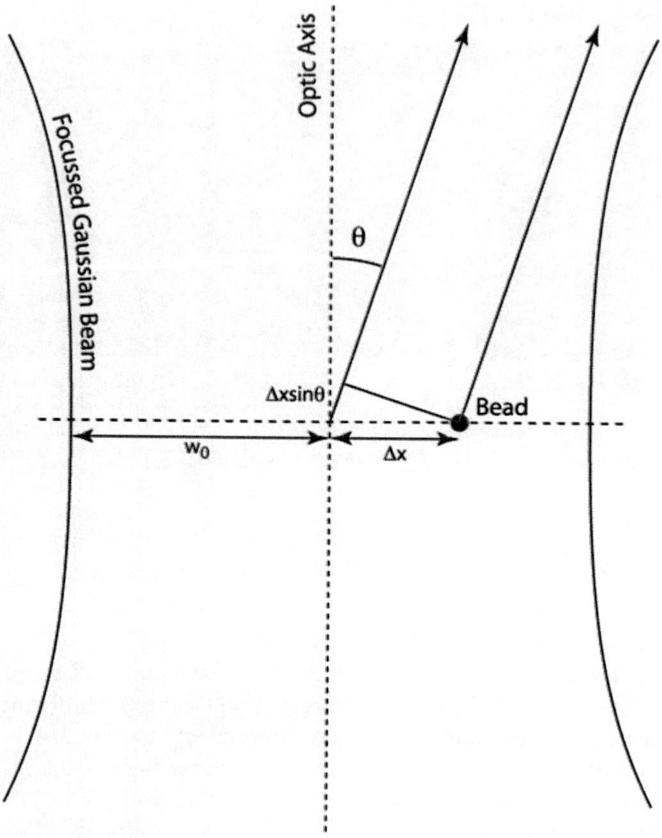

Fig. 9.7 Optical path lengths of a bead on the focal plane of a Gaussian beam. The scattered light from the bead interferes in the far field with the Gaussian beam

of the bead. The interfering fields at the Fourier plane for the particle and the Gaussian beam are

$$\begin{aligned} E &= E_{\text{scat}} + E_0 \\ &= \frac{k\sqrt{S_0}}{f}\left[ka^3\left(\frac{\varepsilon-1}{\varepsilon+2}\right)\exp(-\Delta x^2/2w_0^2)\exp(ik\Delta xx'/f) - \mathrm{i}w_0^2\exp\left(-\frac{2\pi^2 w_0^2}{(f\lambda)^2}x'^2\right)\right]. \end{aligned} \tag{9.12}$$

The intensity is

$$I = \frac{k^2 S_0 w_0^2}{f^2}\left[\exp\left(-\frac{k^2 w_0^2}{f^2}x'^2\right) - 2\exp\left(-\frac{k^2 w_0^2}{2f^2}x'^2\right)\frac{ka^3}{w_0^2}\left(\frac{\varepsilon-1}{\varepsilon+2}\right)\exp(-\Delta x^2/2w_0^2)\sin(k\Delta xx'/f)\right]. \tag{9.13}$$

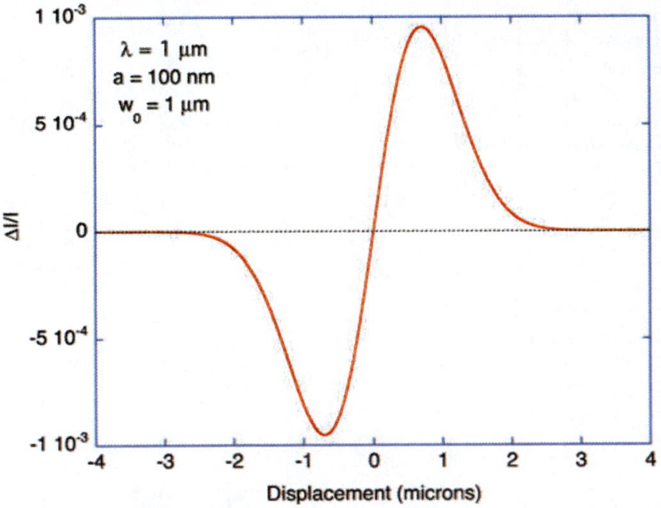

Fig. 9.8 Relative intensity change for a bead in water as a function of displacement off the optic axis of a focused Gaussian beam. The beam radius and wavelength are 1 μm each, and the particle size is 100 nm

This has the form of the original Gaussian beam plus an extra interference term that varies as

$$\Delta I \sim 2\sqrt{I_0 I_{\text{scatt}}}\exp(-k^2 w_0^2 \sin^2\theta)\sin(k\Delta x\ \sin\theta) \tag{9.14}$$

which is an asymmetric function of θ that is linearly proportional to Δx (for Δx much smaller than a wavelength). The asymmetric function can be detected with a split detector at the back focal plane of a collecting lens. The split detector integrates over the two lobes and produces a signal proportional to the difference, which is linearly proportional to the bead displacement. The relative signal is [3,4]

$$\frac{I_+ - I_-}{I_+ + I_-} = \frac{16}{\sqrt{\pi}}\frac{k\alpha}{w_0^2}\frac{x}{w_0}\mathrm{e}^{-x^2/w_0^2} \tag{9.15}$$

with the polarizability

$$\alpha = a^3\frac{n^2-1}{n^2+2} \tag{9.16}$$

and the intensity change is shown as an example in Fig. 9.8. Back-focal-plane interferometry has become especially useful for laser tweezer experiments that confine beads and wish to sense displacements of the bead from the optic axis. By calibrating the optical confinement force and the differential displacement signal, mechanical and dynamical properties of the trapped beads (or what they are attached to) can be extracted quantitatively.

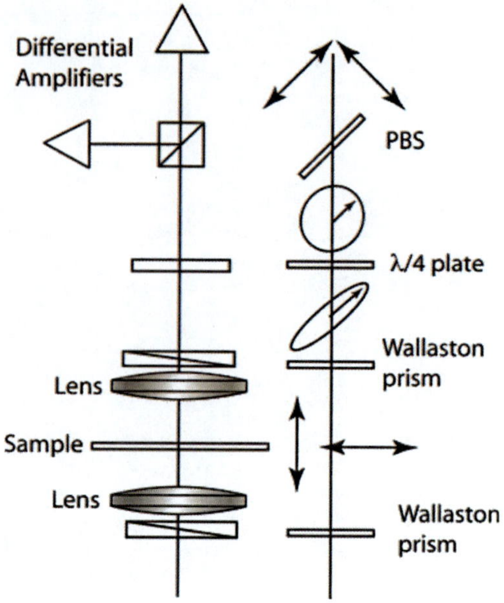

Fig. 9.9 Schematic of a DIC bead displacement interferometer from [6]. Laser illumination is passed through the Nomarski optics to produce two spatially shifted diffraction-limited spots at the focal plane. The displacement of the bead produces a signal proportional to the displacement detected in a balanced photodetector pair

9.3.2 DIC Displacement Measurement

An alternative to back-focal-plane detection uses Nomarski optics to measure the displacement of a bead in the focal spot of a laser trap [5]. The schematic for the interferometer is shown in Fig. 9.9. The chief advantage of the Nomarski approach over the back-focal plane approach is the use of the polarizing beamsplitter and two photodetectors instead of the split detector. The split detector in the back-focal plane configuration requires high mechanical stability. In the Nomarski approach, the detected light is in the zero-order and is not imaged, making it more stable. The optical train in Fig. 9.9 converts a lateral displacement of the particle into a linear (in phase quadrature) intensity. The key element in achieving quadrature is the quarter-wave plate and the polarizing beamsplitter before the balanced photodetectors.

The initial section of the optical train uses the conventional Nomarski process that takes a wave polarized at 45° and converts it into two co-parallel beams with a lateral shift equal to $2d$, shown in Fig. 9.10. The particle is displaced an amount Δx from the bisector. The total field for each beam separately after the sample plane (and already assuming a far-field condition) is

$$E^{-} = \frac{k\sqrt{S_0}}{\sqrt{2}f}\left[k\alpha \exp(-(d-\Delta x)^2/2w_0^2) - \mathrm{i}w_0^2\right],$$
$$E^{+} = \frac{k\sqrt{S_0}}{\sqrt{2}f}\left[k\alpha \exp(-(d+\Delta x)^2/2w_0^2) - \mathrm{i}w_0^2\right], \quad (9.17)$$

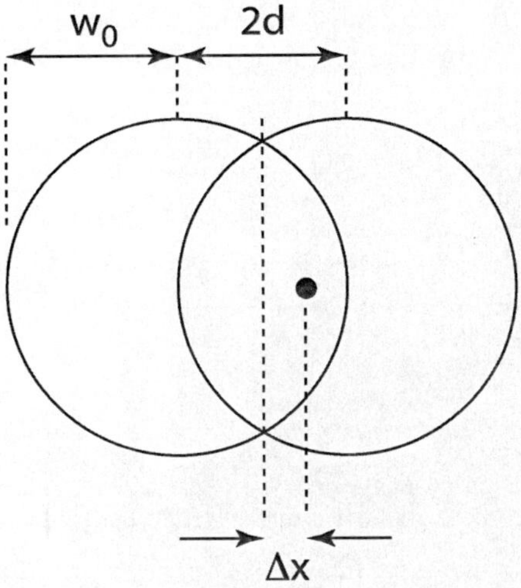

Fig. 9.10 Geometry of the two displaced beams and the particle. The particle is displaced from the center of the two beams by the amount Δx. The spatial shift of the beams is equal to $2d$, and w_0 is the beam radius

where each field is for a separate orthogonal polarization, and where A_d is the detector area that is assumed to be small and on the optic axis. Note that the shift to the right for one beam is a shift to the left relative to the other beam. The total field after the second recombining Wollaston prism is

$$\vec{E} = \hat{x}E^- + \hat{y}E^+, \tag{9.18}$$

where the polarizations are shown explicitly. The combined beams then pass through the quarter wave plate that creates the field

$$\vec{E} = \hat{x}E^- + \hat{y}\mathrm{i}E^+. \tag{9.19}$$

The final step is the polarizing beamsplitter that is oriented at $\pm 45°$. The field projection out each port of the polarizing beamsplitter is

$$\begin{aligned}\vec{E} \cdot \vec{p}^{\pm} &= (\hat{x}E^- + \hat{y}iE^+) \cdot \frac{1}{\sqrt{2}}(\hat{x} \pm \hat{y}) \\ &= \frac{1}{\sqrt{2}}(E^- \pm iE^+).\end{aligned} \tag{9.20}$$

For the + port the detected power for a detector area A_d is

$$I = \frac{1}{2}[|E^-|^2 + |E^+|^2 + \mathrm{i}E^+(E^-)^* - \mathrm{i}(E^+)^*E^-]A_\mathrm{d} \tag{9.21}$$

which is

$$I = \frac{k^2 w_0^2 S_0 A_\mathrm{d}}{2f^2}\left[w_0^2 + 2k\alpha \mathrm{e}^{-d^2/2w_0^2} \sinh\left(\frac{\mathrm{d}\Delta x}{w_0^2}\right)\right]. \tag{9.22}$$

By taking the difference between the two ports, twice the signal is obtained while subtracting common-mode intensity fluctuations. The final signal is

$$\Delta I = 2\frac{k^2 w_0^2 S_0 A_d}{f^2} k\alpha e^{-d^2/2w_0^2} \sinh\left(\frac{d\Delta x}{w_0^2}\right). \tag{9.23}$$

For shot-noise-limited performance, the photon fluctuation noise associated with the intensity S_0 for a detection time $T = 1/\mathrm{BW}$ is

$$\Delta N = \sqrt{\eta \frac{k^2 w_0^4 S_0 A_d}{2h\nu \mathrm{BW} f^2}}, \tag{9.24}$$

where η is the quantum efficiency of the detector, and the signal photon number is

$$N(\Delta x) = \frac{\eta S_0 A_d}{h\nu \mathrm{BW}} \frac{2k^3 \alpha w_0^2}{f^2} e^{-d^2/2w_0^2} \sinh\left(\frac{d\Delta x}{w_0^2}\right). \tag{9.25}$$

Therefore, the signal-to-noise ratio is

$$\mathrm{S/N} = \sqrt{\frac{2\eta S_0 A_d}{h\nu \mathrm{BW}}} \frac{2k^2 \alpha}{f} e^{-d^2/2w_0^2} \sinh\left(\frac{d\Delta x}{w_0^2}\right). \tag{9.26}$$

The noise-equivalent displacement (NED) of the particle is

$$\Delta x_{\mathrm{NED}} = \sqrt{\frac{h\nu \mathrm{BW}}{2\eta S_0 A_d}} \frac{f w_0^2}{2k^2 \alpha d} e^{d^2/2w_0^2}. \tag{9.27}$$

For a detected power of 1 mW, wavelength of 633 nm, focal length of 1 mm, bead radius of 1 μm, beam radius at the sample plane of 2 μm, and beam displacement of 1 μm, the NED is approximately 0.05 pm per root Hertz. This ultra-small displacement sensitivity is typical with these parameters, although in practice the detection bandwidth tends to be large, and picometer displacement can be measured routinely using interferometry in the laboratory.

9.4 Reflection Interference Contrast Microscopy

Phase-contrast microscopy and DIC microscopy rely on diffraction in the first case, and on image shearing in the second case. To these phase-sensitive approaches can be added a more direct form of interferometric imaging that uses in-line common-path geometries that are mechanically stable.

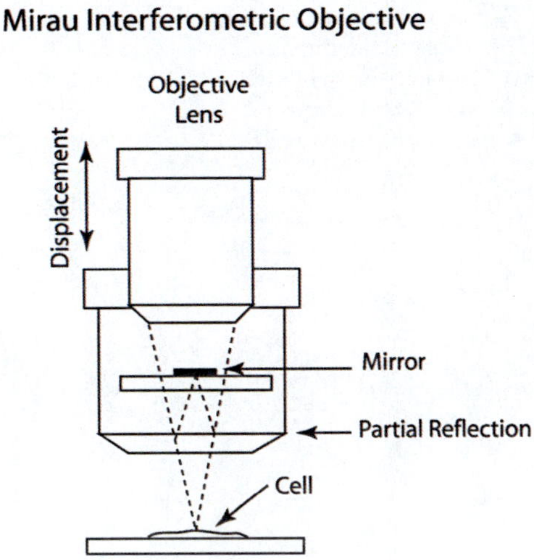

Fig. 9.11 The Mirau interferometric objective uses an in-line common-path design with a normal-incidence beamsplitter and an internal reference mirror

An in-line common-path design that is built directly into the microscope objective lens is the Mirau interferometric objective shown in Fig. 9.11. A normal-incidence beamsplitter is put into the optical path that reflects a reference wave to an internal mirror that is at the same focal distance as the target sample. The reflected reference is recombined with the signal wave at the beamsplitter and directed into the objective lens, which relays the image to a conventional microscope. The topology of the target is referenced to the flat reference mirror and is observed as interference fringes in the image. This interferometer is relatively simple, and can have dynamic phase control using a displacement of the objective lens relative to the beamsplitter. The drawbacks for biological phase imaging are the relatively long working distance, and low fringe contrast if the partial reflection is too large. For these reasons, the Mirau interferometer is usually used for materials metrology of highly reflective surfaces.

An in-line configuration with much shorter working distance, and with higher fringe contrast for weakly reflecting biological samples, is the reflection interference contrast microscope (RICM) shown in Fig. 9.12. The objective lens is modified by a quarter wave plate that helps reduce stray scattered light that would otherwise limit the contrast of the interference fringes. This is called an antiflex objective, because it reduces stray reflectances. Polarized light is sent through the objective lens with a quarter-wave plate between the lens and the immersion oil. The light polarization after the wave plate is circularly polarized. The cover slip performs as a beamsplitter to generate the in-line reference wave, and the biological sample provides the signal wave. These combined waves pass through the quarter wave plate to generate linearly polarized light that is orthogonal to the incident light and is passed by the polarizing beamsplitter. An analyzer filters out background light [7]. The RICM approach is well-suited for cell adhesion and membrane motion studies [8–10].

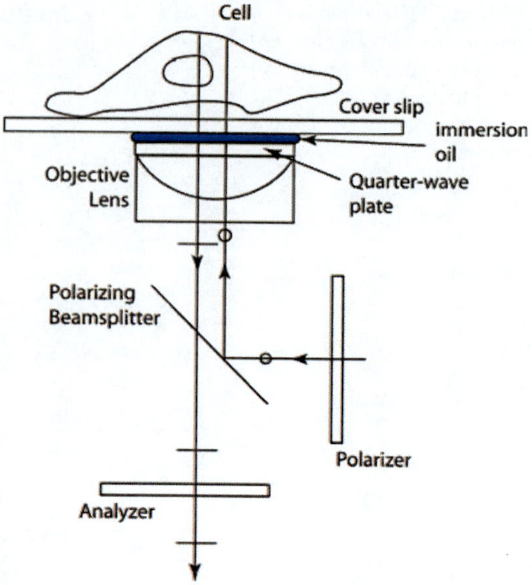

Fig. 9.12 Antiflex objective for reflection interference contrast microscopy (RICM). The polarization control reduces stray reflected light

9.5 Fluorescence Interference Contrast Microscopy

Thin layers that are several hundred nanometers thick or less are effective interference films even for "incoherent" light. This opens the door for sensitive interferometry with arbitrary light sources that do not need long coherence. Therefore, thin nanometer-scale films will show interference phenomena for fluorescent emission from biological samples.

This effect is illustrated in Fig. 9.13. for silicon dioxide layers on silicon. The membrane of a cell is labeled with a fluorophore, and the cell is placed on a tiered oxide. The electric field at the surface of silicon is nearly zero, and hence the fluorescence intensity is very low for the thinnest oxide of 10.9 nm. At a wavelength of 532 nm in silica, the antinode is 92 nm above the silicon surface, and the fluorescent membrane on this thickness of oxide exhibits the largest fluorescence [11,12].

Similar fluorescence interference concepts apply in the the spectral domain [13]. A fluorophore on a spacer above a dielectric mirror interfers with itself in reflection. The reflected light is put into a spectrometer, and interference fringes are observed in the spectral domain analogous to spectral interferometery described in Chap. 1. The fringe spacing is proportional to the optical path length difference between the direct wave and the reflected wave. The substrate configuration and the spectral interferometry are shown in Fig. 9.14. The two spectral interferometry curves are from fluorophores that differ in height by only 20 nm.

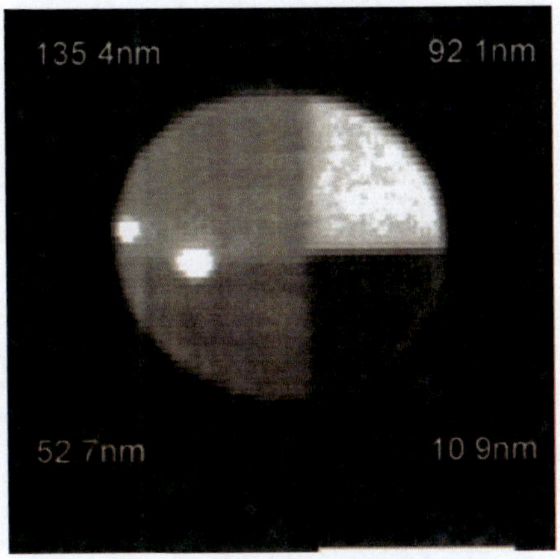

Fig. 9.13 Fluorescence interferometry of a cell using oxide layers of varying thickness. The cell membrane is labeled with a fluorophore. The changing conditions for interference are observed for four oxide thicknesses. With permission from [11,12]

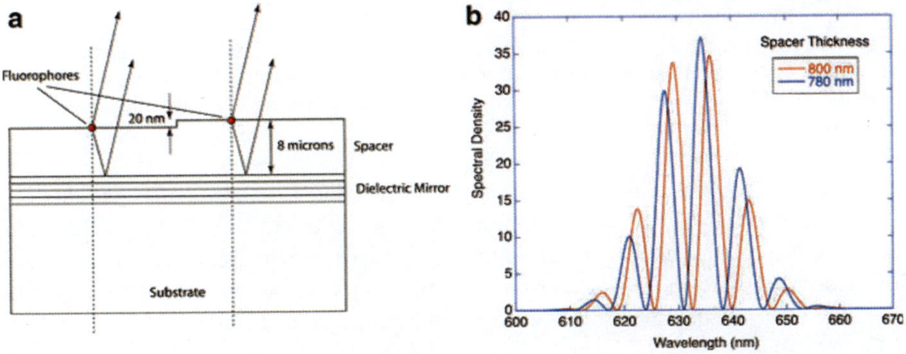

Fig. 9.14 Fluorescence interference in reflection above a dielectric mirror [13]. The structure is shown in (**a**). The fringe spacing in the spectral domain is related to the optical path length difference between the direct and the reflected wave. The two *curves* in (**b**) are from fluorophores that are at different heights that differ by 20 nm

9.6 Angular Scanning Interferometry

The angular dependence of light scattered from living cells and tissues carries information on the size and density of scattering objects. For instance, angular oscillations in scattered light can be related back to the size of the scattering object using Mie scattering theory. It has been suggested that changes in the average size of the nuclei that are too small to see in histology, can be captured by angle-resolved light scattering. Discrimination of the signal from selected depths is accomplished using short-coherence sources.

Two angle-resolved low-coherence interferometry systems are shown in Fig. 9.15. One is a Michelson and the other is a Mach–Zehnder configuration. In each case, a translated lens acts as an angular selective filter that accepts

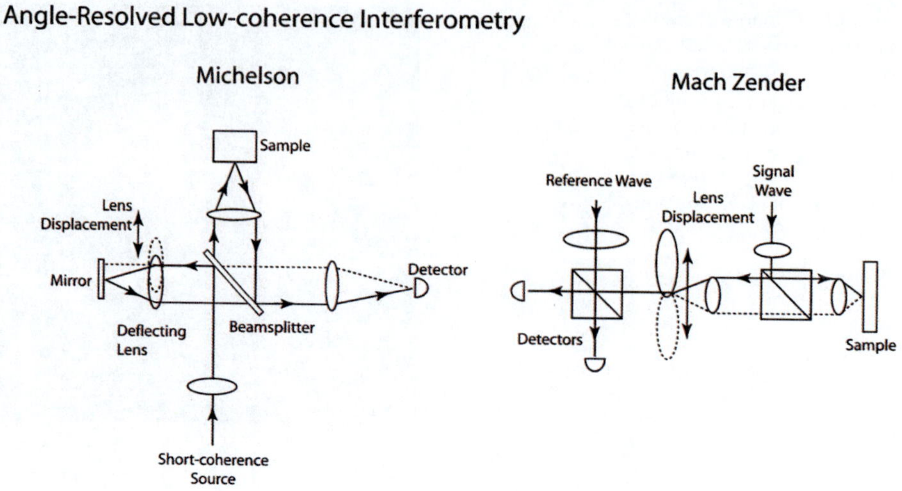

Fig. 9.15 Angle-resolved low-coherence interferometry optical systems [14,15]. The translation of the lens directs light backscattered at different angles to the detector. Two configurations are shown. The Mach–Zehnder has a larger angular range and is detected as a heterodyne at an IF of 10 MHz

backscattered light only from specific angles. The Michelson arrangement is simpler [14], but the Mach–Zehnder has a broader range of angles [15]. The Mach–Zehnder is also operated with heterodyne detection at an intermediate frequency of 10 MHz using an acoustic optic modulator on the reference wave to shift its frequency. The system has fast acquisition and a wide angular range up to 15°.

9.7 Broad-Field Interference Microscopy

An approach complementary to in-line interference microscopy uses conventional Michelson or Mach–Zehnder configurations. These enable phase-shifting interferometric imaging with quantitative phase extraction, but with less stability than the common-path in-line configurations. They also can be combined with low coherence for depth sectioning.

A phase-stepping Michelson microscope is shown in Fig. 9.16 that uses a mirror on a finely adjusted stage (a voltage driven piezoelectric stack known as PZT) that controls the phase [16]. By taking images at four different phases, the optical path length can be quantified. Furthermore, by using low coherence light, the interference is selective to a specific depth (tuned by shifting one of the Michelson mirrors). Images of single cells from a tadpole are shown in Fig. 9.17, from a depth up to 450 μm clearly showing nuclei under high resolution in the cells.

The Michelson short-coherence microscope performs in-line digital holography. The holography is in the image-domain, and the phase information is retrieved by

Fig. 9.16 A Michelson interferometric microscope with phase-stepping control for phase retrieval [16]. A low-coherence source provides depth selectivity

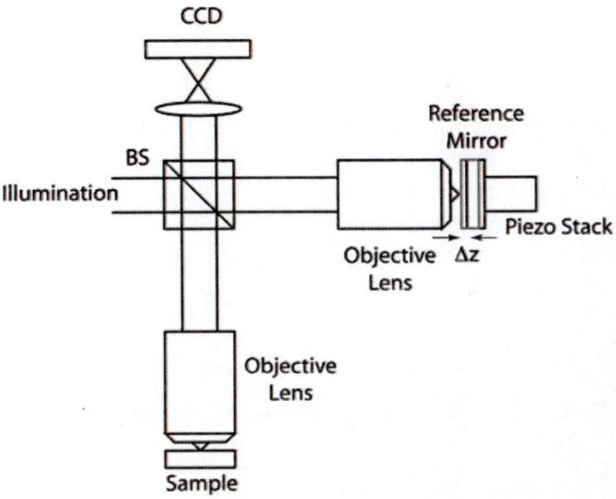

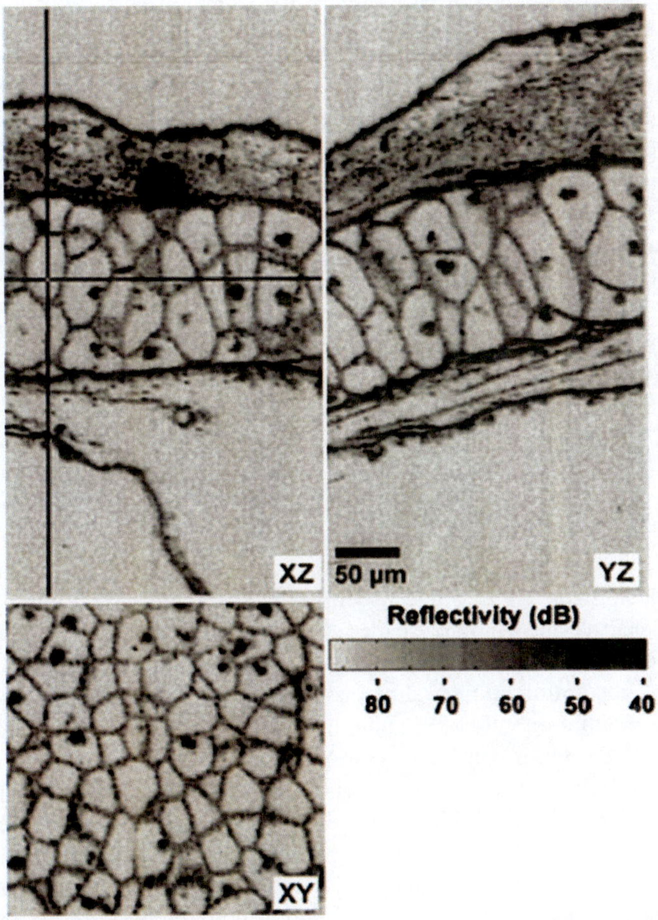

Fig. 9.17 High-resolution optical section images of a frog tadpole, using the Michelson phase-stepping interferometer [16]

the use of the phase-stepping mirror. Because of the use of short coherence to select information from a specific depth, this arrangement is also sometimes known as full-field optical coherence tomography (OCT), although the tomographic part is simply the phase retrieval. We will see in a later chapter that conventional OCT is a laser raster scanning process that builds up three dimensions by scanning. Full-field OCT, on the other hand, captures an entire depth-gated image and thus has a parallel acquisition advantage.

9.8 Digital Holographic Microscopy

The improving bit depth of digital imaging arrays, known as charge-coupled device CCD, opened the opportunity for digital holography under the high background conditions of biological imaging. The principle of digital holography is virtually the same as for traditional holography using photographic recording. Just as interference fringes expose a photographic plate, they expose the wells of a CCD array as buckets of electrons replace silver halide. The difference is in the "read out." While conventional holography uses a laser to diffract off the physical holographic gratings, digital holography performs numerical reconstruction on the electronic data. Digital holography is much more versatile than conventional holography, because there are no disposable materials, and the postrecording numerical manipulations can be as complex as needed to reconstruct information from the scattering target.

The caveats on the comparison of conventional holography with digital holography are on dynamic range and spatial resolution. Silver halide, and other types of holographic recording like the photorefractive effect, can have a larger recording dynamic range because they are not limited to a bit-depth of 12 or 14 bits. In addition, the digital recording resolution is limited to the pixel pitch of the imaging array, which is often on the order of 5 μm or larger. Nyquist sampling requires a fringe spacing equal to twice the pixel pitch, placing the digital holographic fringe spacing around 10 μm. Physical holographic recording media, on the other hand, can record sub-micron fringe spacings. However, the overall ease of use and the power of postprocessing afforded by digital holography make it highly attractive for phase-sensitive recording of scattered light. An example of a digital hologram (a Fresnel-domain hologram not too far from the image plane) is shown in Fig. 9.18 with its high-resolution numerical reconstruction.

An example of a digital holographic microscope is shown in Fig. 9.19 that uses coherent red light that is split by a beamsplitter into a reference wave and an object beam that interacts (in transmission) with a biological sample. The scattered object wave is projected onto a CCD chip and interferes with an off-axis reference wave [18]. A typical interferogram is shown in Fig. 9.19b). The CCD is in a Fresnel regime relative to the image plane of the sample. Therefore, the numerical reconstruction is an inverse Fresnel transform. The power of digital holography arises in the ability to perform dynamic refocusing. By changing the control (distance) parameter in the

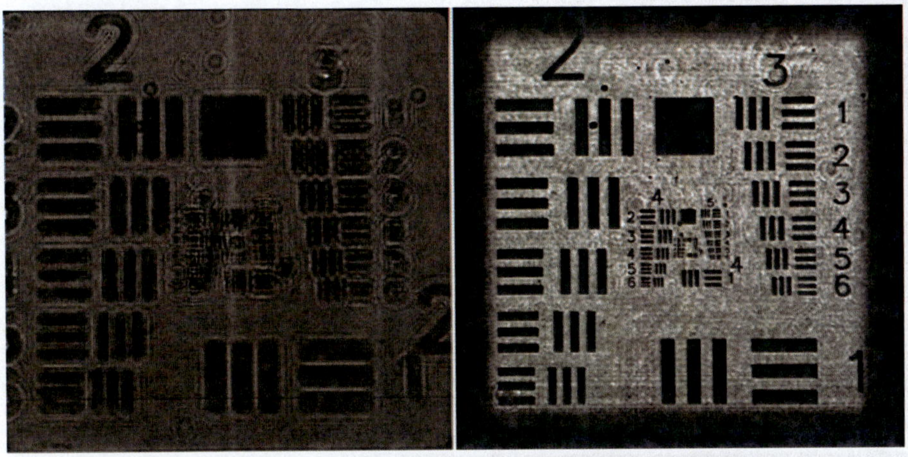

Fig. 9.18 Image-domain digital hologram and its numerical reconstruction [17]

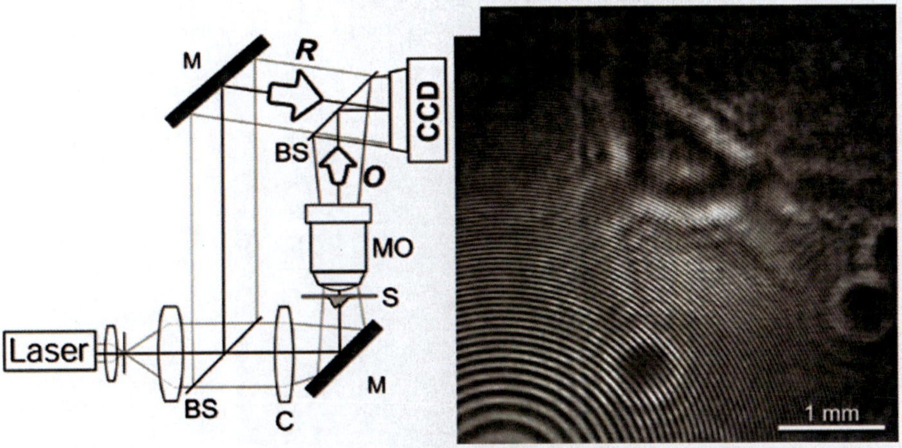

Fig. 9.19 A digital holographic microscope for investigation of biological samples [18]

Fresnel transform, the reconstructed image can be depth-sectioned by numerically focusing at successive depths inside the target. This numerical focusing can provide volumetric data without the need for a coherence (depth) gate as in coherence-domain techniques like OCT. However, the axial selectivity is not as high as in the coherence-domain approaches.

Digital holograms are typically recorded in the Fresnel regime, where slight changes in target distance can be numerically recovered through the inverse Fresnel transform. This aspect of digital holography makes it particularly easy to use without lenses when target and CCD distances are in the range of millimeters. One broad class of applications that can be approached by this regime is cytometry. An example of a lens-free holographic cytometry configuration is shown in Fig. 9.20. Red blood

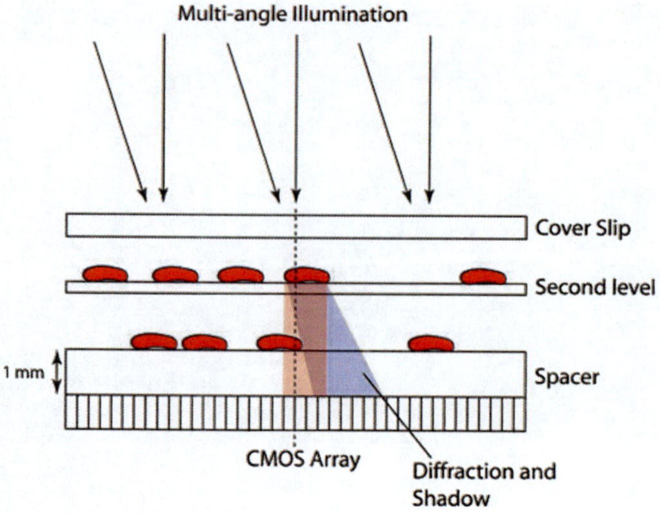

Fig. 9.20 Lens-free holographic recording onto a CMOS pixel array. This "shadow-casting" approach, when combined with multiple illumination angles, allows reconstruction of the cell height to an accuracy of a few microns [19]

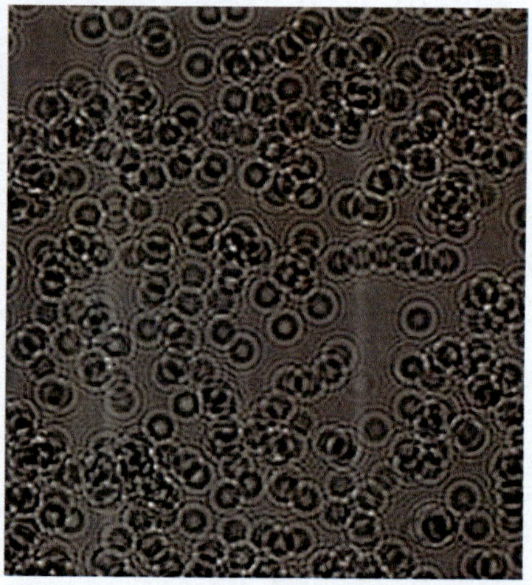

Fig. 9.21 Example of an in-line lens-free hologram of red blood cells [20]

cells are suspended at two different heights (1.025 and 1.080 mm) above the surface of a CMOS pixel array and illuminated with a broad beam at several successive angles. The forward-scattered light interferes with the original wave to form an in-line hologram on the CMOS array. A captured hologram is shown in Fig. 9.21 showing characteristic Airy rings around the individual blood cells.

The continuing improvement of dynamic range and speed of digital pixel arrays (CCD and CMOS) lends itself to increasingly more powerful digital holographic approaches for biology and medicine. Many of these applications are in the field of quantitative phase microscopy, but an increasing ability to use coherence-domain detection also widens the door for digital holography of more complex biological targets such as living tissue. The properties of light propagation through tissue and its detection using coherence techniques are the topics of the next several chapters.

References

1. Goodman, J.W.: Introduction to Fourier Optics. Roberts, Englewood (2004)
2. Popescu, G., Deflores, L.P., Vaughan, J.C., Badizadegan, K., Iwai, H., Dasari, R.R., Feld, M. S.: Fourier phase microscopy for investigation of biological structures and dynamics. Opt. Lett. **29**(21), 2503–2505 (2004).
3. Allersma, M.W., Gittes, F., deCastro, M.J., Stewart, R.J., Schmidt, C.F.: Two-dimensional tracking of ncd motility by back focal plane interferometry. Biophys. J. **74**(2), 1074–1085 (1998)
4. Pralle, A., Prummer, M., Florin, E.L., Stelzer, E.H.K., Horber, J.K.H.: Three-dimensional high-resolution particle tracking for optical tweezers by forward scattered light. Microsc. Res. Tech. **44**(5), 378–386 (1999)
5. Denk, W., Webb, W.W.: Optical measurement of picometer displacements of transparent microscopic objects. Appl. Opt. **29**(16), 2382–2391 (1990)
6. Svoboda, K., Schmidt, C.F., Schnapp, B.J., Block, S.M.: Direct observation of kinesin stepping by optical trapping interferometry. Nature **365**(6448), 721–727 (1993)
7. Weber, I.: Reflection interference contrast microscopy. Biophotonics Pt B Methods Enzymol. 361, 34–47 (2003)
8. Limozin, L., Sengupta, K.: Quantitative reflection interference contrast microscopy (RICM) in soft matter and cell adhesion. Chemphyschem **10**(16), 2752–2768 (2009)
9. Zilker, A., Engelhardt, H., Sackmann, E.: Dynamic reflection interference contrast (Ric-) microscopy – a new method to study surface excitations of cells and to measure membrane bending elastic-moduli. J. Phys. **48**(12), 2139–2151 (1987)
10. Monzel, C., Fenz, S.F., Merkel, R., Sengupta, K.: Probing biomembrane dynamics by dual-wavelength reflection interference contrast microscopy. Chemphyschem **10**(16), 2828–2838 (2009)
11. Braun, D., Fromherz, P.: Fluorescence interference-contrast microscopy of cell adhesion on oxidized silicon. Appl. Phys. A Mater. **65**(4–5), 341–348 (1997)
12. Lambacher, A., Fromherz, P.: Luminescence of dye molecules on oxidized silicon and fluorescence interference contrast microscopy of biomembranes. J. Opt. Soc. Am. B Opt. Phys. **19**(6), 1435–1453 (2002)
13. Marki, I., Bocchio, N.L., Geissbuehler, S., Aguet, F., Bilenca, A., Lasser, T.: Three-dimensional nano-localization of single fluorescent emitters. Opt. Express **18**(19), 20263–20272 (2010)
14. Wax, A., Yang, C.H., Backman, V., Badizadegan, K., Boone, C.W., Dasari, R.R., Feld, M.S.: Cellular organization and substructure measured using angle-resolved low-coherence interferometry. Biophys. J. **82**(4), 2256–2264 (2002)
15. Pyhtila, J.W., Graf, R.N., Wax, A.: Determining nuclear morphology using an improved angle-resolved low coherence interferometry system. Opt. Express **11**(25), 3473–3484 (2003)
16. Dubois, A., Grieve, K., Moneron, G., Lecaque, R., Vabre, L., Boccara, C.: Ultrahigh-resolution full-field optical coherence tomography. Appl. Opt. **43**(14), 2874–2883 (2004)

17. Schnars, U., Juptner, W.P.O.: Digital recording and numerical reconstruction of holograms. Meas. Sci. Technol. **13**(9), R85–R101 (2002)
18. Rappaz, B., Marquet, P., Cuche, E., Emery, Y., Depeursinge, C., Magistretti, P.J.: Measurement of the integral refractive index and dynamic cell morphometry of living cells with digital holographic microscopy. Opt. Express **13**(23), 9361–9373 (2005)
19. Su, T.W., Isikman, S.O., Bishara, W., Tseng, D., Erlinger, A., Ozcan, A.: Multi-angle lensless digital holography for depth resolved imaging on a chip. Opt. Express **18**(9), 9690–9711 (2010)
20. Seo, S., Isikman, S.O., Sencan, I., Mudanyali, O., Su, T.W., Bishara, W., Erlinger, A., Ozcan, A.: High-throughput lens-free blood analysis on a chip. Anal. Chem. **82**(11), 4621–4627 (2010)

Part IV
Interferometry of Biological Tissues

Chapter 10
Light Propagation in Tissue

From the point of view of biomedical applications, the propagation of light in biological tissues is a combination of good news and bad news. The good news is that tissue is highly translucent, allowing light to penetrate far inside. This property has been experienced by anyone who has covered a flashlight with their hand at night and seen it glow red. The bad news is that it is heterogeneous, distorting wave fronts and scattering light in many directions. This is experienced when trying to see a splinter that is lodged just below the surface of the skin where it shows up as a dim blur. This good news/bad news is at the core of a clear split in optical behavior in tissue. Image-bearing light, also called ballistic or coherent light, is attenuated rapidly as light propagates in tissue. Yet scattered photons continue to scatter and transport across large distances. The image-bearing light can be accessed in imaging applications by using a coherence detector (interferometry or holography) to see with micron-scale resolution up to a millimeter inside tissue. On the other hand, diffusing light can be utilized by solving the diffusion equation and reconstructing objects many centimeters inside. Between these two methods, there is a trade-off between depth of penetration and spatial resolution. Coherent light can provide cellular resolution, but only near the surface. Diffusing light provides only coarse imaging, but penetrates deep inside tissue.

Diffusing light is a central feature of light propagation in tissue, and its depth of penetration, combined with low tissue absorption, make it a dominant effect in any optical technique. Most coherence-domain imaging approaches must deal with a high-amplitude diffuse background. It makes all of the coherence approaches intrinsically "high-background" that require extremely high discrimination to extract the small coherent fraction. Although the diffusing light is valuable in its own right as a means to perform deep tomography in tissue, because diffusing light has very low spatial coherence, it is not commonly used for interferometric detection. Therefore, this chapter does not cover the imaging applications of diffuse light, which has been covered extensively elsewhere [1].

Coherent light that is detected in a "coherence gate" is minimally scattered by the tissue. The scattering consists of low-angle forward scattering, which retains much of the coherence properties of the light, plus one or a few high-angle

D.D. Nolte, *Optical Interferometry for Biology and Medicine*, Bioanalysis 1,
DOI 10.1007/978-1-4614-0890-1_10, © Springer Science+Business Media, LLC 2012

scattering events. The number of such events needs to be small enough such that the phase along the optical path is not scrambled too much. Coherent light is a central focus of this book and this chapter.

10.1 Origins of Light Scattering in Tissue

Light is scattered by heterogeneity. Any change in optical property that occurs over a length scale comparable to a wavelength or smaller becomes a source of scattering. The amplitude, angle, and wavelength dependence of such scattering depends on factors such as the size and shape of the scatterer, the density of the scatterers, and the relative refractive index difference between the scatterer and the surrounding medium. The chief question of optical propagation is how much light is scattered, and how much remains unscattered (or weakly scattered) as light penetrates tissue?

Light in uniform tissue is attenuated with an approximately constant probability per length given by the attenuation coefficient μ. This behavior is captured in the general expression

$$\frac{\mathrm{d}I}{\mathrm{d}s} = -\mu I \tag{10.1}$$

which has the simple exponential decay solution

$$I(s) = I_0 \mathrm{e}^{-\mu s} \tag{10.2}$$

where s is the path length and I_0 is the initial intensity. The simple exponential decay is called the Lambert–Beer law with the characteristic decay length given by

$$L = \frac{1}{\mu} \tag{10.3}$$

The expression in (10.1) is generic in that it pertains to many different ways of characterizing light propagation. For instance, the intensity in (10.1) could be the coherent fraction of light that is being attenuated as photons are scattered out of the propagating beam. Or it could be a diffuse fraction, decaying as photons migrate away. It can also be either a scalar or a vector quantity, with the vector quantity describing momentum flux or polarization. The decay of coherence, polarization, or diffuse intensity can all have different attenuation rates μ_i dictated by the different scattering statistics.

10.1.1 Scattering Phase Functions

Before tackling the attenuation of light in optically dense media that has multiple scattering, it helps to look at light attenuation in dilute media that is dominated by single scattering. In this case, the extinction coefficient is related to the scattering cross section of the scatterers C_{sca} through

$$\mu = N_V C_{sca} \tag{10.4}$$

for a volume density N_V of scatterers. However, in addition to total extinction, it is also valuable to understand the likelihood of scattering into an angle θ. The probability function is called $p(\theta)$, and is known as the phase function. The phase function is normalized by

$$\begin{aligned} \frac{1}{4\pi}\int p(\theta)\,\mathrm{d}\Omega &= 1 \\ \frac{1}{2}\int p(\theta)\sin\theta\,\mathrm{d}\theta &= 1 \end{aligned} \tag{10.5}$$

for scattering into 4π solid angle. For a dilute collection of spheres, the phase function is obtained from the Mie Scattering factors

$$\begin{aligned} p(\theta) &= 4\pi\frac{(\mathrm{d}\sigma/\mathrm{d}\Omega)(\theta)}{2\pi\int(\mathrm{d}\sigma/\mathrm{d}\Omega)\sin\theta\,\mathrm{d}\theta} \\ &= \frac{(|S_1|^2+|S_2|^2)}{\frac{1}{2}\int(|S_1|^2+|S_2|^2)\sin\theta\,\mathrm{d}\theta} \end{aligned} \tag{10.6}$$

This is shown in Fig. 10.2 for particles with $n_{\text{rel}} = 1.04$, $\lambda = 633$ nm in a medium with index $n_{\text{m}} = 1.33$ and $a = 1$ μm. Fig. 10.1 shows the log intensity on a polar plot, with a very strong forward scattering lobe. In Fig. 10.2 the 1 μm radius particle scattering phase function $p(\theta)$ is compared with a particle with a half micron radius. Both particles show strong forward scattering with many secondary lobes.

A large particle like the cell nucleus has strong forward scattering, while a small particle has more isotropic scattering. For this reason, forward scattering is dominated by larger organelles like the nucleus, while backscattering is dominated by smaller organelles like the mitochondria. However, the organellar backscatter tends to be structureless, while the nuclear backscatter can show angular dependence similar to that in Fig. 10.2, with periodicity related to the nuclear size [2]. This is used in light scattering spectroscopy (LSS) to detect small changes in nuclear size that are too small to be observed using conventional histology [3, 4]. An example of the backward angular scattering from a particle with a radius of 2.5 μm compared to 2.75 μm (10% larger) is shown in Fig. 10.3. There is a clear change in the oscillating

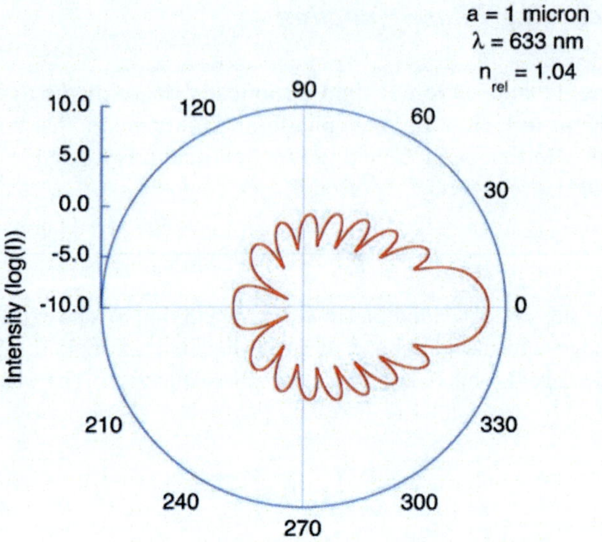

Fig. 10.1 Polar plot of log-intensity as a function of angle for a particle with radius 1 μm in $n_m = 1.33$ with $n_{rel} = 1.04$ at $\lambda = 633$ nm. Even for this small particle, the scattering is strongly in the forward direction

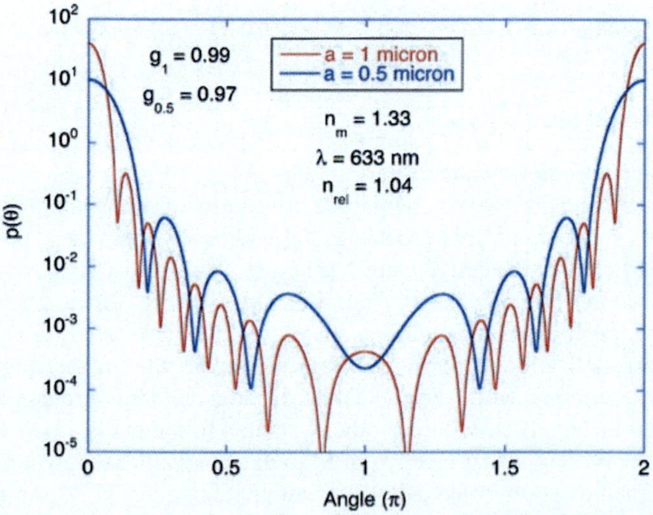

Fig. 10.2 Phase functions at a wavelength of 633 nm for two radii: a particle with a radius of 0.5 μm, a larger particle with a radius of 1 μm. The anisotropy factors are $g = 0.99$ and $g = 0.97$, respectively

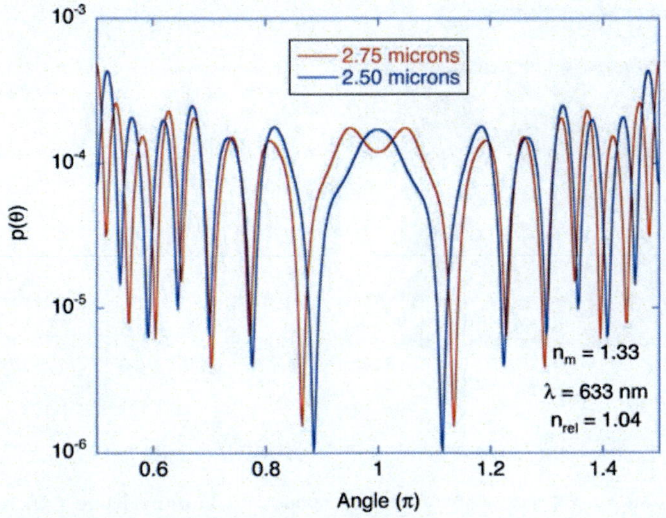

Fig. 10.3 Phase function angular dependence for backscattering from a 5 μm diameter spherical particle compared with a sphere 10% larger. The angular dependence of the backscatter is sensitive to size changes that are below the resolution of histology

pattern with increasing angle, even though this change is on the edge of microscopic resolution. However, biological tissue is very complex, with significant morphology that can affect the interpretation of angular backscattering [5].

Mie backscattering for a given particle exists in a two-dimensional space of angle and wavelength. While LSS explores angular dependence, partial wave spectroscopy (PWS) explores spectral dependence. An example of Mie backscattering is shown in Fig. 10.4a for a 2.5 μm radius particle in water with a relative index of $n_{\mathrm{rel}} = 1.04$. There are clear oscillations in both angle and wavelength. In the direct backscattering direction, a spectrum of the backscattered light is shown in Fig. 10.4b with clear spectral oscillations related to the size of the particle. In microscopy data, which arises from an imaging approach, the Mie backscattering theory captures the general phenomenon, but quantitative details differ that need to be addressed [6]. This PWS approach can be used to detect subtle size changes in nuclear sizes that are indicative of precancerous conditions [7, 8].

An important parameter that characterizes the degree of forward scattering is called the anisotropy factor g, which is the average of cos (θ) over the phase function

$$g = \langle \cos\theta \rangle = \frac{\int p(\theta)\cos\theta \, \mathrm{d}\Omega}{\int p(\theta) \, \mathrm{d}\Omega} \tag{10.7}$$

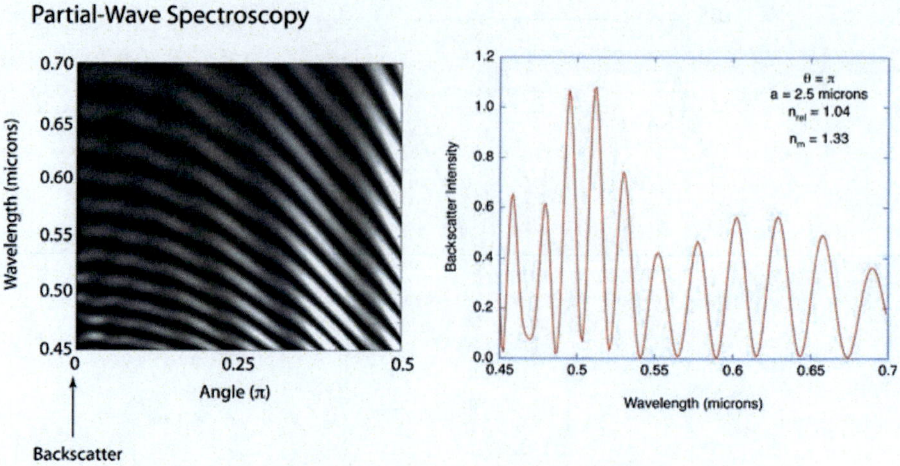

Fig. 10.4 Partial wave spectroscopy of a 2.5 μm radius particle with relative index $n_{rel} = 1.04$ in $n_m = 1.33$. The spectral oscillation period is related to the size of the spherical target

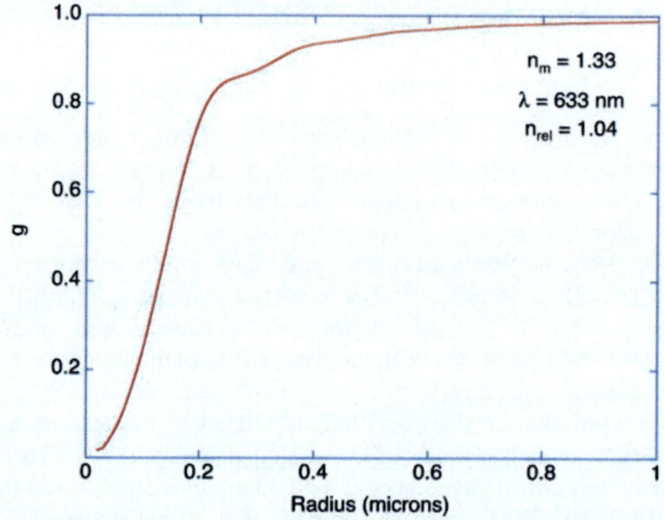

Fig. 10.5 Anisotropy factor as a function of sphere radius for relative index of 1.04 in a medium with $n_m = 1.33$. The anisotropy rises rapidly above a radius of 0.2 μm

The anisotropy factor g as a function of sphere radius is plotted in Fig. 10.5 for a relative index of 1.04. The anisotropy rises quickly around 0.3 μm radius to $g = 0.9$. Many tissue types have anisotropy factors around $g = 0.9$. However, light propagation through tissue cannot be described as single-scattering from spherical particles. Instead, more general principles need to be developed to understand tissue anisotropy.

10.1.2 Henyey–Greenstein Phase Function

Tissue cannot be represented as a dilute target. Cells are densely packed in tissue, and are filled with a high density of organelles. Therefore, Mie coefficients are not necessarily the most appropriate means to describe the probability distribution. A probability function that has matched experimental data relatively well is the Henyey–Greenstein function

$$p(\theta) = \frac{1 - g^2}{(1 + g^2 - 2g\ \cos\theta)^{3/2}}, \tag{10.8}$$

where g is the average cosine. Henyey–Greenstein distributions are shown in Fig. 10.6a for $g = 0.7$, 0.8, and 0.9. A comparison of the Henyey–Greenstein function with Mie scattering for $g = 0.99$ on a particle with $a = 1$ μm in water with a relative index $n_{\mathrm{rel}} = 1.04$ is shown in Fig. 10.6b. Many tissues have high anisotropy with g greater than 0.8. For instance bovine muscle, pig brain, and white chicken muscle have g in the range 0.94–0.97, for human dermis it is 0.82, and for the gray and white matter of the brain it is 0.96 and 0.83, respectively, although these tissues can have large variability.

The Henyey–Greenstein function is expressed as an expansion of Legendre polynomials through

$$p(\theta) = \sum_{n=0}^{\infty} (2n + 1) x_n P_n(\cos\theta) \tag{10.9}$$

where the coefficients x_n are

$$x_n = \frac{1}{2} \int p(\theta) P_n(\cos\theta) \sin\theta\, \mathrm{d}\theta \tag{10.10}$$

For large anisotropy ($g > 0.7$) the coefficients in the expansion of $p(\theta)$ have approximately the simple form

$$x_n \sim \exp(-(1 - g)n) \tag{10.11}$$

which causes the low-n (forward-scattering) expansion to dominate.

10.1.3 Absorption, Scattering, and Extinction

When a medium has absorption in addition to scattering, the extinction is the combination of both

$$\mu_t = \mu_a + \mu_s \tag{10.12}$$

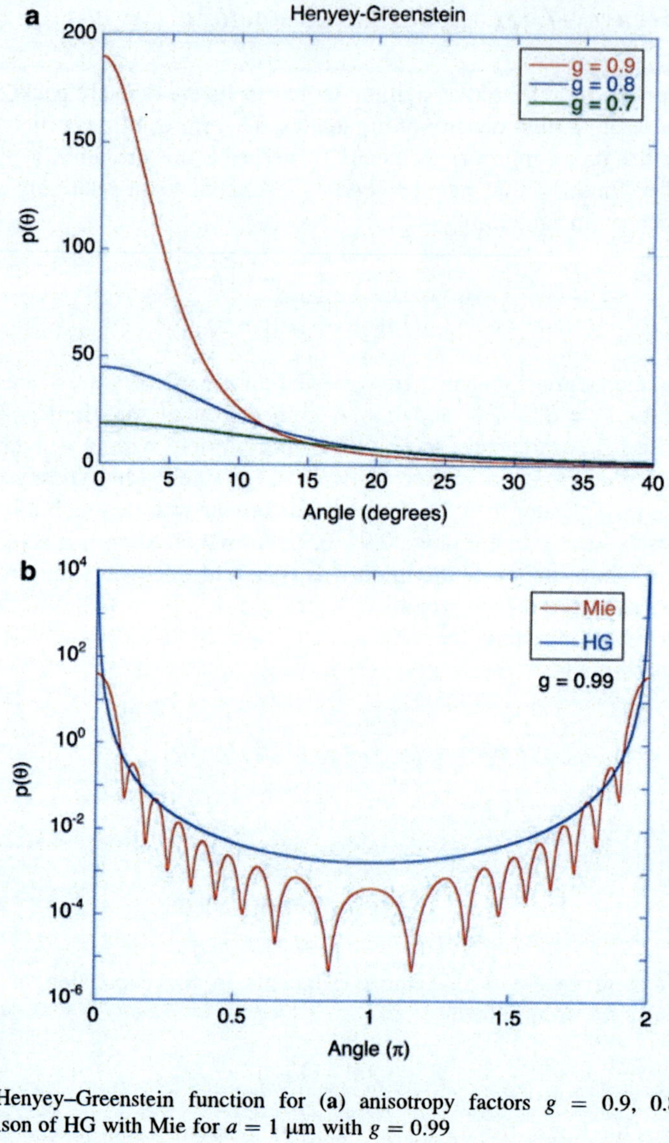

Fig. 10.6 Henyey–Greenstein function for (**a**) anisotropy factors $g = 0.9$, 0.8, and 0.7. (**b**) Comparison of HG with Mie for $a = 1$ μm with $g = 0.99$

and the mean free path is

$$L_t = \frac{1}{\mu_a + \mu_s} = \frac{1}{\mu_t} \tag{10.13}$$

These are dimensional quantities. Dimensionless quantities that relate to light propagation in the presence of both absorption and scattering are the albedo

$$a = \frac{\mu_s}{\mu_t} = \frac{\mu_s}{\mu_a + \mu_s} \tag{10.14}$$

and the optical thickness

$$\tau = \int_0^d \mu_t \mathrm{d}s \tag{10.15}$$

A system with high scattering relative to absorption has a high albedo, while high absorption means a low albedo. The optical thickness is the thickness of a sample (or a depth inside the sample) in units of scattering lengths.

10.2 Photon Transport

Tissue is optically thick, which means that typical biological samples have physical scales that are many times the photon scattering length. Therefore, any light that penetrates to a significant depth inside tissue experiences multiple scattering. In this situation, the photons "diffuse," and one can define average intensities within the tissue through the effect of scattering on energy and momentum flux.

The transport equation for a flux J under the condition of multiple scattering is

$$\frac{\mathrm{d}J(\vec{r},\hat{s})}{\mathrm{d}s} = -\mu_t J(\vec{r},\hat{s}) + \frac{\mu_s}{4\pi}\int_{4\pi} p(\hat{s}',\hat{s})J(\vec{r},\hat{s}')\,\mathrm{d}\Omega' \tag{10.16}$$

where $J(\vec{r},\vec{s})$ is the irradiance per solid angle propagating in the direction $\hat{s}$, and the phase function is $p(\hat{s}',\hat{s}) = p(\theta)$ for azimuthally isotropic scattering with $\cos\theta = \hat{s}'\cdot\hat{s}$. The irradiance is related to the intensity through

$$I(\vec{r}) = \int_{4\pi} J(\vec{r},\hat{s})\,\mathrm{d}\Omega \tag{10.17}$$

An approximate solution to the transport equation of (10.16) can be made using the small-angle approximation (SAA). This approximation is appropriate for high anisotropy factor g in which most scattering is small-angle forward scattering. The flux in (10.16) is replaced by an angular intensity and is divided by μ_t to yield

$$\frac{\mathrm{d}I(\tau,\theta)}{\mathrm{d}\tau} = -I(\tau,\theta) + \frac{a}{4\pi}\int_{4\pi} p(\theta',\theta)I(\tau,\theta')\,\mathrm{d}\Omega' \tag{10.18}$$

where the phase function is

$$p(\theta', \theta) = \sum_{n=0}^{\infty} x_n P_n(\cos\theta) P_n(\cos\theta') \tag{10.19}$$

and the coefficients x_n are given in (10.10). The solution of the differential equation in (10.18) has the form

$$I(\tau, \theta) = \sum_{n=0}^{\infty} b_n(\tau) P_n(\cos\theta) \tag{10.20}$$

When this is plugged into (10.18) the expression for the b_n is obtained as

$$\frac{db_n(\tau)}{d\tau} = -c_n b_n(\tau) \tag{10.21}$$

where

$$c_n = 1 - a x_n \tag{10.22}$$

and a is the albedo. The solutions to (10.21) are single exponentials

$$b_n(\tau) = A_n \exp(-c_n \tau), \tag{10.23}$$

where the values A_n are found from the initial conditions. The intensity is then

$$I(\tau, \theta) = \sum_{n=0}^{\infty} A_n e^{-c_n \tau} P_n(\cos\theta) \tag{10.24}$$

For plane-wave illumination at normal incidence,

$$A_n = \frac{2n+1}{4\pi} \tag{10.25}$$

and therefore

$$I(\tau, \theta) = \frac{I_0}{2\pi} \sum_{n=0}^{\infty} \frac{2n+1}{2} e^{-c_n \tau} P_n(\cos\theta) \tag{10.26}$$

The diffuse angular scattering in the SAA is obtained by subtracting the coherent component from the total intensity. The diffuse intensity is thus

$$I_d(\tau, \theta) = \frac{I_0}{2\pi} \sum_{n=0}^{\infty} \frac{2n+1}{2} (e^{-c_n \tau} - e^{-\tau}) P_n(\cos\theta) \tag{10.27}$$

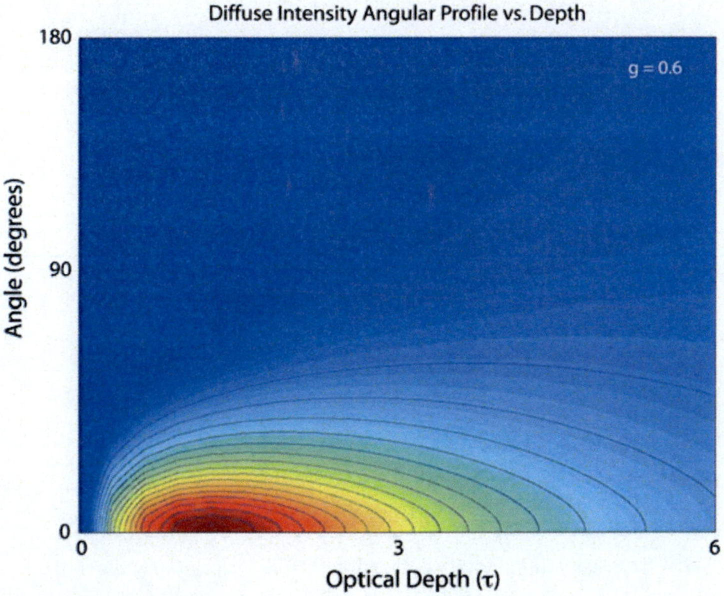

Fig. 10.7 Angular profile of the diffuse intensity as a function of optical depth for $g = 0.6$. Most of the light forward scatters, with a peak in the diffuse intensity in the forward direction at an optical depth of approximately 1.5

The diffuse intensity is plotted in Fig. 10.7 as a function of optical depth and propagation angle from 0 to 180° for $g = 0.6$. The diffuse light has a strong forward character with increasing backscattering at increasing depth. The forward scattered diffuse light increases with increasing depth to about $\tau = 1.5$ and then decays for deeper optical depths. The forward-scattered diffuse light $I_d(\tau, 0°)$ is plotted in Fig. 10.8 for $g = 0.9$ and $g = 0.8$. The forward scattering is stronger for higher anisotropy.

The solution in (10.27) is for a plane-wave illumination with no spatial profile. The total flux in the propagation direction is the fluence that penetrates ballistically to depth into the tissue. This net forward fluence is obtained by projecting the fluence along the propagation axis and integrating over all angles. The result is

$$\begin{aligned} I(\tau) &= \int_{4\pi} I(\tau,\theta)\cos\theta \, \mathrm{d}\Omega \\ &= I_0 \int_0^{\pi} \sum_{n=0}^{\infty} \frac{2n+1}{2} \mathrm{e}^{-c_n \tau} P_n(\cos\theta)\cos\theta \, \sin\theta \, \mathrm{d}\theta \\ &= I_0 \exp(-(\mu_a + (1-g)\mu_s)z) \\ &= I_0 \exp(-(\mu_a + \mu_s')z) \end{aligned} \tag{10.28}$$

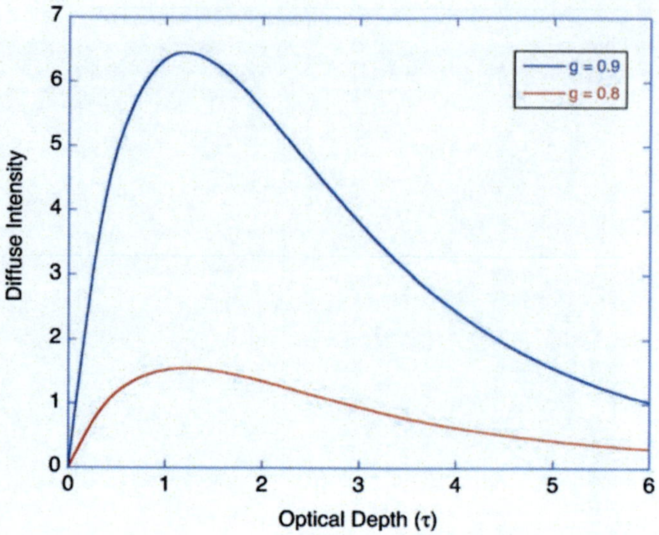

Fig. 10.8 Diffuse intensity in the forward direction $\theta = 0°$ for $g = 0.9$ and $g = 0.8$

where $\tau = (\mu_a + \mu_s)z$ and

$$\mu_s' = \mu_s(1 - g) \tag{10.29}$$

is the reduced scattering coefficient. The reduced scattering coefficient is an important feature of photon transport in highly anisotropic scattering media (that have large values of the anisotropy factor g), such as in living tissue. Most photons scatter in the forward direction, which preserves much of the directionality of the initial plane wave, even in the presence of multiple scattering. For large anisotropies, light penetrates much deeper than $1/\mu_s$. The results of the integration in (10.28) are shown in Fig. 10.9 for $g = 0.6$ as the curve marked with μ' which has a noticeably slower decay than the curve marked with μ. The depths achieved by optical coherence tomography are made possible by coherent detection of light that is multiply scattered in the forward direction, retaining partial coherence, and aided by spatial filtering provided by the tight focus of the incident probe beam and collection optics.

10.2.1 Diffuse Surface Reflectance

The small-angle scattering approximation (SAA) neglects backscatter from the surface of the tissue, which is clearly not physical. Indeed, diffuse reflectance is one of the most important experimental techniques in tissue optics [9, 10].

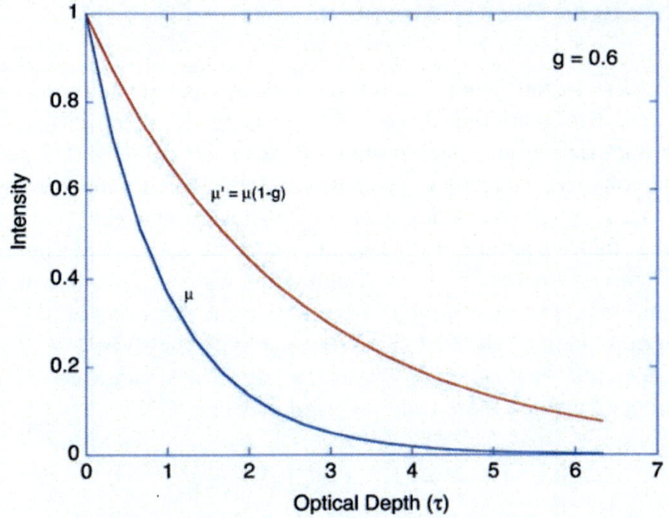

Fig. 10.9 Intensity vs. optical depth for the ballistic fraction as well as for the total forward fluence integrated over theta

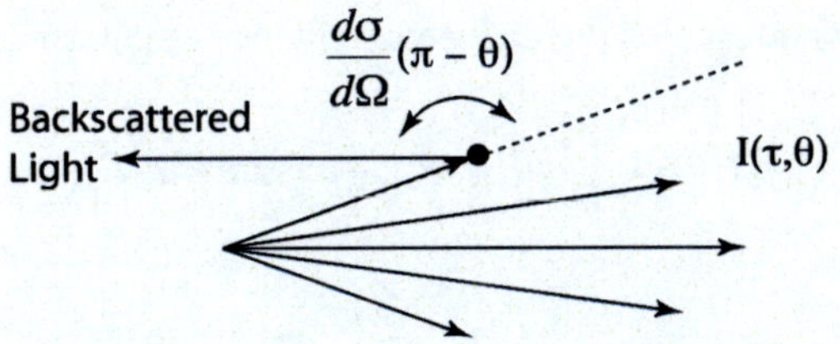

Fig. 10.10 Small-angle approximation with a single backscattering event. The backscattered photon is forward scattered back to the sample surface where it contributes the diffuse reflectance of the sample

This deficiency in the SAA can be partially remedied by considering small-angle scattering with a single high-angle scattering event. The backscattered intensity is

$$I_b(\tau) = 2\pi\mu_s N_V e^{-(1-g)\tau} \int_0^{\pi} I(\tau,\theta) \left[\frac{d\sigma}{d\Omega}(\pi-\theta)\right] \sin\theta \, d\theta \tag{10.30}$$

where $I(\tau,\theta)$ is from (10.26) and N_V is the number density of particles. This expression integrates backscattering from all internal angles and assumes propagation back to the surface with a reduced extinction coefficient. The resultant backscattered intensity as a function of optical depth is approximately

$$I_b(\tau) = I_0 N_A(\tau)\sigma_b e^{-(1-g)2\tau} \tag{10.31}$$

where σ_b is the total backscattering cross-section and $N_A(\tau)$ is the effective areal density of scatterers (Fig. 10.10).

10.3 Enhanced Backscattering

Enhanced backscattering (EBS) occurs when coherent light illuminates a random medium. In the backscattering direction, there is an enhanced diffuse reflectance that has an angular width of approximately $\Delta\theta \approx \lambda/\ell'$, where ℓ' is the transport length. This enhanced reflectance is caused by constructive interference of time-reversed paths. The light-scattering geometry is illustrated in Fig. 10.11. Consider a single ray that follows a multiply scattered path and emerges from the surface at a distance $\Delta\vec{r}$ with a k-vector $\vec{k}^{\text{out}}$. If the illuminating wave is a plane wave, then there must be a time-reverse path in which a ray enters at the exit point of the first ray, and exits at entrance point of the first ray to emerge with the same $\vec{k}^{\text{out}}$. The phase of both paths are equal and interfere constructively in the backward direction. The phase difference between these time-reversed paths is

$$\Delta\phi = (\vec{k}^{out} + \vec{k}^{in})(\vec{r}_2 - \vec{r}_1) \tag{10.32}$$

For perfect backscattering, the phase difference vanishes. With increasing angle of diffuse reflectance, and for larger deflections of the entering and exiting paths, the phase difference grows, and the constructive enhancement decreases to the incoherent background.

The angular dependence of the diffuse reflectance is determined by

$$I(\theta) = \int\!\!\int p(\vec{r})\exp(-ik\vec{r}\,\sin\theta)d^2r \tag{10.33}$$

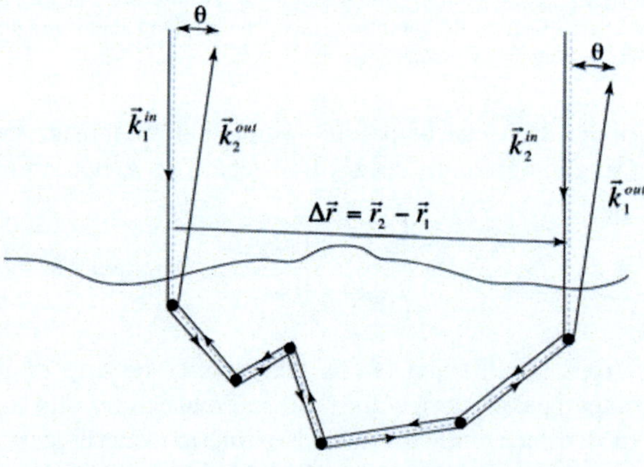

Fig. 10.11 An incident plane wave can execute two multiply scattered time-reversed paths. The backscattered waves have the same phase and interfere constructively to produce an enhanced backscatter signal

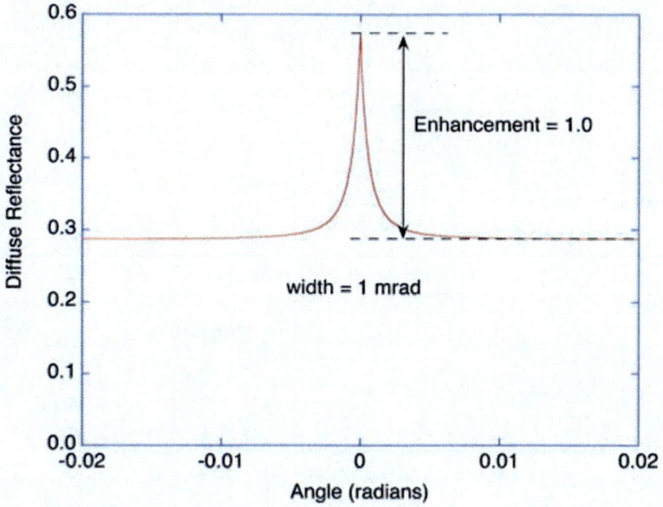

Fig. 10.12 Enhanced backscatter (EBS) angular distribution from a diffusely scattering medium. The full-width at half maximum is approximately 1 millirad

where $p(\vec{r})$ is the probability of a backscattered photon exiting at a distance r from where it entered the sample. In the diffusion approximation, for which the sample thickness is much larger that a mean transport length ℓ', the angular dependence of the diffuse reflectance is [11]

$$I_{\text{EBS}}(\theta) \approx \frac{3}{8\pi}\left[1 + \sqrt{2} + \frac{1}{(1 + k\theta\ell')^2}\left(1 + \frac{1 - \exp\sqrt{2}k\theta\ell'}{k\theta\ell'}\right)\right] \tag{10.34}$$

which has a full-width at half maximum

$$\Delta\theta \approx \frac{1.8}{k\ell'} \approx \frac{\lambda}{\pi\ell'} \tag{10.35}$$

which is extremely small, typically in the milliradian range for biological tissues. An example of the EBS is shown in Fig. 10.12. The enhancement is a factor of two in this ideal case. The narrow angular width of the EBS makes it difficult to observe this effect in practice, although it has been accomplished for relatively dilute diffusive media [12].

A significant advance on the detection of EBS was achieved by the use of probe light with short spatial coherence [13]. The use of partial coherence modifies (10.33) by modulating $p(\vec{r})$ by the second-order degree of coherence $g^{(2)}(\vec{r})$

$$I(\theta) = \iint p(\vec{r})g^{(2)}(\vec{r})\exp(-ik\vec{r}\sin\theta)d^2r \tag{10.36}$$

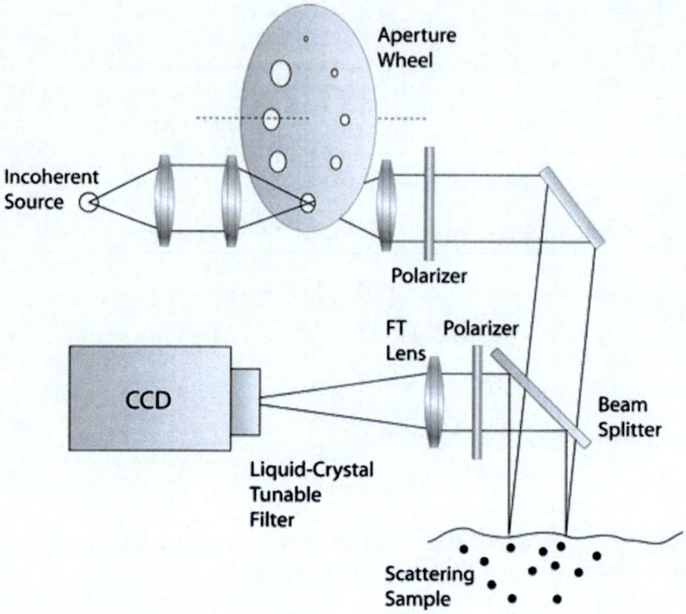

Fig. 10.13 Low-coherence enhanced backscattering (LEBS) experimental arrangement. The aperture wheel controls the spatial coherence at the sample surface. The Fourier-transform (FT) lens maps the angular distribution of intensity to the surface of the CCD chip. The liquid-crystal tunable filter selects wavelengths for light-scattering spectroscopy (LSS) [14]

with the requirement that the spatial coherence length must be much greater than the transport length ℓ' so that the diffusion approximation to $p(\vec{r})$ may be used. Equation (10.36) is a Fourier transform, so that the diffuse reflectance for low-coherence enhanced backscattering (LEBS) becomes a convolution [14]

$$I_{\mathrm{LEBS}}(\theta) = \mathrm{FT}(p(\vec{r}) \cdot g^{(2)}(\vec{r})) = I_{\mathrm{EBS}}(\theta) * G^{(2)}(\theta) \tag{10.37}$$

with $G^{(2)}(\theta)$ as the angular Fourier transform of $g^{(2)}(\vec{r})$. The angular diffuse reflectance is the convolution of $I_{\mathrm{EBS}}(\theta)$ (which is extremely narrow) with the Fourier transform of the second-order coherence function, which can be much broader when short coherence is used. Therefore, the convolution broadens the enhanced backscattered peak, which can be adjusted by controlling the spatial coherence length in the experiment.

An experimental arrangement for low-coherence enhanced backscatter (LEBS) is shown in Fig. 10.13. An incoherent source is spatially filtered by the aperture wheel to control the spatial coherence length at the sample surface. The scattered light is passed through a Fourier transform lens to the surface of the CCD chip. The polarizers and liquid-crystal tunable filter select polarizations and wavelengths for LSS analysis. The shorter coherence length significantly suppresses speckle.

When the coherence length ξ_{sc} is smaller than the transport length ℓ', the angular width of the enhanced backscatter peak is

$$\Delta\theta \approx \frac{\lambda}{\xi_{sc}} \tag{10.38}$$

although the details depend on the form of the second-order coherence function and on $p(\vec{r})$.

Low-coherence enhanced backscatter has potential for biomedical optical diagnosis applications. The sensitivity of the backscattered diffuse reflectance to the morphological details of the scattering sample exceeds the ability of conventional histology to detect sub-wavelength changes in the structure of nuclei in cells in tissue. The low-coherence backscattering approach has been applied to the early detection of colorectal cancer [15] and pancreatic cancer [16, 17].

10.4 Multiple Dynamic Light Scattering

Dynamic light scattering is a powerful characterization tool for remotely extracting dynamic information from inside media, such as internal displacements, diffusivities and velocities [18, 19]. In the dilute limit, in which scattered light is only scattered once, dynamic light scattering is called quasi-elastic light scattering (QELS) [20]. In biological applications, QELS is useful for optically thin systems, such as cell monolayer cultures, in which it can provide information on diffusion and directed motion in the cytosol [21, 22] and nucleus [23] and undulations of the plasma membrane [24–28]. However, as the optical thickness of a sample increases beyond one optical transport length, multiple scattering becomes significant, and the diffusive character of light begins to dominate [29–31]. The challenge, and the goal, is to continue to be able to extract meaningful dynamics from inside living tissues.

10.4.1 Diffusing Wave Spectroscopy

Even in the presence of multiple scattering, dynamic light scattering information can still be extracted [32, 33] using diffusing wave spectroscopy (DWS) [34, 35] and an equivalent formulation of multi-scattering dynamic light scattering called diffuse correlation spectroscopy (DCS) [1, 36–38]. In biological applications, DWS has been used to assess tissue response to burns [39], brain activity [40], to monitor blood flow [1, 41], and to monitor tissue structure [42].

There are two basic principles underpinning DWS. First, each scattering event along a path contributes stochastically to the phase shift of the photon following that path. In this way, the stochastic phase modulations accumulate in the phase of the photon, and hence can be expressed as a product of stochastic exponentials. Second, all the possible paths that a photon can take must be considered in the total correlation function. The density of these possible paths are found as a solution to the photon diffusion equation.

The complex field of a single path consisting of N_k scattering events is

$$E_k(t) = \mathrm{e}^{-\mathrm{i}\omega t} \prod_{n=1}^{N_k} a_n \mathrm{e}^{\mathrm{i}\bar{q}_n \bar{r}_n(t)} \tag{10.39}$$

The total field is the combination of all paths

$$E(t) = \sum_k E_k \tag{10.40}$$

The first-order correlation function for the field fluctuations is

$$\begin{aligned} g_k^{(1)}(\tau) &= \mathrm{e}^{-\mathrm{i}\omega t} \left\langle \prod_{n=1}^{N_k} |a_n|^2 \mathrm{e}^{\mathrm{i}\bar{q}_n \Delta \bar{r}_n(\tau)} \right\rangle \\ &= \mathrm{e}^{-\mathrm{i}\omega t} \langle |a_n|^2 \rangle \exp(-\langle \bar{q}^2 \rangle \langle \Delta r^2(\tau) \rangle N_k/6) \end{aligned} \tag{10.41}$$

where the cross terms in the correlation function vanish because scattering events are assumed to be uncorrelated (an assumption that can break down in some circumstances). Because the scattering vector q_n varies with each scattering event, a mean value can be assigned to it as

$$\langle \bar{q}^2 \rangle = 2k^2 \frac{l}{\ell'} \tag{10.42}$$

where l is the scattering mean free path, and ℓ' is the transport mean free path. The number of scattering events for each path is

$$N_k = s_k / l \tag{10.43}$$

where s_k is the total path length. The kth path correlation function is then

$$g_k^{(1)}(\tau) = \mathrm{e}^{-\mathrm{i}\omega t} \langle |a_n|^2 \rangle \exp(-k^2 \langle \Delta r(\tau) \rangle s_k / 3l^*) \tag{10.44}$$

The combined correlation function of all paths is

$$g^{(1)} = \sum_k P(k) g_k^{(1)}(\tau) \tag{10.45}$$

For a continuous distribution of possible paths this is

$$g^{(1)}(\tau) = \int_0^\infty \exp\left(-\frac{k^2 s}{3l^*} \langle \Delta r^2(\tau) \rangle \right) \rho(s)\, \mathrm{d}s \tag{10.46}$$

where $\rho(s)$ is the probability density of possible paths that is obtained by solving the photon diffusion equation subject to the boundary geometry of the sample and the intensity distribution of the incident light [1].

The autocorrelation function can be reexpressed as

$$g^{(1)}(\tau) = \int_0^\infty \exp\left(-2 \frac{\tau}{\tau_0} \frac{s}{l^*} \right) \rho(s)\, \mathrm{d}s \tag{10.47}$$

where the argument $2(\tau/\tau_0)(s/l^*)$ of the exponential is the single-scattering case, multiplied by the quantity s/l^*, which is the average number of scattering events along the path. Note that longer paths lead to faster decorrelation times because more scattering events add to scramble the phase. For this reason, transmission experiments have much faster decay rates than backscattering experiments in the multiple scattering regime [34].

Selected Bibliography

Boas, D.A, Dunn A.K.: Laser speckle contrast imaging in biomedical optics. J. Biomed. Opt. **15**(1), 011109 (2010) (A recent review on speckle contrast, with an emphasis on blood flow imaging)

Durduran, T., Choe R., et al.: Diffuse optics for tissue monitoring and tomography. Rep. Prog. Phys. **73**(7), 076701 (2010) (A recent review of diffusing wave optics in tissues)

Kokhanovsky A.A.: Optics of Light Scattering Media: Problems and Solutions. Springer, New York (2001) (A good source for the small-angle approximation and Legendre polynomial expansion)

Pine, D.J., Weitz D.A., et al.: Diffusing-wave spectroscopy. Phys. Rev. Lett. **60**(12), 1134–1137 (1988) (The paper that introduced diffusing-wave spectroscopy, which is a fluctuation spectroscopy)

Wang L.V., Wu H.: Biomedical Optics: Principles and Imaging. Wiley, Hoboken (2007) (This is becoming the standard textbook of biomedical optics)

References

1. Durduran, T., Choe, R., Baker, W.B., Yodh, A.G.: Diffuse optics for tissue monitoring and tomography. Rep. Prog. Phys. **73**(7), 076701 (2010)
2. Perelman, L.T., Backman, V., Wallace, M., Zonios, G., Manoharan, R., Nusrat, A., Shields, S., Seiler, M., Lima, C., Hamano, T., Itzkan, I., Van Dam, J., Crawford, J.M., Feld, M.S.: Observation of periodic fine structure in reflectance from biological tissue: a new technique for measuring nuclear size distribution. Phys. Rev. Lett. **80**(3), 627–630 (1998)

3. Backman, V., Wallace, M.B., Perelman, L.T., Arendt, J.T., Gurjar, R., Muller, M.G., Zhang, Q., Zonios, G., Kline, E., McGillican, T., Shapshay, S., Valdez, T., Badizadegan, K., Crawford, J.M., Fitzmaurice, M., Kabani, S., Levin, H.S., Seiler, M., Dasari, R.R., Itzkan, I., Van Dam, J., Feld, M.S.: Detection of preinvasive cancer cells. Nature **406**(6791), 35–36 (2000)
4. Fang, H., Ollero, M., Vitkin, E., Kimerer, L.M., Cipolloni, P.B., Zaman, M.M., Freedman, S. D., Bigio, I.J., Itzkan, I., Hanlon, E.B., Perelman, L.T.: Noninvasive sizing of subcellular organelles with light scattering spectroscopy. IEEE J. Sel. Top. Quantum Electron. **9**(2), 267–276 (2003)
5. Lau, C., Scepanovic, O., Mirkovic, J., McGee, S., Yu, C.C., Fulghum, S., Wallace, M., Tunnell, J., Bechtel, K., Feld, M.: Re-evaluation of model-based light-scattering spectroscopy for tissue spectroscopy. J. Biomed. Opt. **14**(2), 024031 (2009)
6. Liu, Y., Li, X., Kim, Y.L., Backman, V.: Elastic backscattering spectroscopic microscopy. Opt. Lett. **30**(18), 2445–2447 (2005)
7. Subramanian, H., Pradhan, P., Liu, Y., Capoglu, I.R., Li, X., Rogers, J.D., Heifetz, A., Kunte, D., Roy, H.K., Taflove, A., Backman, V.: Optical methodology for detecting histologically unapparent nanoscale consequences of genetic alterations in biological cells. Proc. Natl. Acad. Sci. U.S.A. **105**(51), 20118–20123 (2008)
8. Subramanian, H., Roy, H.K., Pradhan, P., Goldberg, M.J., Muldoon, J., Brand, R.E., Sturgis, C., Hensing, T., Ray, D., Bogojevic, A., Mohammed, J., Chang, J.S., Backman, V.: Nanoscale cellular changes in field carcinogenesis detected by partial wave spectroscopy. Cancer Res. **69** (13), 5357–5363 (2009)
9. Farrell, T.J., Patterson, M.S., Wilson, B.: A diffusion-theory model of spatially resolved, steady-state diffuse reflectance for the noninvasive determination of tissue optical-properties invivo. Med. Phys. **19**(4), 879–888 (1992)
10. Kienle, A., Lilge, L., Patterson, M.S., Hibst, R., Steiner, R., Wilson, B.C.: Spatially resolved absolute diffuse reflectance measurements for noninvasive determination of the optical scattering and absorption coefficients of biological tissue. Appl. Opt. **35**(13), 2304–2314 (1996)
11. Akkermans, E., Wolf, P.E., Maynard, R.: Coherent backscattering of light by disordered media - analysis of the peak line-shape. Phys. Rev. Lett. **56**(14), 1471–1474 (1986)
12. Wolf, P.E., Maret, G., Akkermans, E., Maynard, R.: Optical coherent backscattering by random-media - an experimental-study. J. Phys. **49**(1), 63–75 (1988)
13. Kim, Y.L., Liu, Y., Turzhitsky, V.M., Roy, H.K., Wali, R.K., Backman, V.: Coherent backscattering spectroscopy. Opt. Lett. **29**(16), 1906–1908 (2004)
14. Turzhitsky, V., Rogers, J.D., Mutyal, N.N., Roy, H.K., Backman, V.: Characterization of light transport in scattering media at subdiffusion length scales with low-coherence enhanced backscattering. IEEE J. Sel. Top. Quantum Electron. **16**(3), 619–626 (2010)
15. Roy, H.K., Turzhitsky, V., Kim, Y., Goldberg, M.J., Watson, P., Rogers, J.D., Gomes, A.J., Kromine, A., Brand, R.E., Jameel, M., Bogovejic, A., Pradhan, P., Backman, V.: Association between rectal optical signatures and colonic neoplasia: potential applications for screening. Cancer Res. **69**(10), 4476–4483 (2009)
16. Liu, Y., Brand, R.E., Turzhitsky, V., Kim, Y.L., Roy, H.K., Hasabou, N., Sturgis, C., Shah, D., Hall, C., Backman, V.: Optical markers in duodenal mucosa predict the presence of pancreatic cancer. Clin. Cancer Res. **13**(15), 4392–4399 (2007)
17. Turzhitsky, V., Liu, Y., Hasabou, N., Goldberg, M., Roy, H.K., Backman, V., Brand, R.: Investigating population risk factors of pancreatic cancer by evaluation of optical markers in the duodenal mucosa (vol 25, pg 313, 2008). Dis. Markers **27**(5), 253–253 (2009)
18. Berne, B.J., Pecora, R.: Dynamic Light Scattering: With Applications To Chemistry, Biology, And Physics. Dover, New York (2000)
19. Weitz, D.A., Pine, D.J.: In: Brown, W. (ed.) Dynamic Light Scattering: The Method and Some Applications, vol. 49. Monographs on the Physics and Chemistry of Material, pp. 652–720. Oxford University Press, Oxford (1993)

20. Bloomfield, V.A.: Quasi-elastic light-scattering applications in biochemistry and biology. Annu. Rev. Biophys. Bioeng. **10**, 421–450 (1981)
21. Mason, T.G., Weitz, D.A.: Optical measurements of frequency-dependent linear viscoelastic moduli of complex fluids. Phys. Rev. Lett. **74**(7), 1250–1253 (1995)
22. Joo, C., Evans, C.L., Stepinac, T., Hasan, T., de Boer, J.F.: Diffusive and directional intracellular dynamics measured by field-based dynamic light scattering. Opt. Express **18**(3), 2858–2871 (2010)
23. Suissa, M., Place, C., Goillot, E., Freyssingeas, E.: Internal dynamics of a living cell nucleus investigated by dynamic light scattering. Eur. Phys. J. E **26**(4), 435–448 (2008)
24. Kramer, L.: Theory of light scattering from fluctuations of membranes and monolayers. J. Chem. Phys. **55**(5), 2097 (1971)
25. Hirn, R., Bayer, T.M., Radler, J.O., Sackmann, E.: Collective membrane motions of high and low amplitude, studied by dynamic light scattering and micro-interferometry. Faraday Discussions **111**, 17–30 (1998)
26. Haidekker, M.A., Stevens, H.Y., Frangos, J.A.: Cell membrane fluidity changes and membrane undulations observed using a laser scattering technique. Ann. Biomed. Eng. **32**(4), 531–536 (2004)
27. Amin, M.S., Park, Y., Lue, N., Dasari, R.R., Badizadegan, K., Feld, M.S., Popescu, G.: Microrheology of red blood cell membranes using dynamic scattering microscopy. Opt. Express **15**(25), 17001–17009 (2007)
28. Tishler, R.B., Carlson, F.D.: A study of the dynamic properties of the human red-blood-cell membrane using quasi-elastic light-scattering spectroscopy. Biophys. J. **65**(6), 2586–2600 (1993)
29. Wax, A., Yang, C.H., Dasari, R.R., Feld, M.S.: Path-length-resolved dynamic light scattering: modeling the transition from single to diffusive scattering. Appl. Opt. **40**(24), 4222–4227 (2001)
30. Carminati, R., Elaloufi, R., Greffet, J.J.: Beyond the diffusing-wave spectroscopy model for the temporal fluctuations of scattered light. Phys. Rev. Lett. **92**(21) (2004)
31. Lemieux, P.A., Vera, M.U., Durian, D.J.: Diffusing-light spectroscopies beyond the diffusion limit: The role of ballistic transport and anisotropic scattering. Phys. Rev. E **57**(4), 4498–4515 (1998)
32. Maret, G., Wolf, P.E.: Multiple light-scattering from disordered media – the effect of brownian-motion of scatterers. Zeitschrift Fur Physik B-Condensed Matter **65**(4), 409–413 (1987)
33. Stephen, M.J.: Temporal fluctuations in wave-propagation in random-media. Phys. Rev. B **37** (1), 1–5 (1988)
34. Pine, D.J., Weitz, D.A., Chaikin, P.M., Herbolzheimer, E.: Diffusing-wave spectroscopy. Phys. Rev. Lett. **60**(12), 1134–1137 (1988)
35. Maret, G.: Diffusing-wave spectroscopy. Curr. Opin. Colloid Interface Sci. **2**(3), 251–257 (1997)
36. Ackerson, B.J., Dougherty, R.L., Reguigui, N.M., Nobbmann, U.: Correlation transfer – application of radiative-transfer solution methods to photon-correlation problems. J. Thermophys. Heat Transf. **6**(4), 577–588 (1992)
37. Dougherty, R.L., Ackerson, B.J., Reguigui, N.M., Dorrinowkoorani, F., Nobbmann, U.: Correlation transfer – development and application. J. Quant. Spectrosc. Radiat. Transf. **52** (6), 713–727 (1994)
38. Boas, D.A., Campbell, L.E., Yodh, A.G.: Scattering and imaging with diffusing temporal field correlations. Phys. Rev. Lett. **75**(9), 1855–1858 (1995)
39. Boas, D.A., Yodh, A.G.: Spatially varying dynamical properties of turbid media probed with diffusing temporal light correlation. J. Opt. Soc. Am. A Opt. Image Sci. Vis. **14**(1), 192–215 (1997)

40. Li, J., Dietsche, G., Iftime, D., Skipetrov, S.E., Maret, G., Elbert, T., Rockstroh, B., Gisler, T.: Noninvasive detection of functional brain activity with near-infrared diffusing-wave spectroscopy. J. Biomed. Opt. **10**(4) (2005)
41. Boas, D.A., Dunn, A.K.: Laser speckle contrast imaging in biomedical optics. J. Biomed. Opt. **15**(1) (2010)
42. Zimnyakov, D.A., Tuchin, V.V.: Optical tomography of tissues. Quantum Electron. **32**(10), 849–867 (2002)

Chapter 11
Optical Coherence Tomography

Optical coherence tomography (OCT) has emerged as one of the most important and successful applications of interferometry in the biomedical sciences. This is due, in part, to its simplicity, but also because its advantages are well matched to applications on transparent or translucent media, particularly in ophthalmology. OCT is a form of laser ranging, somewhat like radar. But where radar seeks to isolate distant reflections based on *time of flight*, OCT isolates reflections based on *distance of flight*, while rejecting the predominant noncoherent diffuse light that arises in dense turbid media. As we saw in the previous chapter, light propagating in tissue retains partial coherence as unscattered or minimally scattered photons, which can be reflected by structures inside the tissue and returned to the surface. OCT uses coherence-domain detection, namely interferometry, to isolate these coherent returns and to measure how far the reflected light has traveled. By scanning across a biological sample, three-dimensional maps of reflective structure are created up to about 1–2 mm inside. There are two principle modes of short-coherence interference that are Fourier transforms of each other. One is in the time domain, actively tuning optical path length changes in the reference arm to adjust the detection gate to a specific depth. The other is in the frequency domain, which uses spectral interference fringes to identify reflections coming from many different depths simultaneously. Time-domain OCT was the first configuration that was investigated [1], but the need to actively adjust the depth gate made the acquisition time-consuming. Fourier-domain OCT was developed about 10 years later [2], bringing with it great advantages of parallelism and speed.

11.1 Coherence Gating

Coherence gating refers to the use of short-coherence light to spatially localize the interference of two waves. In its simplest manifestation, using ultrafast pulsed lasers, the coherence length of a laser pulse is equal to $l_c = c\Delta t_p$, where Δt_p is the pulse duration. For $\Delta t_p = 10$ fs, the coherence length is $l_c = 3$ μm, which is

D.D. Nolte, *Optical Interferometry for Biology and Medicine*, Bioanalysis 1,
DOI 10.1007/978-1-4614-0890-1_11,

approximately the level of axial resolution that is possible using laser ranging with ultrafast light sources. However, coherence gating does not need to be performed using pulses. Virtually any broadband light source (with good spatial coherence), even a continuous wave (CW) source, with a spectral width $\Delta\lambda$ may be used to perform coherence gating. The key concept is the matching of optical path lengths between a signal path and a reference path. When the optical path lengths match, or at least are within a coherence length, then interference will occur.

One of the key aspects in OCT performance is the axial resolution that can be achieved for a given source spectrum. It is common to define widths in OCT applications as full-width at half maximum (FWHM). The axial coherence function of a short-coherence source is

$$I_1(z) = \langle E^*(0)E(z)\rangle = E_0^2 \exp\left[-4\ln 2\left(\frac{z}{\Delta z}\right)^2\right] \tag{11.1}$$

in which Δz is the FWHM of the coherence function. The Fourier transform of the spatial coherence function is

$$\begin{aligned} I_1(q) = \mathrm{FT}[I_1(z)] &= E_0^2\sqrt{\frac{\pi}{4\ln 2}}\Delta x \exp\left[\frac{-q^2\Delta z^2}{16\ln 2}\right] \\ &= E_0^2\sqrt{\frac{\pi}{4\ln 2}}\Delta x\ \exp\left[-4\ln 2\frac{q^2}{\Delta q^2}\right] \end{aligned} \tag{11.2}$$

from which the assignment can be made

$$\Delta z\Delta q = 8\ln 2 \tag{11.3}$$

that relates Δz to the Δq. OCT is performed almost exclusively in the backscatter configuration for which $q = 2k$. Therefore,

$$\Delta z = \frac{8\ln 2}{\Delta(4\pi/\lambda)} = \left(\frac{2\ln 2}{\pi}\right)\frac{\lambda^2}{\Delta\lambda} \tag{11.4}$$

This equation defines the full-width half-maximum axial resolution of OCT. Note that the resolution length is half of the coherence length of the light source. This factor of 2 comes from $q = 2k$ in the backscatter configuration. The axial resolution as a function of bandwidth is shown in Fig. 11.1 for two center wavelengths at 800 nm and 1.33 μm. Bandwidths of 50 nm are typical, with 5 μm resolution at 800 nm. Ultra-high resolution toward 1 μm requires bandwidths greater than 150 nm.

Axial resolution in OCT is improved by decreasing the mean wavelength or by expanding the spectral bandwidth of the source. An example of ultra-high OCT resolution, with a bandwidth of $\Delta\lambda = 350$ nm at a center wavelength of $\lambda = 800$ nm, has an axial resolution of 1 μm. This resolution is sufficient to see single cells and even nuclei within the cells. *In vivo* ultra-high resolution OCT is shown in Fig. 11.2 for an African tadpole (*Xenopus laevis*) in which nuclei are clearly visible.

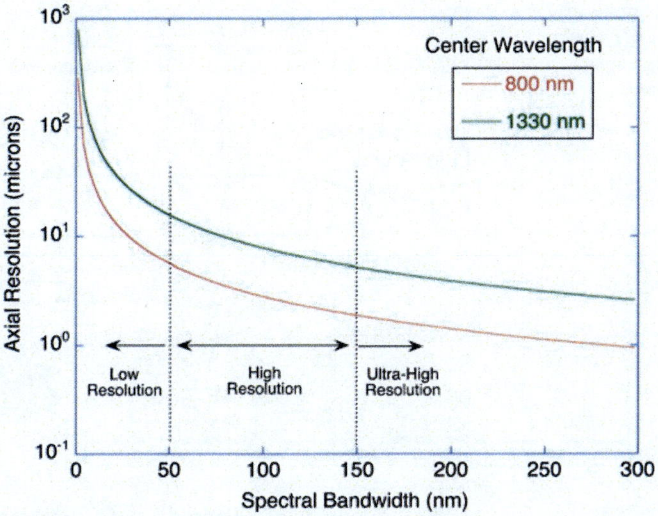

Fig. 11.1 Axial resolution vs. light source bandwidth for center wavelengths $\lambda = 800$ nm and $\lambda = 1.33$ nm

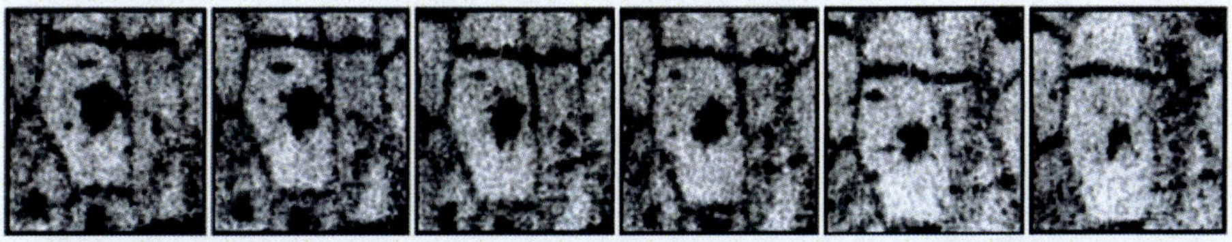

Fig. 11.2 Ultra-high resolution of living cells in the African tadpole (*Xenopus laevis*) with 1 μm axial resolution achieved using a bandwidth of 350 nm [3]

11.2 Time-Domain OCT

Time-domain OCT is the simplest, conceptually, to configure in the laboratory and to analyze. It consists of a conventional Michelson interferometer with a delay line to tune the optical path length of the reference arm to match a chosen depth in the sample. Examples of a free-space and a fiber-optic-based system are illustrated in Fig. 11.3. The reference-arm mirror in these examples is on translation stages with piezoelectric stacks for rapid scanning of short distances. The range of motion should be up to a millimeter, which is beyond piezoelectric displacements. Fast linear displacements are possible with voice-coil actuators, but again with limited range of motion. To scan through a full millimeter of tissue (approximately the limit of the penetration depth of OCT) requires a faster and larger displacement approach that can be provided by Fourier optics.

Fast delay lines can be constructed using Fourier-transform techniques that were originally developed for pulse-shaping applications [5]. The concept is

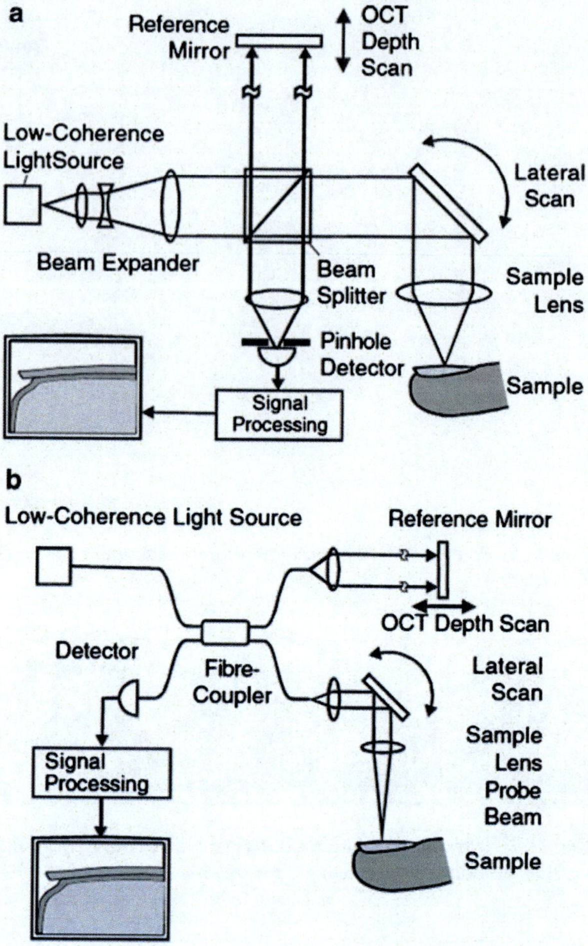

Fig. 11.3 Examples of time-domain OCT systems. (**a**) A free-space system. (**b**) A fiber-optics system [4]

illustrated in Fig. 11.4. A collimated beam is incident on the diffraction grating that disperses the wavelengths into a spectrum that is recollimated by the Fourier-transform lens. At the Fourier plane of the lens, a rotational galvanometer imparts a small angle θ on the spectrum that is reflected and transformed back onto the grating. The spectral components of the beam are recombined by the grating into an output beam that has a path delay proportional to θ.

The path delay on the output beam originates in the Fourier-transform relationship between time delay in the time domain and phase in the Fourier domain. A real-valued Gaussian spectral amplitude given by

$$s(\omega) = \frac{1}{\sqrt{\sqrt{\pi}\sigma_\omega}} \exp(-(\omega - \omega_0)^2/2\sigma_\omega^2) \tag{11.5}$$

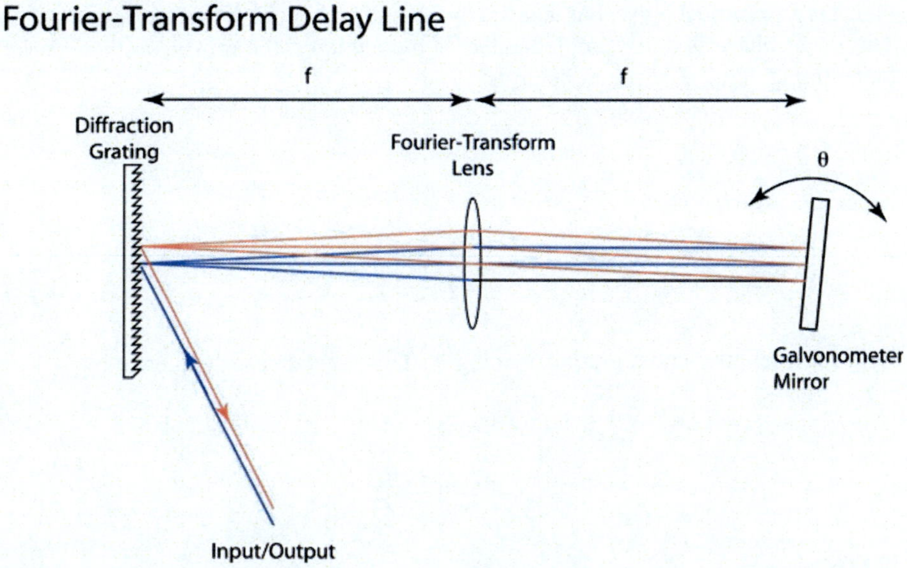

Fig. 11.4 Fourier transform rapid scanning delay line. The reference wave is dispersed spectrally and Fourier transformed to a mirror with a rapidly cycling tilt angle. The reflected and retransformed wave has an optical path length delay proportional to the mirror angle

is transformed into the time domain as

$$
\begin{aligned}
E(t) &= \frac{1}{\sqrt{\pi\sqrt{\pi}\sigma_\omega}} \int_0^\infty \exp(-(\omega-\omega_0)^2/2\sigma_\omega^2)\mathrm{e}^{-\mathrm{i}\omega t}\,\mathrm{d}\omega \\
&= \frac{\sqrt{2\sigma_\omega}}{\pi^{1/4}}\,\mathrm{e}^{-\mathrm{i}\omega_0 t}\mathrm{e}^{-t^2/2\Delta t^2}
\end{aligned}
\tag{11.6}
$$

which is a Gaussian pulse of center frequency ω_0 and pulse duration

$$
\Delta t = \frac{1}{\sigma_\omega}. \tag{11.7}
$$

Now, if a linear phase is imposed on the spectral amplitude as

$$
s(\omega) = \frac{1}{\sqrt{\sqrt{\pi}\sigma_\omega}} \exp(-(\omega-\omega_0)^2/2\sigma_\omega^2)\exp[\mathrm{i}(\omega-\omega_0)\tau] \tag{11.8}
$$

this is transformed to

$$
\begin{aligned}
E(t) &= \frac{1}{\sqrt{\pi\sqrt{\pi}\sigma_\omega}} \int_0^\infty \mathrm{e}^{-(\omega-\omega_0)^2/2\sigma_\omega^2}\mathrm{e}^{-\mathrm{i}\omega t}\mathrm{e}^{\mathrm{i}(\omega-\omega_0)\tau}\mathrm{d}\omega \\
&= \frac{\sqrt{2\sigma_\omega}}{\pi^{1/4}}\mathrm{e}^{-\mathrm{i}\omega_0(t+\tau)}\mathrm{e}^{-(t-\tau)^2/2\Delta t^2}
\end{aligned}
\tag{11.9}
$$

which is an identical pulse, but delayed by an amount τ relative to the original pulse. The linear phase is easily implemented as the angle on the mirror on the Fourier plane in Fig. 11.4. The relationship is

$$\phi = 2ky \sin\theta, \tag{11.10}$$

where the position y is

$$y = f\frac{\Delta\lambda}{\Lambda} \tag{11.11}$$

for a focal length f and a grating spacing Λ. This is related to frequency by

$$\begin{aligned} \Delta\omega &= c\Delta k \\ &= -c2\pi\frac{\Delta\lambda}{\lambda^2} \end{aligned} \tag{11.12}$$

to give

$$\begin{aligned} \phi &= 2kf\frac{\Delta\lambda}{\Lambda}\sin\theta \\ &= -\frac{2(\omega-\omega_0)}{c}\frac{f\lambda}{\Lambda}\sin\theta. \end{aligned} \tag{11.13}$$

The time delay is then

$$\tau = \frac{2f\lambda}{c\Lambda}\sin\theta. \tag{11.14}$$

As an example, for a grating spacing of 1 μm, a focal length of 10 cm at a center wavelength of 840 nm, this is a delay of 2.5 ps per degree. The delay for a 1 mm coherence depth is about 10 ps. Therefore, only a few degrees of deflection for the mirror is sufficient to scan to the full depth in tissue.

The application of Fourier domain and spectral approaches to rapid delay lines has clear advantages. These advantages can be utilized even more directly through the principle of spectral interferometry for more sensitive and faster modes of OCT.

11.3 Fourier-Domain OCT

Time-domain OCT is a serial acquisition process in which one voxel value is obtained at a time as the time delay is scanned in z for a given position (x, y). The voxels are compiled one by one as each dimension is scanned individually.

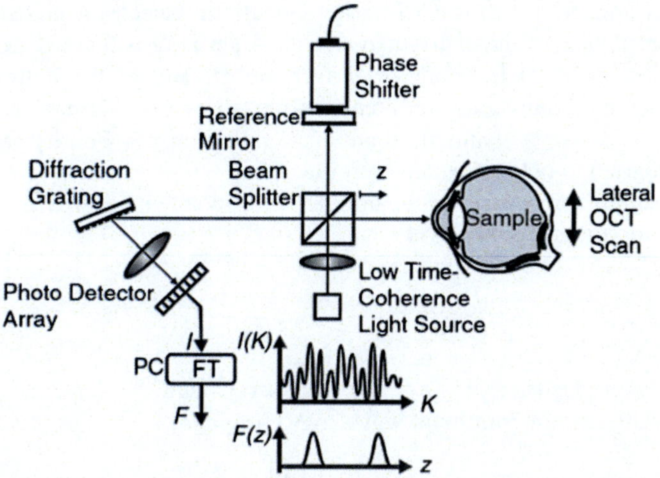

Fig. 11.5 Spectral-domain OCT system. Scattered light with multiple time-of-flights is combined with a single reference pulse and interferes at the output of the spectrometer grating [4]

Multiplex approaches that detect multiple data points simultaneously have an obvious advantage of speed and signal-to-noise. Simultaneous measurements of all depths (within a range defined by the operational parameters of the system) can be made at a given position (x, y) using Fourier-domain techniques such as spectral interferometry or by using a swept source.

11.3.1 Spectral-Domain OCT

Spectral interferometry was introduced in Chap. 1. There is a somewhat counterintuitive effect that two well-separated temporal pulses entering a spectrometer will cause interference fringes to appear in the measured spectrum. The fringe spacing $\Delta\lambda$ is related to the delay between the two pulses.

$$\Delta t = \frac{\lambda^2}{c\Delta\lambda}. \tag{11.15}$$

The depth resolution is the same as for time-domain OCT and is based on the spectral width of the source. These same relations hold for a continuous-wave short-coherence source, such as a superluminescent diode. Delay times are simply converted to optical path length, and the same form of spectral interferometry continues to hold.

An example of a spectral-domain OCT system is shown in Fig. 11.5. A short-coherence source is used in a Michelson interferometer configuration in which a reference wave and a distributed time-of-flight signal are combined and sent into a spectrometer. A linear CCD array detects the spectrum that includes the spectral interference of the distributed time-of-flight returns that interfere with the reference. The spectral interference fringes are already in the Fourier domain, containing all the combined interferences from all of the return reflections from the sample. To extract depth information requires only a Fourier transform of the interferogram to obtain the depth profile.

The maximum range is set by the minimum resolvable interference $\Delta\lambda_{\text{min}}$ on the linear detector array. The range is

$$\text{range} = \frac{c\Delta t_{\text{max}}}{2} = \frac{\lambda^2}{4\Delta\lambda_{\text{min}}} \tag{11.16}$$

for a minimum measurable $\Delta\lambda_{\text{min}}$. For an array with N elements detecting a spectrum width $\Delta\lambda$, the minimum wavelength fringe is

$$\Delta\lambda_{\text{min}} = 2\frac{\Delta\lambda}{N}, \tag{11.17}$$

where the factor of 2 is for Nyquist sampling. The range is then

$$\text{range} = \frac{N\lambda^2}{8\Delta\lambda}, \tag{11.18}$$

where $\Delta\lambda$ is the $e^{-1/2}$ bandwidth.

11.3.2 Swept-Source and In-Line OCT

Short-coherence light sources tend to have low powers and uncollimated spatial modes that require special engineering solutions to be useful in OCT applications. Tunable lasers, on the other hand, can have high powers and excellent collimation. Even though the coherence length of tunable lasers tends to be very large, they can still be used as Fourier-domain OCT sources because of the Fourier-transform relationship between phase and position. For instance, if there are N distinct reflections distributed in depth d_i, then the net field arising from these reflections is

$$E_s = \sum_{n=1}^{N} E_n e^{i(\omega/c)2nd_i}, \tag{11.19}$$

where ω is the instantaneous angular frequency of a long-coherence source. If the frequency is swept [6–8], and detected relative to a stable reference, then the interferogram is

$$I_{SS}(\omega) = I_R + E_r \sum_{n=1}^{N} |E_n| \cos\left(\frac{\omega}{c}2nd_n - \phi_n\right). \tag{11.20}$$

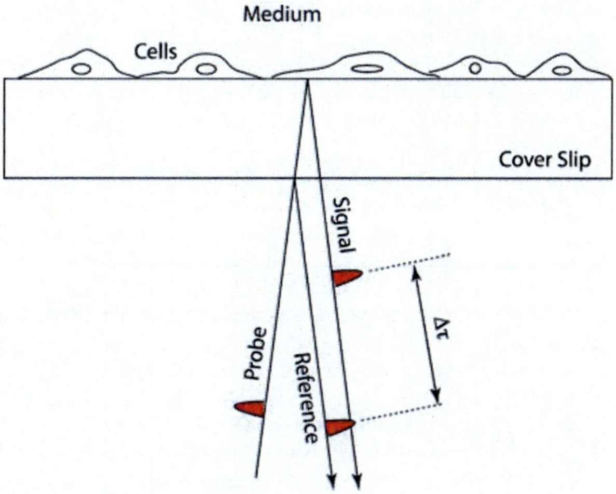

Fig. 11.6 In-line spectral-domain OCT. The front reflection from the coverslip provides a stable reference for spectral interferometry performed with high phase stability. The displacement sensitivity is sub-nanometer

When the swept-source spectrum is transformed back into real space, delta-functions occur at the locations of the reflectors

$$I_{SS}(x) = E_r \sum_{n=1}^{N} |E_n| \delta(2nd_n). \tag{11.21}$$

The resolution of swept-source OCT is set by the tunability range $\Delta\lambda$, and the depth range is set by the minimum resolvable wavelength during tuning $\Delta\lambda_{min}$. The advantage in signal-to-noise over time-domain OCT is typically 20–30 dB. Swept-source OCT has become one of the favored OCT approaches because of the advantages on speed, signal-to-noise, and easy optical implementation.

Fourier-domain OCT makes possible ultra-stable reference configurations that enable phase-sensitive detection. An in-line common-path configuration is shown in Fig. 11.6 that consists of a coverslip carrying cells. The coverslip provides an in-line reference reflection for the Fourier-domain system. Because the coverslip has no mechanical instability, the reference wave is stable, providing high axial sensitivity to sub-nanometer displacements. This approach can be used to study systems with small dynamic displacements [9–11].

References

1. Huang, D., Swanson, E.A., Lin, C.P., Schuman, J.S., Stinson, W.G., Chang, W., Hee, M.R., Flotte, T., Gregory, K., Puliafto, C.A., Fujimoto, J.G.: Optical coherence tomography. Science **254**, 1178 (1991)
2. de Boer, J.F., Cense, B., Park, B.H., Pierce, M.C., Tearney, G.J., Bouma, B.E.: Improved signal-to-noise ratio in spectral-domain compared with time-domain optical coherence tomography. Opt. Lett. **28**(21), 2067–2069 (2003)

3. Drexler, W.: Ultrahigh-resolution optical coherence tomography. J. Biomed. Opt. **9**(1), 47–74 (2004)
4. Fercher, A.F., Drexler, W., Hitzenberger, C.K., Lasser, T.: Optical coherence tomography – principles and applications. Rep. Prog. Phys. **66**(2), 239–303 (2003)
5. Weiner, A.M.: Femtosecond optical pulse shaping and processing. Prog. Quant. Electron. **19**, 161–237 (1995)
6. Fercher, A.F., Hitzenberger, C.K., Kamp, G., Elzaiat, S.Y.: Measurement of intraocular distances by backscattering spectral interferometry. Opt. Commun. **117**(1–2), 43–48 (1995)
7. Chinn, S.R., Swanson, E.A., Fujimoto, J.G.: Optical coherence tomography using a frequency-tunable optical source. Opt. Lett. **22**(5), 340–342 (1997)
8. Choma, M.A., Sarunic, M.V., Yang, C.H., Izatt, J.A.: Sensitivity advantage of swept source and Fourier domain optical coherence tomography. Opt. Express **11**(18), 2183–2189 (2003)
9. Choma, M.A., Ellerbee, A.K., Yang, C.H., Creazzo, T.L., Izatt, J.A.: Spectral-domain phase microscopy. Opt. Lett. **30**(10), 1162–1164 (2005)
10. Ellerbee, A.K., Creazzo, T.L., Izatt, J.A.: Investigating nanoscale cellular dynamics with cross-sectional spectral domain phase microscopy. Opt. Express **15**(13), 8115–8124 (2007)
11. McDowell, E.J., Ellerbee, A.K., Choma, M.A., Applegate, B.E., Izatt, J.A.: Spectral domain phase microscopy for local measurements of cytoskeletal rheology in single cells. J. Biomed. Opt. **12**(4) (2007)

Chapter 12
Holography of Tissues

Living biological tissue is an unlikely candidate as a target for holographic imaging. It is strongly aberating, strongly scattering, strongly heterogeneous, nonstationary, anisotropic, and volumetric in extent. Coherent light propagating through it becomes spatially diverse (many scattered wave vectors) and temporally diverse (many scattered frequencies and phases). Holographic imaging, which requires a steady relative phase condition between a signal and a reference wave, is difficult under these conditions. Nonetheless, holographic techniques are increasingly being applied to probe the physical and chemical properties of living tissue. For instance, for short penetration through translucent targets, in-line phase-shifting interferometry has provided full-field coherence-gated imaging at microscopic resolutions [1, 2]. Alternatively, off-axis holography has probed structural variations at broader tissue scales [3], and can be used to measure dynamic processes such as intracellular motion [4]. The advent of digital holography [5, 6], in which holographic interference fringes are captured directly on a CCD chip, has made holography easily accessible and affordable for an expanding number of biological applications.

The main character of all holographic imaging of tissue is the presence of speckle. Speckle is ubiquitous in all coherent imaging systems, and is guaranteed when the target is as structurally and dynamically diverse as living tissue. Therefore, all tissue holography is speckle holography. In this sense, speckle mitigation is the opposite of what is required for hologram recording, although speckle reduction *after* image reconstruction is still useful when striving for high-resolution imaging. The properties of speckle fields, their dependence on the optical properties of the imaging system, and the effect they have on spatial and temporal coherence were discussed in detail in Chap. 3. Speckle holography is performed when a coherent off-angle reference is added to the speckle field, as discussed in Sect. 3.5. In this chapter, the optical configurations for speckle holography are described, and the recording of speckle holograms is explained, along with specific applications to tissue holography.

D.D. Nolte, *Optical Interferometry for Biology and Medicine*, Bioanalysis 1,
DOI 10.1007/978-1-4614-0890-1_12,

12.1 Dynamic Holography

Tissue holography is fundamentally dynamic holography. Tissue is a living and moving target, with intracellular motions that are fast enough to produce Doppler shifts above 20–50 Hz. The speckle field that is returned from a target sample displays flickering and shimmering. To capture the instantaneous speckle field holographically (with high fringe contrast) requires short hologram exposure times down to 10 ms, and perhaps faster. Dynamic holography is simply high-speed holography, in which the holographic reconstruction (or at least the holographic recording) is performed at high speeds with short hologram refresh times. Dynamic holograms can be physical entities, such as optical changes induced in physical media, like photorefractive crystals, or they can be detected digitally on a planar detector array (CCD or CMOS digital camera).

12.1.1 Photorefractive Holography

The photorefractive effect [7] is one of the most sensitive forms of physical hologram recording. Photorefractive materials convert a spatial variation in intensity (e.g., interference fringes) into a matched spatial variation in optical properties (e.g., refractive index). If the interference fringes arise from the interference of a signal wave and a coherent reference wave, then the refractive index changes inside the material constitute a physical hologram. Because photorefractive materials are dynamic, with hologram storage times that last only as long as the dielectric relaxation time, it is common to read the hologram out at the same time it is recorded by the same reference wave. This simultaneous readout (diffraction) gives the photorefractive effect the character of a third-order nonlinear optical material, known as a $\chi^{(3)}$ material, in which waves mix through two-wave mixing and four-wave mixing.

In two-wave mixing, two coherent waves with k-vectors $\vec{k}_1$ and $\vec{k}_2$ cross inside a material and produce interference fringes with a grating vector $\vec{K} = (\vec{k}_2 - \vec{k}_1)$. The incident waves are scattered (diffracted) by the grating into $\vec{k}_{1,\mathrm{scatt}} = \vec{k}_1 + \vec{K} = \vec{k}_2$ and $\vec{k}_{2,\mathrm{scatt}} = \vec{k}_2 - \vec{K} = \vec{k}_1$. In other words, each wave diffracts into the direction of the other. The two output waves are each coherent superpositions of the two input waves. This two-wave mixing process is shown at the top of Fig. 12.1. The details of the grating formation in the material and the diffraction efficiency determine how much of each wave is mixed in the output, and determines their relative phase. Two-wave mixing can be used as an adaptive beam combiner that is insensitive to mechanical vibrations in an adaptive interferometer [8], which was the basis of the adaptive BioCD application [9].

In four-wave mixing, shown at the bottom of Fig. 12.1, two coherent waves with k-vectors $\vec{k}_1$ and $\vec{k}_2$ again write a grating in the photorefractive material, but there is a counter-propagating wave $\vec{k}_3 = -\vec{k}_2$ that is coherent and counter-propagates

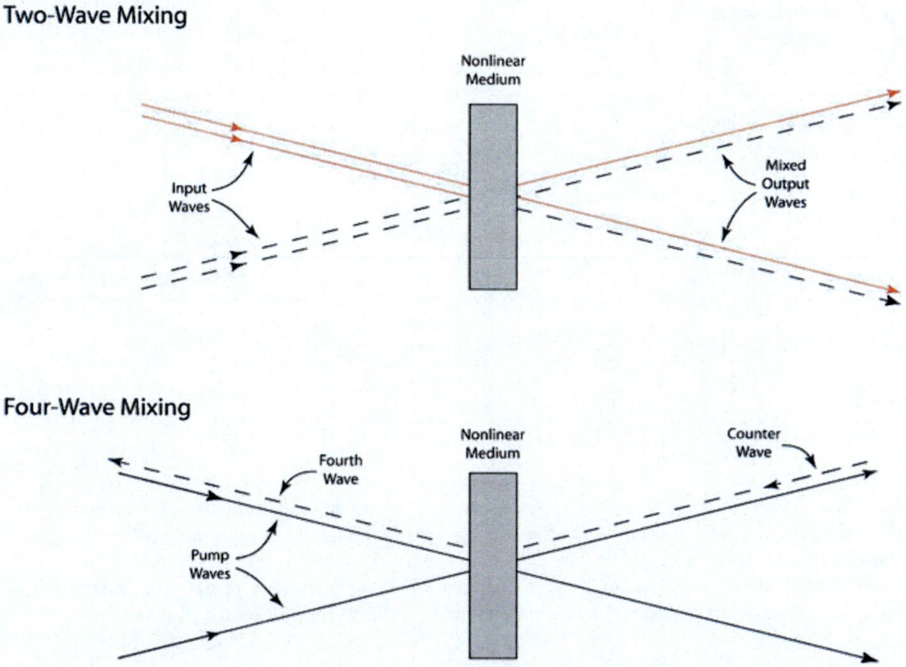

Fig. 12.1 Two-wave mixing and four-wave mixing geometries in dynamic media. In two-wave mixing, the waves write gratings that cause self-diffraction that is phase matched to each wave. In four-wave mixing two waves write a diffraction grating that diffracts a probe wave

against the direction of one of the pump waves. This counter wave is diffracted into a new wave that counter-propagates in the direction of the other pump wave $\vec{k}_{3,\text{scatt}} = -\vec{k}_2 + \vec{K} = -\vec{k}_1$. This "fourth wave" can have special properties, known as phase conjugation [10–12], that have applications for distortion-free imaging in biological samples [13].

There are several classes of photorefractive materials, with trade-offs between diffraction efficiency and response rate. The strongest photorefractive gratings are generated in ferroelectrics such as $LiNbO_3$ and $BaTiO_3$, but the weak transport in these materials causes them to respond slowly to changing fringe patterns giving them slow update rates [14, 15]. These materials are ideal for storage and self-organizing memory applications, but are not ideal for fast image processing. The fastest photorefractive materials are high-mobility semiconductors such as GaAs, InP, and CdTe [16]. The dielectric relaxation time in these materials can be as short as a microsecond under high illumination because of the high carrier mobility. These materials are good choices for real-time image processing applications, but have relatively low diffraction efficiencies. A highly sensitive class of photorefractive semiconductors are photorefractive quantum-wells [17, 18]. These are thin films of multiple layers of GaAs and AlGaAs grown by molecular beam epitaxy that use quantum-confinement effects to enhance the optical properties.

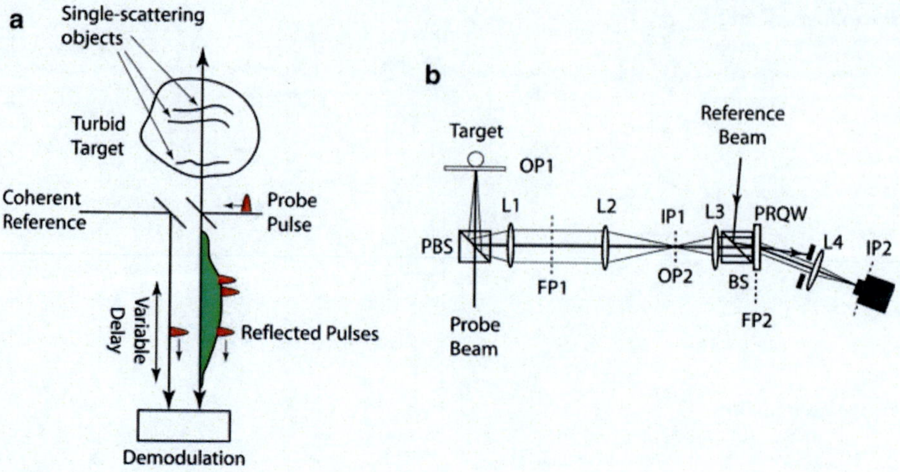

Fig. 12.2 Photorefractive coherence gating of light backscattered from tissue. The concept is illustrated in (**a**) using a pulsed source that creates distributed time-of-flight "echoes" from a turbid sample. When these reflections are coincident with a coherent reference pulse inside the photorefractive material, the interference fringes create a diffraction grating that diffracts the reference wave, shown in the optical schematic in (**b**), that is imaged onto a digital camera

Photorefractive quantum-well devices are the most sensitivity dynamic holographic media with the highest speeds, but diffraction efficiency remains a trade-off. A material that occupies the middle ground between efficiency and speed are photorefractive polymers [19, 20]. Despite being polymers, these can have fairly good carrier transport, while special chromophores can provide fairly large electro-optic effects. Speeds in the range of milliseconds are achievable with diffraction efficiencies approaching 100%. These photorefractive materials have found applications in holographic imaging of tissue [3, 21–24].

12.1.2 Holographic Coherence-Gating

Coherence-domain detection of light scattered from tissue requires a short-coherence gate to distinguish path-matched photons from non-path-matched photons. In conventional optical coherence tomography (OCT) this is provided by a coherent reference on the point detector, yielding a temporal heterodyne signal that is demodulated to extract signal vs. depth. Topography is recorded by scanning the point detection over the sample. Off-axis short-coherence holography also constitutes coherence-domain detection, with the added advantage that it uses spatial heterodyne (spatial interference fringes) and thus is a full-frame approach that captures x–y images at a gated depth z.

The principles of short-coherence holographic coherence-gating are illustrated in Fig. 12.2. The coherent laser ranging aspects are identical to OCT. However,

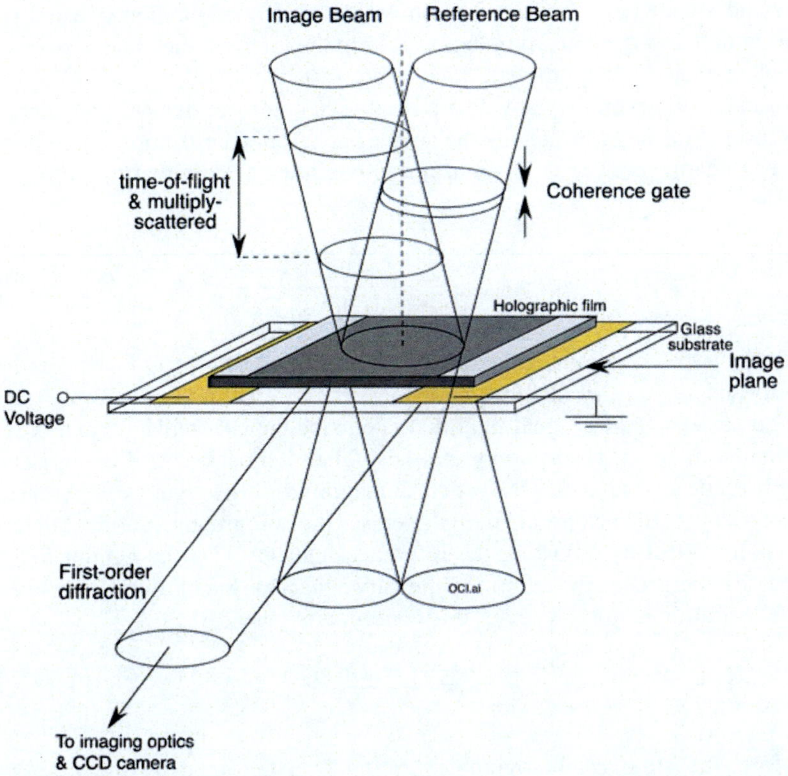

Fig. 12.3 A photorefractive quantum-well device used as a holographic coherence gate. The photorefractive device is a thin film that diffracts the reference wave into a background-free direction where it is imaged directly, with no computed tomography, onto a camera plane

instead of using a temporal heterodyne, in photorefractive coherence-gating the reference wave is off-axis, generating a spatial heterodyne signal – the hologram. The hologram is only created by light that has the same optical path length as the reference, which is swept through the volume of the target sample. At each depth selected by the reference optical path, only coherently reflected waves write a hologram. This hologram diffracts the reference wave into an imaging system that reconstructs the depth-selected reflectances on a digital camera.

An important aspect of holographic coherence-gating using dynamic media, such as photorefractive quantum-well devices, is the background-free nature of the coherence detection. The signal wave returning from a translucent volumetric sample is a complicated combination of distributed time-of-flights from coherent reflectors as well as a very high background of multiply scattered light, illustrated in Fig. 12.3. However, only the coherent part writes the hologram. The background passes through the holographic film and is not recorded. The hologram, which is only of the coherent structure at the selected depth, diffracts the reference wave into a diffracted direction that has no other background and can be detected directly using imaging lenses and a camera. A Fourier-domain holographic configuration is usually used, requiring only a single transform lens to reform the image at the camera plane [25].

Photorefractive quantum-well devices are fast, with dynamic recording speeds up to 1 kHz. However, they are absorption-based nonlinear devices and hence have diffraction efficiencies limited by the device background absorption. The image and reference beams produce an internal refractive index grating that has a magnitude equal to

$$n_K = n_{max} \frac{2\sqrt{I_{sig} I_{ref}}}{I_{sig} + I_{ref} + I_{background}}, \tag{12.1}$$

where I_{sig} is the coherent component of the returned light, I_{ref} is the reference intensity and $I_{background}$ is all other intensities, mainly arising from multiply scattered light, but also including signal light that is outside the coherence length. A similar equation holds for an absorption grating α_K. The high background that arises from multiple scattering can be several orders of magnitude larger than the coherent signal intensity. This high background partially erases the holographic grating, but this can be partially offset by adjusting the reference intensity. The maximum hologram amplitude occurs when the reference intensity equals the background intensity $I_{ref} = I_{back}$. The grating amplitude under this condition is given by

$$n_K = \frac{n_{max}}{\sqrt{I_{back}/I_{sig}}}. \tag{12.2}$$

Therefore, the strongest holographic grating is determined by the ratio of the background intensity to the signal intensity. The diffraction efficiency from the holographic grating depends on the grating amplitude as

$$\eta = \left[\left(\frac{\pi n_K L}{\lambda} \right)^2 + \left(\frac{\alpha_K L}{4} \right)^2 \right] e^{-\alpha_0 L}, \tag{12.3}$$

where L is the thickness of the active region of the device. The exponential absorption term decreases the overall diffracted intensity. Under optimal performance, the diffraction efficiency of the photorefractive quantum-well devices approach several percent. Although the efficiency is smaller under high background tissue imaging conditions, the square-root dependence in (12.2) enables observable holograms to be recorded even up to relatively high background levels.

12.1.3 *Multicellular Tumor Spheroids*

There is a significant benefit to probing three-dimensional tissues, as opposed to two-dimensional monolayer cultures, when seeking to uncover the collective behavior of cells in tissues. While microscopic imaging of cellular motility and motility-related gene expression is well-established in two dimensions [26], cells in

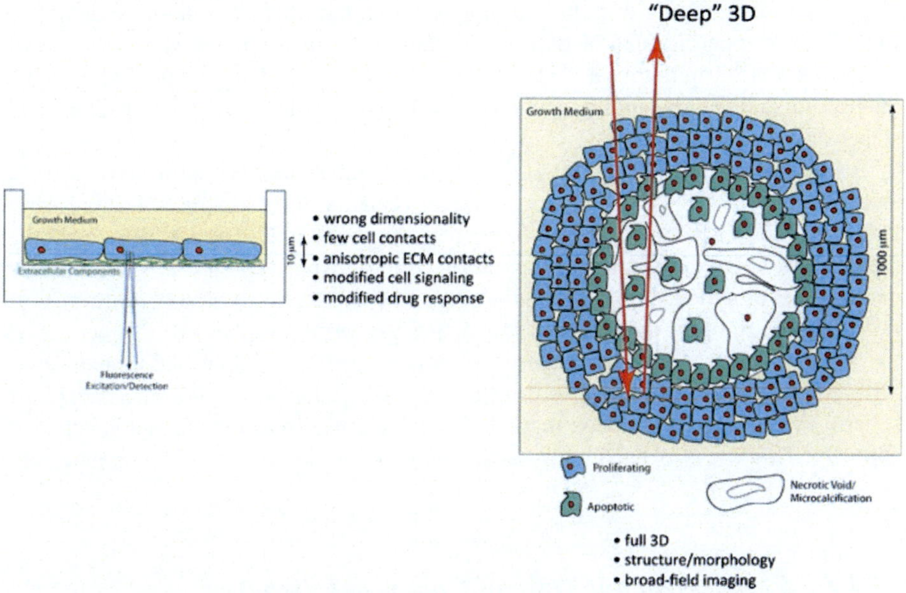

Fig. 12.4 Monolayer culture compared with a three-dimensional multicellular spheroid. Cells behave differently in two dimensions relative to three dimensions, that captures the natural cellular environment

contact with flat hard surfaces do not behave the same as cells embedded in the extracellular matrix [27, 28]. A major conceptual and experimental shift has occurred in the past 10 years away from 2D cell cultures to 3D tissue cultures [29–32]. Cancer morphogenesis progression and metastasis, in particular, is strongly dependent on a three-dimensional environment [33–35]. Supporting this experimental trend are models of the extracellular matrix (ECM) in which different ECM models have different properties, with the goal of including the relevant three-dimensional biology. Special attention has been paid lately to mechanical properties [36–38] and cell-ECM adhesions [27, 39, 40].

Recent work has raised the dimensionality of cellular motility imaging from 2D to 3D, using microscopies such as confocal fluorescence [41, 42], two-photon [43, 44], optical projection tomography (OPT) [45], and single-plane illumination projection (SPIM) [46]. Three-dimensional videography of motile metastatic cells in ECM [28, 47] has been performed primarily with confocal microscopy [48, 49] and light-sheet-illumination approaches [29, 46]. Lateral resolution in these cases is diffraction-limited at the surface of the 3D matrix but degrades the deeper the sectioned plane penetrates into the target sample. Although structured illumination [50] and stimulated emission [51] approaches can beat that limit under special circumstances, these also suffer significant degradation in resolution with increasing probe depth, limiting access to motility information from deep inside the sample where it is far from the artificial influence of surfaces.

Multicellular spheroids of normal cells or neoplastic cells (tumor spheroids) are ideal targets for optical coherence imaging (OCI) applications (Fig. 12.4).

The spheroids are balls of cells that may be cultured up to 1 mm in size in vitro [52–57]. The spheroids can be used to simulate the optical properties of a variety of tissues such as the epidermis and various epithelial tissues and may be used to simulate the histological and metabolic features of small nodular tumors in the early avascular stages of growth [57].

Beyond a critical size (about 100 μm), most spheroids develop a necrotic core surrounded by a shell of viable, proliferating cells, with a thickness varying from 100 to 300 mm. The development of necrosis has been linked to deficiencies in the metabolites related to energy generation and transfer. The limiting factor for necrosis development is oxygen – the oxygen consumption and oxygen transport reflecting the status of the spheroid [53, 58]. Early work on spheroids [59] launched the study of therapeutic strategies for cancer using the spheroid response to different drugs. The response to drug therapy was quantified from an analysis of spheroid volume growth delay, increase in the necrotic area, and change in survival capacity. This work focused on hypoxia and its induction by chemical agents [60].

12.1.4 Photorefractive Optical Coherence Imaging

Photorefractive OCI is a full-frame coherence-domain volumetric imaging approach that shares much in common with OCT, but performs direct imaging inside highly scattering media and needs no computed tomography [61, 62]. Holographic coherence-gating is background free, and uses conventional optics to image directly onto a camera or video recorder. In this sense, it is more direct than full-frame OCT, which still requires phase stepping and image reconstruction [1, 63].

A selected set of pseudo B-scans obtained using holographic OCI for a representative tumor spheroid is shown in Fig. 12.5. The 800 μm diameter tumor is resting on a Petri dish. The grayscale is the logarithm of reflected intensity. The strongest reflectors are from inside the necrotic core of the spheroid, while the outer shell is optically more homogeneous and scatterers less. The full three-dimensional data set is shown in Fig. 12.6 for two thresholds. On the left, the threshold captures all reflected signals. On the right of the figure, only the brightest reflections are plotted. The healthy shell is stripped away and the necrotic core is revealed, including its "shadow" on the Petri dish.

The reflected intensities from a tumor spheroid are histogrammed in Fig. 12.7 at several selected radii from the center of the spheroid. The histograms are fit by single-decay exponentials over many orders of magnitude, with the brightest reflections coming from the necrotic core, and weaker reflections from the homogeneous healthy shell. The speckle contrast in all cases is near unity, signifying fully developed speckle caused by multiple scattering and channel cross talk driven by the full-field illumination [65]. Despite the lack of spatial detail in the reconstructed tissue holograms, the temporal fluctuations carry significant dynamical information, which is discussed in Sect. 12.3.

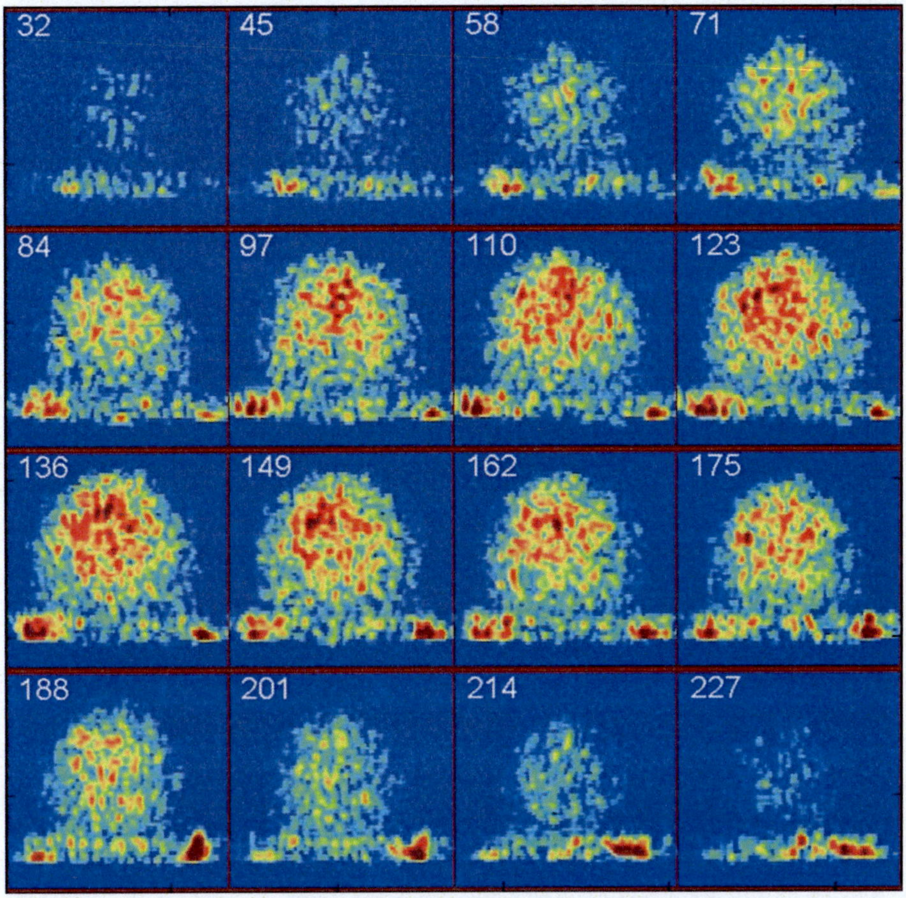

Fig. 12.5 Holographic optical coherence imaging (OCI) pseudo B-scans (y–z plane at selected x-values) of a 800-μm diameter tumor spheroid resting on a Petri dish. From ref. [64]

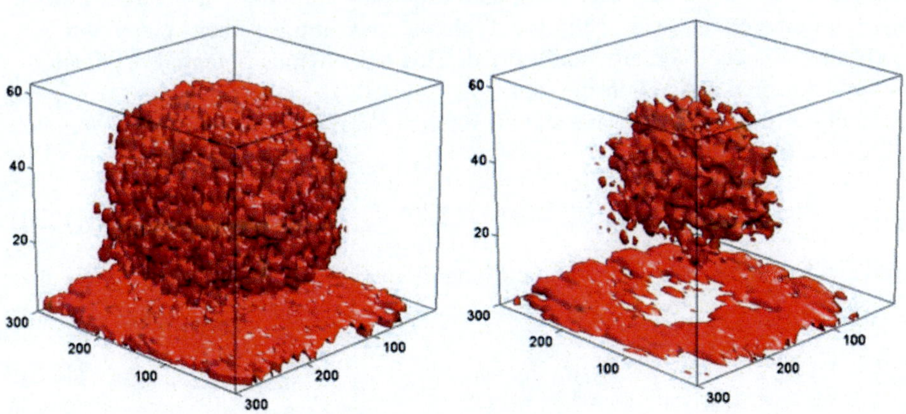

Fig. 12.6 Volumetric reconstruction of a tumor spheroid resting on a Petri dish. The image on the *left* shows the volumetric reflectances. The image on the *right* is thresholded to show only the highest reflectances, uncovering the necrotic core and its "shadow" on the Petri dish. From ref. [64]

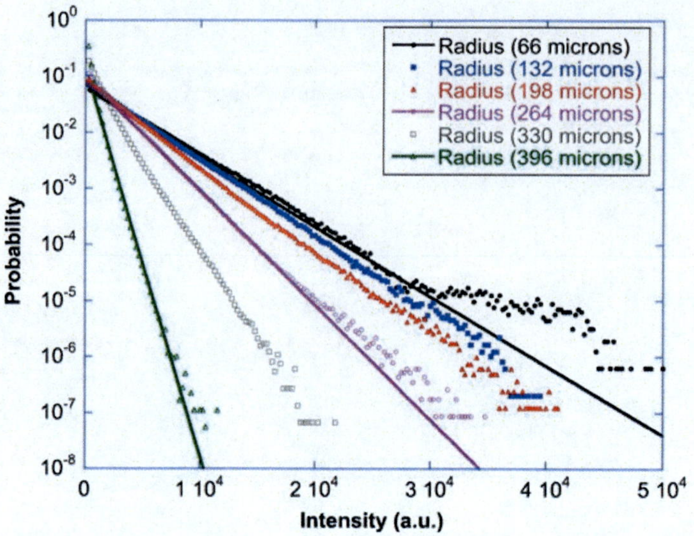

Fig. 12.7 Intensity histograms for tissue at increasing radii from the center of the spheroid. The intensity distributions are Poissonian across several orders of magnitude, with the brightest intensities found at the center of the necrotic core

12.1.5 Phase-Conjugate Imaging

Dynamic holography in photorefractive materials provides a unique opportunity to compensate directly the aberrations that degrade image resolution deep inside tissue. The four-wave mixing geometry shown in Fig. 12.8 uses a back-propagating reference wave to read out the hologram written by the two forward waves. This backward-diffracted wave that counter-propagates back along the direction of the original object wave has interesting and important properties: it behaves as if it were time-reversed, with conjugated phases that retrace their paths through aberrating objects to emerge undistorted. This has obvious potential applications in distortion-free image transmission.

In Fig. 12.8, the signal wave and the reference wave generate a refractive index grating with the component

$$\Delta n \propto S^* R e^{iKy}, \tag{12.4}$$

where S and R are the complex field amplitudes. The index grating carries the conjugated phase of the signal wave S^*. The grating vector is

$$\vec{K} = \vec{k}_R - \vec{k}_S. \tag{12.5}$$

When the back-propagating wave scatters from the grating, it creates a wave that acquires the conjugated phase of the original signal wave

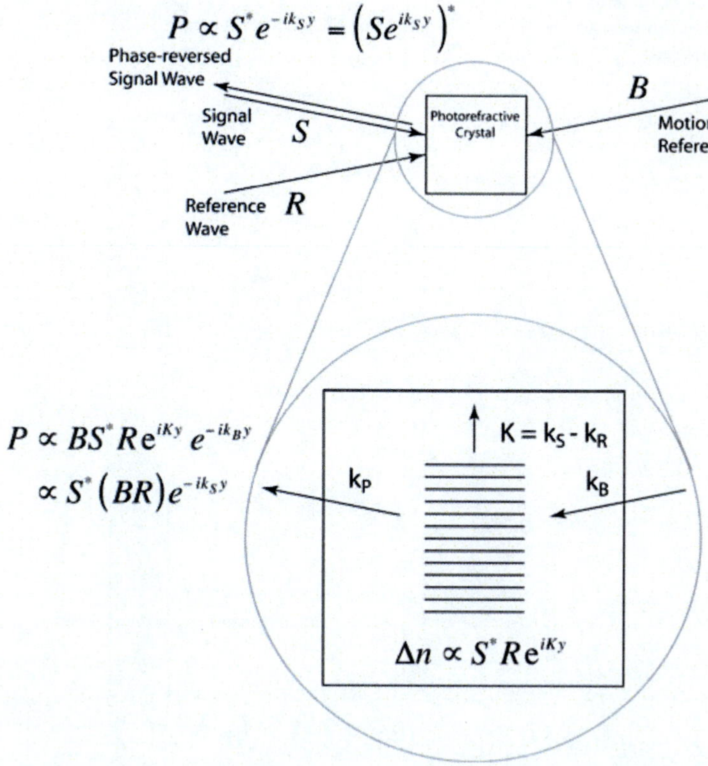

Fig. 12.8 Phase-conjugate scattering geometry. The refractive grating carries information on the conjugated phase of the signal wave. The backpropagating wave scatters from this grating and acquires the conjugated phase

$$\begin{aligned} P &\propto B\mathrm{e}^{-\mathrm{i}k_B y}\Delta n \\ &\propto S^* \mathrm{e}^{-\mathrm{i}k_B y} R \mathrm{e}^{\mathrm{i}Ky} \\ &\propto S^* (BR)\mathrm{e}^{-\mathrm{i}k_S y}. \end{aligned} \tag{12.6}$$

Because the reference wave is structureless (a plane wave), the reference and back wave amplitudes are constant. Therefore, the phase-conjugated wave is

$$\begin{aligned} P &\propto S^* \mathrm{e}^{-\mathrm{i}k_S y} \\ &\propto (S\mathrm{e}^{\mathrm{i}k_S y})^* \end{aligned} \tag{12.7}$$

which is the phase-conjugate of the original signal wave.

Time-reversal is a powerful concept in physics that helps to unify phenomena and to establish correspondences, such as the Stokes relations for transmission and reflection [66]. Time-reversal becomes a powerful tool in optics through its ability

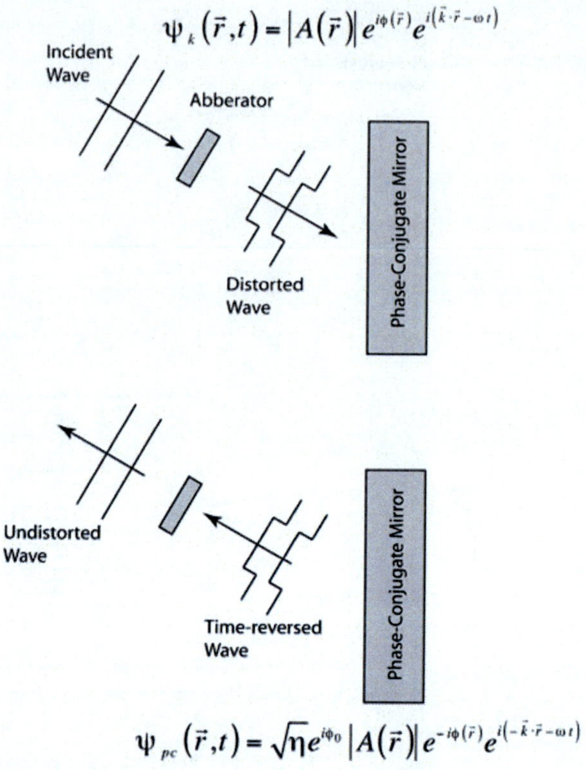

Fig. 12.9 Phase-conjugate aberration correction using a time-reversal mirror

to compensate for refractive aberrations that occur during wave propagation through heterogeneous media. The correspondence of phase conjugation with time-reversal is made by identifying the phase-conjugate wave as

$$\begin{aligned} P &\propto (S\mathrm{e}^{\mathrm{i}k_S y - \mathrm{i}\omega t})^* \\ &\propto (S\mathrm{e}^{\mathrm{i}k_S y + \mathrm{i}\omega(-t)})^* \end{aligned} \tag{12.8}$$

in which the scattered wave is complex-conjugated with a negative time argument (-t). The operation of complex conjugation with negative time is the definition of the time-reversal operator on a complex wave Ψ

$$T\Psi = \Psi^*(-t). \tag{12.9}$$

Therefore, the phase-conjugated wave is also a time-reversed wave. This property allows phase-conjugated waves to retrace their paths through aberrating objects to emerge undistorted.

The behavior of a phase-conjugate mirror is shown in Fig. 12.9 for an incident wave that is distorted by a phase aberration; in the figure it is a dielectric obstacle. The distorted wave is one of the two pump waves on a photorefractive crystal.

The other pump wave is a plane wave that is reflected to counter-propagate against the direction of the incident pump wave. This wave is diffracted through four-wave mixing into the direction of the incident distorted wave. Because the diffracted wave picks up a phase modulation that is the phase-conjugate of the incident wave phase aberrations, when it counter-propagates back through the distorting object, the phase distortion is nulled. What emerges is a plane wave reflected back, despite the interposed distorting object. The reflected phase-conjugated wave is time-reversed – like running the movie backward, as all the phases are compensated and the wave emerges as a plane wave again (bottom of figure). This principle can be used to create an imaging approach that compensates for the aberrating properties of tissue.

Phase conjugation has been applied to biological tissues [13] using the photovoltaic material $LiNbO_3$ as the photorefractive crystal. An image passes through aberrating tissue onto the photorefractive crystal that acts as a phase-conjugating mirror. The conjugate reference beam is diffracted, passing back through the aberrating tissue and is imaged onto a CCD camera. This technique can be used to suppress the turbidity of tissue [67]. However, the dynamic nature of living tissue, relative to the slow speed of the phase-conjugate mirror, limits the fidelity of the phase-conjugate readout [68].

12.2 Digital Holography

The rapidly advancing development of CCD chips and digital cameras has provided low-cost and high-performance digital recording platforms for holographic applications. The hologram writing processes that were discussed in Chap. 2 and in Sect. 12.1 apply directly to hologram recording on CCD camera planes. What changes is the form of readout. In physical holograms, the readout is by diffraction of an illuminating laser. In digital holograms, the readout is performed numerically. Not only does this eliminate the physical hologram and the readout optics, but digital holography has the additional advantage that the quantitative phase of the object is computed directly. The optical arrangement for transmission digital holography is the same as for conventional holography. A schematic is shown in Fig. 12.10 for a thin object and for lens-free holography.

12.2.1 *Free-Space Propagation*

The electric field $\mathbf{E}_s(\mathbf{r})$ at the hologram plane is related to the field $\mathbf{E}_0(\mathbf{r}_0)$ just after the object by the free-space propagator $H(\mathbf{r}, z_d)$ through

$$\mathbf{E}_s(\mathbf{r}) = \int H(\mathbf{r}, z_d)\mathbf{E}_0(\mathbf{r}_0)\, d^2\mathbf{r}_0 \tag{12.10}$$

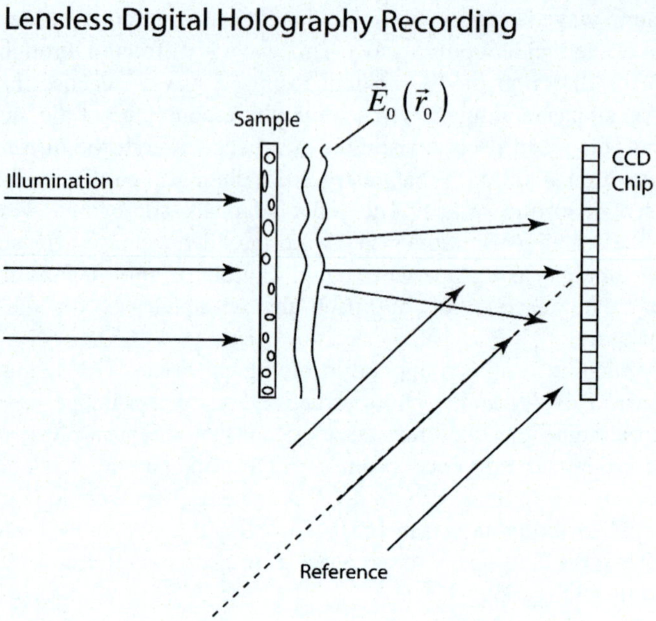

Fig. 12.10 Lens-free holographic recording with a CCD chip. This specific schematic is for a transmission hologram with off-axis reference. The reference angle is highly exaggerated

which is the expression (2.66) in Sect. 2.2.1. The free-space propagator is, in general, a Fresnel propagator, given previously in (2.68)

$$H(\mathbf{r}) = -\mathrm{i}\frac{k}{2\pi z_\mathrm{d}}\mathrm{e}^{\mathrm{i}kz_\mathrm{d}}\mathrm{e}^{\mathrm{i}(k/2z_\mathrm{d})r^2} \tag{12.11}$$

but it also can be any free-space propagator, including that for imaging through a lens, or for performing an optical Fourier transform.

At the hologram plane the signal field combines with the reference field to produce the hologram field

$$\mathbf{E}_\mathrm{H} = \mathbf{E}_\mathrm{r} + \mathbf{E}_\mathrm{s} \tag{12.12}$$

with the hologram intensity

$$I_\mathrm{H} = I_\mathrm{r} + I_\mathrm{s} + E_\mathrm{r}E_\mathrm{s}^* + E_\mathrm{r}^*E_\mathrm{s}. \tag{12.13}$$

Because the reference field is usually a simple wave (like a plane wave or a spherical wave), the amplitude and phase associated with the reference field may be a constant or smoothly varying. Therefore, the amplitude and phase variations of

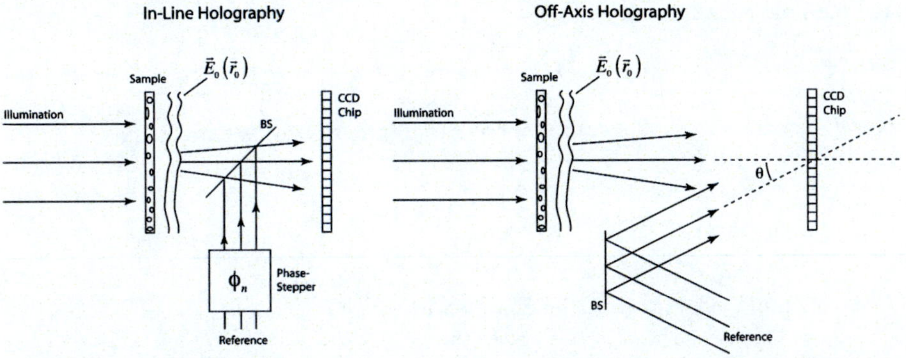

Fig. 12.11 In-line and off-axis holographic recording configurations. The in-line configuration uses a discrete phase stepper to control the relative phase for post-processing phase extraction. The off-axis configuration is spatial heterodyning, with the interference fringes acting as the carrier frequency

the fourth term in the above equation arise from the signal field E_s, which was directly related to the field at the object by

$$I_{Hs} = E_r^* E_s = E_r^* \iint H(\mathbf{r}, z_d)\mathbf{E}_0(\mathbf{r}_0)\, d^2\mathbf{r}_0. \qquad (12.14)$$

The goal of digital holography is therefore accomplished in two steps: (1) extraction of the complex function I_{Hs} from the hologram intensity; and (2) reconstruction of the object phase from the complex function I_{Hs}.

12.2.2 Phase Extraction

The complex function I_{Hs} must be extracted from an intensity distribution on the hologram plane that is purely real. Only by isolating I_{Hs} separately from the third term in (12.13) is it possible to quantify the phase of the object field. There are in general two ways to isolate I_{Hs}: (1) by phase-stepping interferometry; and (2) by off-axis holography (Fig. 12.11).

12.2.2.1 Phase-Stepping Interferometry

In phase-stepping interferometry, the constant phase of the reference wave is stepped through discrete values ϕ_n, and multiple hologram images are recorded, one for each reference phase value. The hologram field is then

$$\mathbf{E}_H = \mathbf{E}_r e^{i\phi_n} + \mathbf{E}_s \qquad (12.15)$$

and the hologram intensity is

$$I_H(\phi_n) = I_r + I_s + e^{i\phi_n} E_r E_s^* + e^{-i\phi_n} E_r^* E_s. \tag{12.16}$$

For N discrete phases, the complex hologram intensity I_{Hs} is extracted as

$$I_{Hs} = \frac{1}{N} \sum_{n=1}^{N} e^{i\phi_n} I_H(\phi_n). \tag{12.17}$$

As an example, when $N = 4$, the complex intensity is extracted as

$$I_{Hs}(x, y) = \frac{1}{4}[(I_H(0) - I_H(\pi)) + i(I_H(\pi/2) - I_H(3\pi/2))]. \tag{12.18}$$

The resulting function $I_{Hs}(x, y)$ is a complex-valued function of pixel position (x,y) on the CCD camera that contains the information of the phase of the object.

12.2.2.2 Off-Axis Holography

Phase-stepping interferometry requires multiple exposures to extract the complex function I_{Hs}. An alternative approach is to tilt the axis of the reference wave relative to the optic axis of the signal wave. This creates a spatial phase that modulates the reference wave as

$$\mathbf{E}_H = \mathbf{E}_r e^{-ikx \sin\theta} + \mathbf{E}_s \tag{12.19}$$

with the resulting interference intensity pattern

$$I_H(x, y) = I_r + I_s + e^{ikx \sin\theta} E_r E_s^* + e^{-ikx \sin\theta} E_r^* E_s. \tag{12.20}$$

This intensity pattern represents a hologram spatially modulated by a carrier wave. An example is shown in Fig. 12.12. The full-frame hologram is shown in Fig. 12.12a with a magnified portion in Fig. 12.12b. The periodic fringes modulate the speckled hologram in the x-direction. A line section through the data of Fig. 12.12a is shown in Fig. 12.13a, which shows the combination of speckle and the fringe pattern. When the trace in Fig. 12.13a is Fourier transformed, it produces the spectrum in Fig. 12.13b that consists of the Fourier transform of I_{Hs}, plus its complex conjugate, in addition to the zero-order field.

From Fig. 12.13b it is seen that the carrier wave and subsequent transformation have spatially separated the multiple terms in (12.20). This is the consequence of taking the Fourier transform that gives

$$\mathrm{FT}(I_H(x, y)) = i_r \delta(k_x)\delta(k_y) + i_s(k_x, k_y) + e_r e_s^*(k_x + k \sin\theta, k_y) + e_r^* e_s(k_x - k \sin\theta, k_y), \tag{12.21}$$

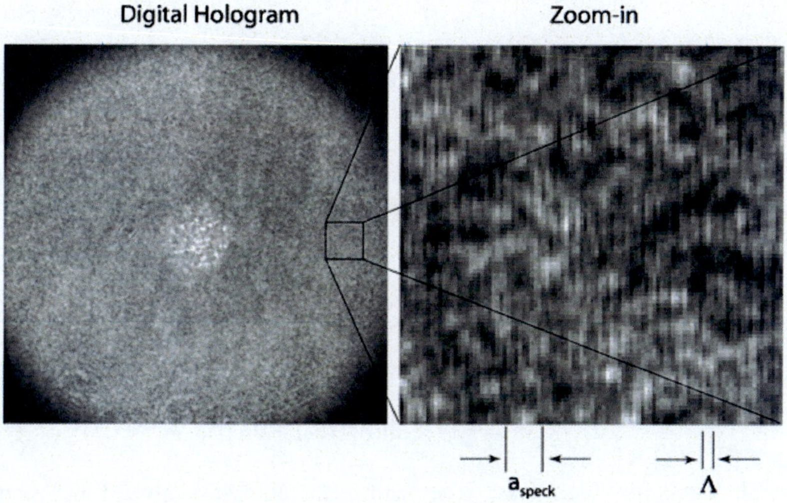

Fig. 12.12 Digital Fourier-domain hologram (*left*) and magnified showing the interference fringes (*right*). From ref. [3]

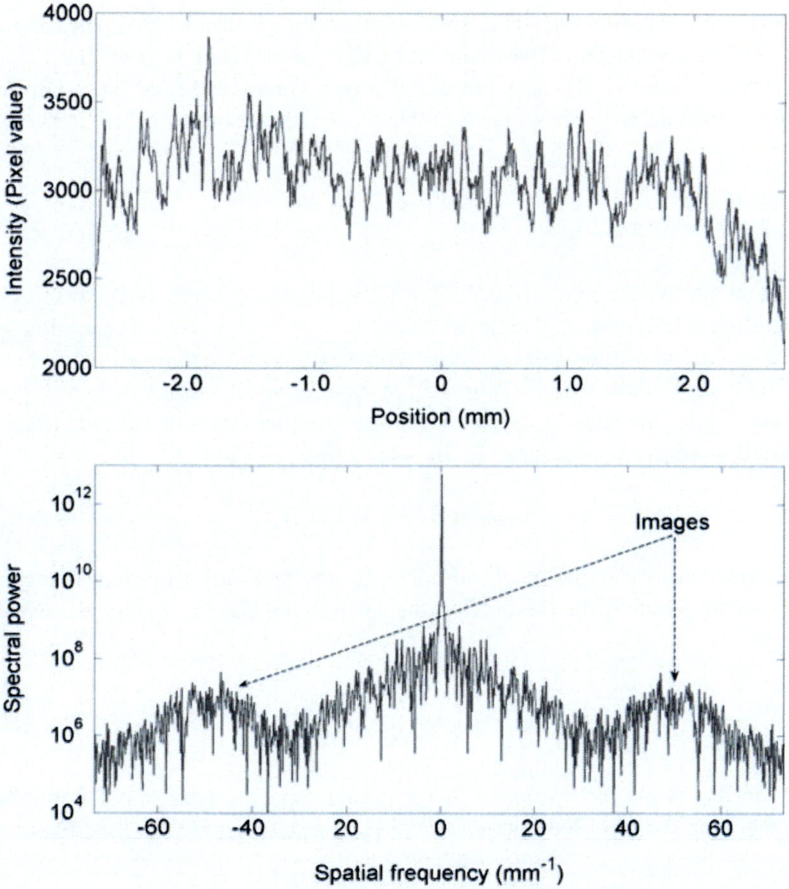

Fig. 12.13 Digital Fourier-domain hologram (*top*) and reconstructed image (*bottom*). From ref. [3]

where i and e represent the Fourier transforms of intensity I and field E, respectively. The resulting transformation consists of a delta function at the center from the reference wave, a zero-order field from the signal, and two displaced complex fields, one centered on $-k \sin \theta$ and the other at $k \sin \theta$. These complex fields have only the spatial variation from E_s, because E_r is assumed uniform.

The retrieval of I_{Hs} is accomplished by isolating the complex values around $k = -k\sin \theta$, recentering them at the origin of the Fourier plane, and transforming back to the real-space plane. This procedure is equivalent to single-sideband demodulation, because the off-axis configuration is equivalent to spatial heterodyne. The complex intensity is then retrieved as

$$I_{Hs}(x, y) = \mathrm{FT}^{-1}(\mathrm{Shift}(\mathrm{HP}(\mathrm{FT}(I_H)))), \tag{12.22}$$

where HP stands for "high-pass." The result is the complex-valued function that is to be related back to the object field $\mathbf{E}_0(\mathbf{r}_0)$.

Off-axis digital holography using coherence-gated depth selection was applied to multicellular tumor spheroids and to the eye of a mouse [3]. Cross-sections of a tumor are shown in Fig. 12.14, which can be compared with Fig. 12.5. The digital reconstructions have finer speckle structure than the photorefractive structures, but the same differentiation between the optically heterogeneous core and the more homogeneous outer shell are captured. The reconstructed digital hologram of the eye of a mouse is shown in Fig. 12.15, showing the cornea, lens, and iris.

12.2.2.3 Resolution Limits

The discreteness of the pixels of a CCD or CMOS array set limits on the resolution of off-axis digital holography. The interference fringes need to be well sampled by the discrete pixels, as shown in Fig. 12.16. Full resolution is predicted using Nyquist sampling theory, with two pixels per fringe spacing Λ, and two fringes per speckle. However, a general rule-of-thumb for practical implementation is three pixels per fringe and three fringes per speckle. In other words

$$a_{\mathrm{speck}} \approx 3\Lambda \approx 3(3\Delta x_{\mathrm{pix}}). \tag{12.23}$$

These values are evaluated on the Fourier plane for Fourier-domain holography. This recording condition establishes the spatial resolution of the reconstructed image as

$$\mathrm{Res} \approx \frac{9}{N}\mathrm{FOV}, \tag{12.24}$$

where the Res is the reconstructed image-domain speckle size, N is the number of pixels that span the recorded Fourier-domain hologram, and FOV is the field-of-view

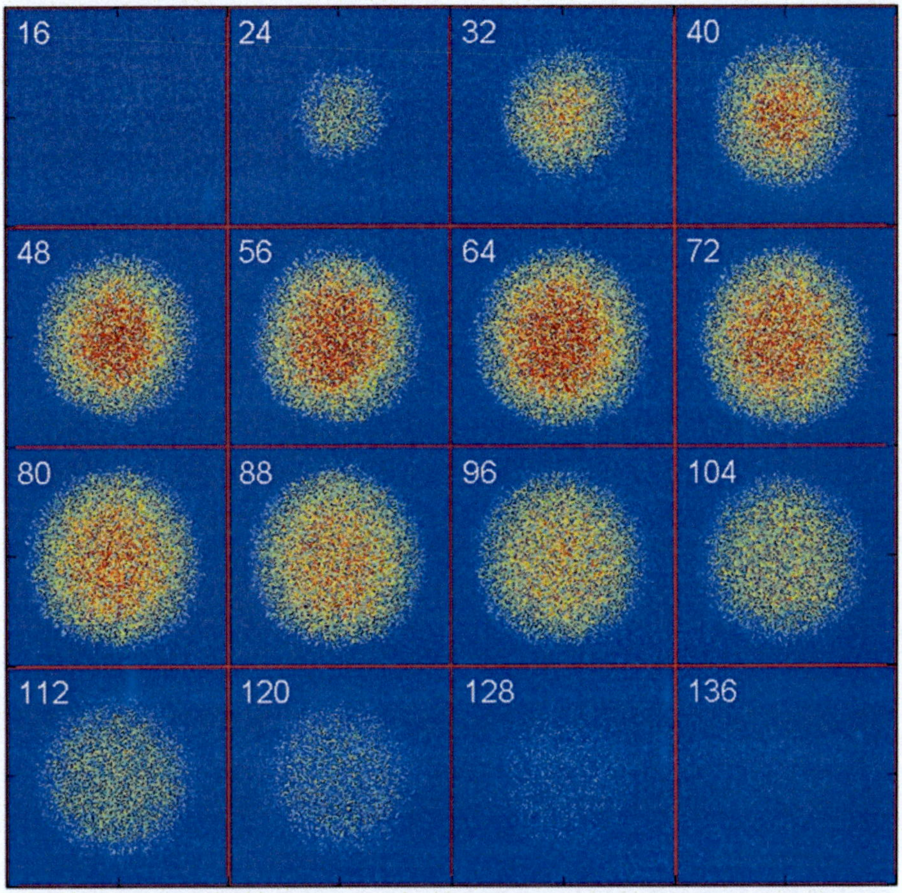

Fig. 12.14 Reconstructed digital holograms from OCI of a tumor spheroid shown at selected coherence-gated depths from *top* to *bottom*. The bright reflections from the necrotic core are clearly distinguishable from the weaker reflections of the homogeneous shell

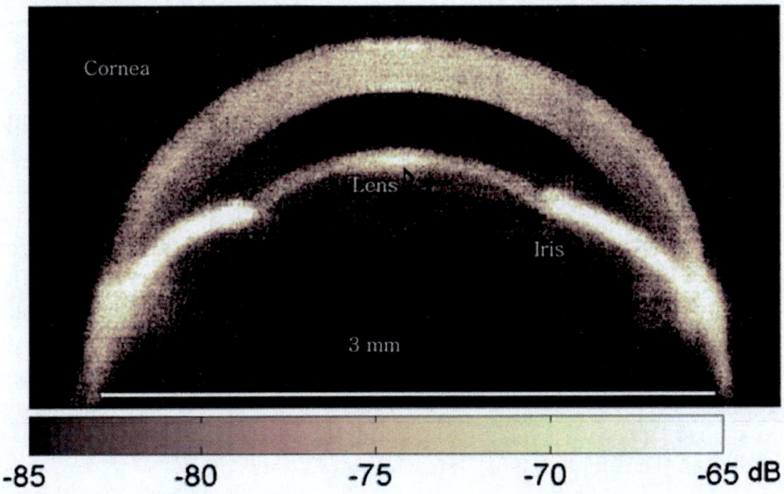

Fig. 12.15 Reconstructed digital holograms from OCI of a mouse eye. From ref. [3]

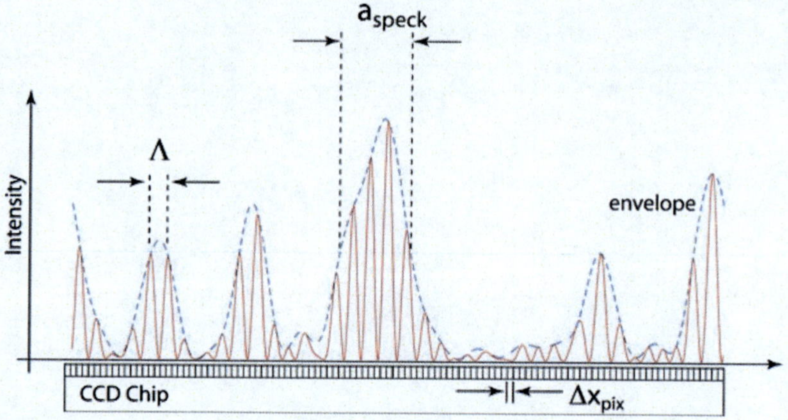

Fig. 12.16 Relationships of sizes for well-resolved speckle holograms. There should be several pixels per fringe, and several fringes per speckle

of the object. As an example, for a field-of-view of 1 mm and a hologram that spans 300 pixels, the resolution is 30 μm.

12.3 Motility Contrast Imaging and Tissue Dynamics Spectroscopy

The active internal motions of cells were described in Sect. 8.4 of Chap. 8, comprised of organelle transport, membrane undulations, and cell shape changes. These different processes occur on different spatial and time scales, with the general trend of large spatial scales occurring on long time scales and small objects moving on fast time scales. When light scatters from living tissue, the internal motions contribute strongly to the dynamic evolution of the light-scattering phases and hence to the speckle. The speckle fields returned from living tissue are highly dynamic, consisting of intensity fluctuations in both space and time. For the case of broad illumination, the spatial variability is not as important as the temporal variability. By measuring the fluctuating intensities, and analyzing their statistical properties, information can be extracted related to the degree of activity of living tissue and the relative contributions from different types of intracellular motions.

Coherence gating can be applied to dynamic light-scattering to provide three-dimensional localization of the intracellular motions of cells [69]. This is the principle of motility contrast imaging (MCI) which has been applied to living tissue and used to study the effects of antimitotic cytoskeletal drugs on tissue dynamics [70]. Coherence-gated dynamic light-scattering lies between the regimes of single and multiple scattering. The coherence gate preferentially selects singly scattered ballistic photons [71–73], but multiply scattered photons may also be path-matched to the reference [74, 75].

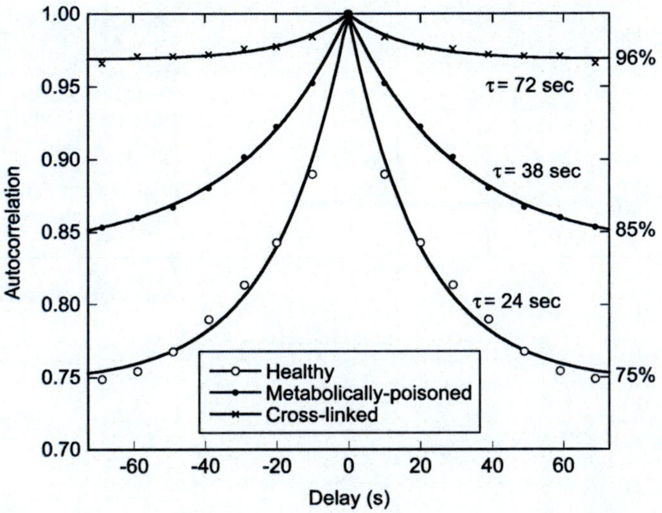

Fig. 12.17 Speckle intensity autocorrelation traces of healthy tissue, metabolically poisoned tissue and cross-linked tissue

12.3.1 Motility Contrast Imaging

Holographic OCI applies a holographic coherence gate on fully developed speckle, capturing dynamic speckle fluctuations from a selected depth inside tissue. An example of intensity fluctuation autocorrelations is shown in Fig. 12.17 for healthy tissue, metabolically poisoned tissue and cross-linked tissue. The variance of the healthy tissue, and the characteristic correlation time, set it apart from cross-linked tissue. Metabolically poisoned tissue has a behavior that falls between these extremes. The statistical differences between healthy vs. necrotic tissue, and between healthy vs. treated tissues provides the basis for a new kind of imaging contrast, called MCI. Examples of MCI, and applications for studying drug–response, are shown in Fig. 12.18.

The basic optical setup in Fig. 12.18a uses light backscatter with an off-axis holographic coherence gate on a CCD camera. The camera is on a Fourier plane, capturing Fourier-domain holograms of the tissue. At a fixed depth, multiple sequential holograms are recorded and reconstructed numerically. The varying intensities of each pixel from frame-to-frame carry information about the dynamical motions that occur at that depth and at that *x*–*y* location. The normalized standard deviation (NSD) is color-coded and plotted in Fig. 12.18b for several selected depths inside an 800-μm diameter tumor spheroid. The proliferating shell shows highly dynamic behavior (red colors) surrounding the necrotic core (blue colors). The volumetric reconstruction is shown in Fig. 12.18c as a cut-away. This was the first volumetric motility contrast image of living tissue [70]. The dynamical motion of

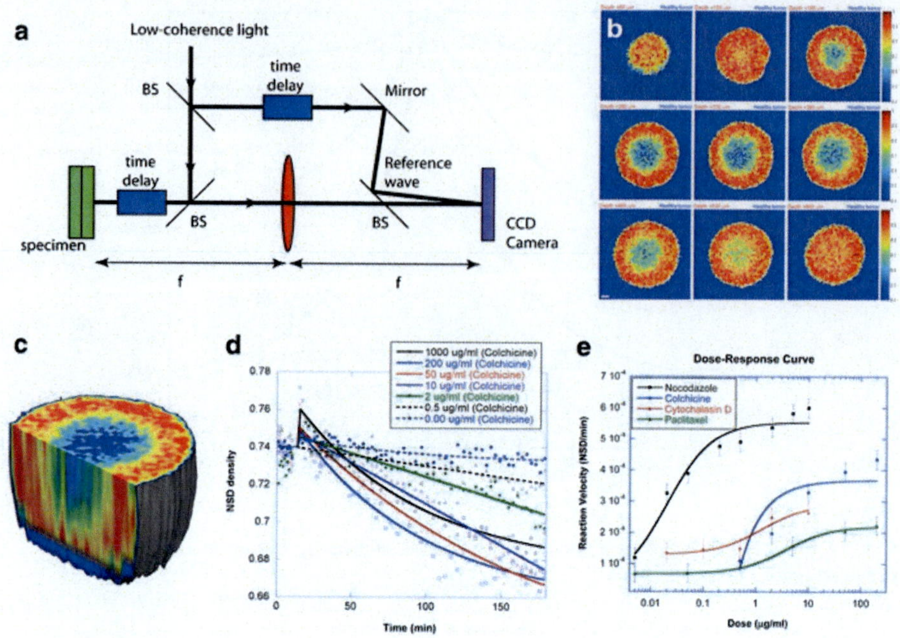

Fig. 12.18 Examples of motility contrast imaging (MCI). (**a**) Schematic of the holographic coherence gating optics. (**b**) Motility contrast image sections of a tumor, with false color proportional to the degree of motion. The healthy shell is clearly contrasted with the necrotic core. (**c**) A volumetric reconstruction based on intracellular motility. (**d**) Response curves for increasing concentrations of colchicine. (**e**) Dose–response curves for nocodazole, colchicine, cytochalasin D, and taxol

the living tissue provides a natural (endogenous) image contrast for function imaging.

In addition to providing an imaging contrast, the dynamical motion is also a measure of the internal dynamics of the tissue, and can be used to track changes caused by changes in the tissue environment or the tissue response to applied drugs. The NSD motility metric can be tracked in time after a dose is applied. The time-course of the motility is shown in Fig. 12.18d for several doses of the antimitotic drug colchicine. Higher doses act faster and have a stronger suppression of the intracellular motions. Colchicine acts by inhibiting microtubule polymerization and hence inhibits mitosis, but also suppresses the motions of organelles that rely on the microtubules as their pathway. A different antimitotic drug, taxol, has the opposite effect and stabilizes microtubules. This also inhibits mitosis, but the microtubules remain available for other natural functions of the cell. The dose-response of several antimitotic drugs is shown in Fig. 12.18e, for tubulin destabilizers colchicine and nocodazole, tubulin stabilizer taxol, and actin destabilizer cytochalasin D. The use of the motility metric to track tissue response to drugs is performed nondestructively, without labels, and in real-time.

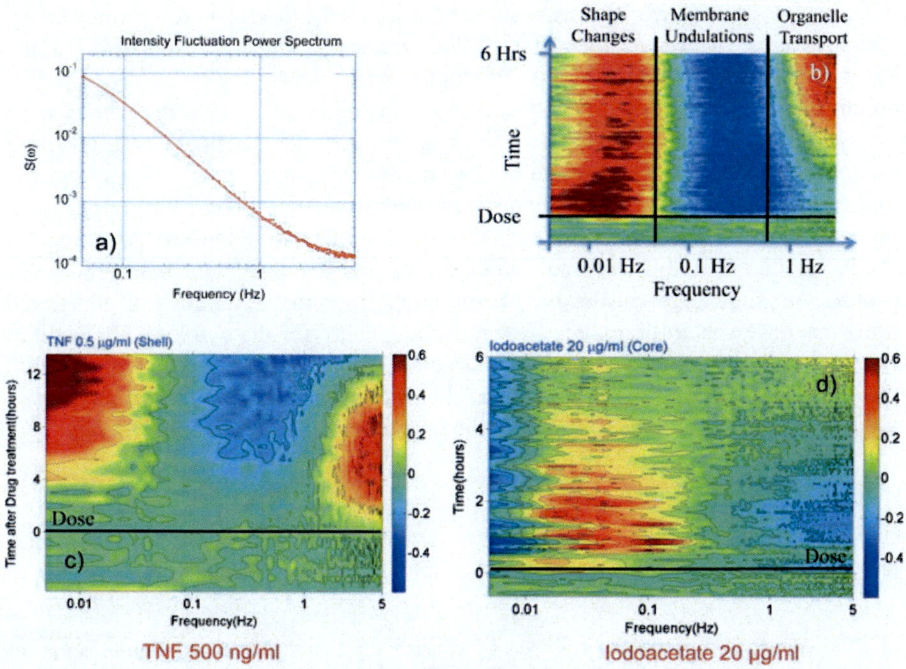

Fig. 12.19 Examples of tissue dynamics spectroscopy (TDS). (**a**) Power spectrum of speckle intensity fluctuations. (**b**) A drug–response spectrogram, showing relative spectral changes with frequency along the x-axis and time along the y-axis. The drug–dose is applied at $t = 0$. (**c**) Drug–response spectrogram for the cytokine TNF. (**d**) Drug–response spectrogram for iodoacetate

12.3.2 *Tissue Dynamics Spectroscopy (TDS)*

MCI captures the total motions of living tissue, but does not differentiate the many types of motion that were discussed in Chap. 8. The key to identifying different types of motion using coherence gating is to analyze the frequency contributions to dynamic speckle intensity fluctuations returned from living tissue. For instance, the spectral power density from healthy tissue is shown in Fig. 12.19a. This power spectrum is constructed by finding the different frequency contributions to the fluctuating speckle. The spectrum has a frequency roll-off above 0.1 Hz and falls with an exponent around $n = 1.7$ as it approaches the noise floor above 5 Hz. The spectrum is essentially featureless, without any significant spectral features to identify possibly different contributions. However, changes in the power spectrum caused by environmental perturbations or applied drugs can show more specific signatures.

The differential relative spectral power density is defined as

$$D(\omega, t) = \frac{S(\omega, t) - S(\omega, t_0)}{S(\omega, t_0)}, \tag{12.25}$$

where $S(\omega,t)$ is the power spectrum at time t, and t_0 is the time used for normalization (prior to perturbation of the tissue). The frequency-time representation of the spectral changes constitutes a tissue response spectrogram for an applied drug or condition [4]. An example for nocodazole is shown in Fig. 12.19b. Frequency is on the x-axis and time is on the y-axis. The drug dose is applied at time $t = 0$. The drug induces enhanced spectral content at low and high frequencies, and suppresses the mid frequencies. There are approximately three frequency bands. The upper band above 0.5 Hz is associated with organelle transport. The mid band from 0.05 to 0.5 Hz is associated with membrane undulations, and the low band below 0.05 Hz captures gross changes in the membrane and cell shape. Examples of additional drug–response spectrograms are shown in Fig. 12.19c for the cytokine TNF and in Fig. 12.19d for the metabolic poison iodoacetate. Each drug has its own "fingerprint" that captures how the drug affects intracellular motions. These fingerprints may be useful for drug candidate screening in the drug development pipeline.

References

1. Beaurepaire, E., Boccara, A.C., Lebec, M., Blanchot, L., Saint-Jalmes, H.: Full-field optical coherence microscopy. Opt. Lett. **23**, 244 (1998)
2. Dubois, A., Grieve, K., Moneron, G., Lecaque, R., Vabre, L., Boccara, C.: Ultrahigh-resolution full-field optical coherence tomography. Appl. Opt. **43**(14), 2874–2883 (2004)
3. Jeong, K., Turek, J.J., Nolte, D.D.: Fourier-domain digital holographic optical coherence imaging of living tissue. Appl. Opt. **46**, 4999–5008 (2007)
4. Jeong, K., Turek, J.J., Nolte, D.D.: Speckle fluctuation spectroscopy of intracellular motion in living tissue using coherence-domain digital holography. J. Biomed. Opt. **15**(3), 030514 (2010)
5. Cuche, E., Marquet, P., Depeursinge, C.: Simultaneous amplitude-contrast and quantitative phase-contrast microscopy by numerical reconstruction of Fresnel off-axis holograms. Appl. Opt. **38**(34), 6994–7001 (1999)
6. Schnars, U., Juptner, W.P.O.: Digital recording and numerical reconstruction of holograms. Meas. Sci. Technol. **13**(9), R85–R101 (2002)
7. Nolte, D.D. (ed.): Photorefractive effects and materials. Kluwer Academic Publishers, Dordrecht (1995)
8. Nolte, D.D., Cubel, T., Pyrak-Nolte, L.J., Melloch, M.R.: Adaptive beam combining and interferometry using photorefractive quantum wells. J. Opt. Soc. Am. B **18**, 195–205 (2001)
9. Peng, L., Varma, M.M., Cho, W., Regnier, F.E., Nolte, D.D.: Adaptive interferometry of protein on a BioCD. Appl. Opt. **46**, 5384–5395 (2007)
10. Zeldovich, B.Y., Pilipetsky, N.F., Shkunov, V.V.: Principles of phase conjugation. Optical sciences. Springer, Berlin (1985)
11. Sakai, J.-I.: Phase conjugate optics. McGraw-Hill, New York (1992)
12. Gower, M., Proch, D. (eds.): Optical phase conjugation. Springer, Berlin (1994)
13. Yaqoob, Z., Psaltis, D., Feld, M.S., Yang, C.H.: Optical phase conjugation for turbidity suppression in biological samples. Nat. Photon. **2**(2), 110–115 (2008)
14. Glass, A.M., von der Linde, D., Negran, T.J.: High-voltage bulk photovoltaic effect and the photorefractive process in LiNbO3. Appl. Phys. Lett. **25**, 233 (1974)
15. Feinberg, J.: Photorefractive nonlinear optics. Phys. Today **41**, 46 (1988)

16. Glass, A.M., Johnson, A.M., Olson, D.H., Simpson, W., Ballman, A.A.: Four-wave mixing in semi-insulating InP and GaAs using the photorefractive effect. Appl. Phys. Lett. **44**(10), 948–950 (1984)
17. Nolte, D.D., Melloch, M.R.: Photorefractive quantum wells and thin films. In: Nolte, D.D. (ed.) Photorefractive Effects and Materials. Kluwer Academic Publishers, Dordrecht (1995)
18. Nolte, D.D.: Semi-insulating semiconductor heterostructures: optoelectronic properties and applications. J. Appl. Phys. **85**, 6259 (1999)
19. Ducharme, S., Scott, J.C., Twieg, R.J., Moerner, W.E.: Observation of the photorefractive effect in a polymer. Phys. Rev. Lett. **66**(14), 1846–1849 (1991)
20. Meerholz, K., Volodin, B.L., Sandalphon, Kippelen, B., Peyghambarian, N.: A photorefractive polymer with high optical gain and diffraction efficiency near 100-percent. Nature **371**(6497), 497–500 (1994)
21. Hyde, S.C.W., Barry, N.P., Jones, R., Dainty, J.C., French, P.M.W., Klein, M.B., Wechsler, B. A.: Depth-resolved holographic imaging through scattering media by photorefraction. Opt. Lett. **20**, 1331 (1995)
22. Hyde, S.C.W., Jones, R., Barry, N.P., Dainty, J.C., French, P.M.W., Kwolek, K.M., Nolte, D. D., Melloch, M.R.: Depth-resolved holography through turbid media using photorefraction. IEEE J. Sel. Top. Quantum Electron. **2**(4), 965–975 (1996)
23. Yu, P., Mustata, M., French, P.M.W., Turek, J.J., Melloch, M.R., Nolte, D.D.: Holographic optical coherence imaging of tumor spheroids. Appl. Phys. Lett. **83**, 575 (2003)
24. Salvador, M., Prauzner, J., Kober, S., Meerholz, K., Jeong, K., Nolte, D.D.: Depth-resolved holographic optical coherence imaging using a high-sensitivity photorefractive polymer device. Appl. Phys. Lett. **93**(23) (2008)
25. Jeong, K., Peng, L.L., Nolte, D.D., Melloch, M.R.: Fourier-domain holography in photorefractive quantum-well films. Appl. Opt. **43**(19), 3802–3811 (2004)
26. Pollard, T.D., Borisy, G.G.: Cellular motility driven by assembly and disassembly of actin filaments. Cell **112**(4), 453–465 (2003)
27. Cukierman, E., Pankov, R., Stevens, D.R., Yamada, K.M.: Taking cell-matrix adhesions to the third dimension. Science **294**(5547), 1708–1712 (2001)
28. Webb, D.J., Horwitz, A.F.: New dimensions in cell migration. Nat. Cell Biol. **5**(8), 690 (2003)
29. Keller, P.J., Pampaloni, F., Stelzer, E.H.K.: Life sciences require the third dimension. Curr. Opin. Cell Biol. **18**(1), 117–124 (2006)
30. Pampaloni, F., Reynaud, E.G., Stelzer, E.H.K.: The third dimension bridges the gap between cell culture and live tissue. Nat. Rev. Mol. Cell Biol. **8**(10), 839–845 (2007)
31. Voytik-Harbin, S.L.: Three-dimensional extracellular matrix substrates for cell culture. Meth. Cell Biol. **63**, 561 (2001)
32. Lee, J., Cuddihy, M.J., Kotov, N.A.: Three-dimensional cell culture matrices: state of the art. Tissue Eng. B Rev. **14**(1), 61–86 (2008)
33. Bissell, M.J., Myers, C., Lee, G., Lee, E., Rizki, A., Mian, S., Gray, J., Radisky, D.: A breast cancer progression model: the importance of three-dimensional tissue architecture and metalloproteinases. Breast Cancer Res. **7**, S6–S6 (2005)
34. Johnson, K.R., Leight, J.L., Weaver, V.M.: Demystifying the effects of a three-dimensional microenvironment in tissue morphogenesis. In: Wang, Y.L., Discher, D.E. (eds.) Cell Mechanics. Methods in Cell Biology, vol. 83, pp. 547–583. Elsevier Academic Press, San Diego (2007)
35. Lee, G.Y., Kenny, P.A., Lee, E.H., Bissell, M.J.: Three-dimensional culture models of normal and malignant breast epithelial cells. Nat. Methods **4**(4), 359–365 (2007)
36. Kenny, P.A.: Three-dimensional extracellular matrix culture models of EGFR signalling and drug response. Biochem. Soc. Trans. **35**, 665–668 (2007)
37. Pizzo, A.M., Kokini, K., Vaughn, L.C., Waisner, B.Z., Voytik-Harbin, S.L.: Extracellular matrix (ECM) microstructural composition regulates local cell-ECM biomechanics and fundamental fibroblast behavior: a multidimensional perspective. J. Appl. Physiol. **98**(5), 1909–1921 (2005)

38. Roeder, B.A., Kokini, K., Sturgis, J.E., Robinson, J.P., Voytik-Harbin, S.L.: Tensile mechanical properties of three-dimensional type I collagen extracellular matrices with varied microstructure. J. Biomech. Eng. **124**(2), 214–222 (2002)
39. Pedersen, J.A., Swartz, M.A.: Mechanobiology in the third dimension. Ann. Biomed. Eng. **33** (11), 1469–1490 (2005)
40. Voytik-Harbin, S.L., Rajwa, B., Robinson, J.P.: Three-dimensional imaging of extracellular matrix and extracellular matrix-cell interactions. Meth. Cell Biol. **63**, 583 (2001)
41. Moerner, W.E., Fromm, D.P.: Methods of single-molecule fluorescence spectroscopy and microscopy. Rev. Sci. Instrum. **74**(8), 3597–3619 (2003)
42. Webb, R.H.: Confocal optical microscopy. Rep. Prog. Phys. **59**(3), 427–471 (1996)
43. Cahalan, M.D., Parker, I., Wei, S.H., Miller, M.J.: Two-photon tissue imaging: Seeing the immune system in a fresh light. Nat. Rev. Immunol. **2**(11), 872–880 (2002)
44. Konig, K.: Multiphoton microscopy in life sciences. J. Microsc. **200**, 83–104 (2000)
45. Sharpe, J., Ahlgren, U., Perry, P., Hill, B., Ross, A., Hecksher-Sorensen, J., Baldock, R., Davidson, D.: Optical projection tomography as a tool for 3D microscopy and gene expression studies. Science **296**(5567), 541–545 (2002)
46. Huisken, J., Swoger, J., Del Bene, F., Wittbrodt, J., Stelzer, E.H.K.: Optical sectioning deep inside live embryos by selective plane illumination microscopy. Science **305**(5686), 1007–1009 (2004)
47. Grinnell, F., Rocha, L.B., Iucu, C., Rhee, S., Jiang, H.M.: Nested collagen matrices: a new model to study migration of human fibroblast populations in three dimensions. Exp. Cell Res. **312**(1), 86–94 (2006)
48. Friedl, P., Brocker, E.B.: The biology of cell locomotion within three-dimensional extracellular matrix. Cell. Mol. Life Sci. **57**(1), 41–64 (2000)
49. Friedl, P., Zanker, K.S., Brocker, E.B.: Cell migration strategies in 3-D extracellular matrix: Differences in morphology, cell matrix interactions, and integrin function. Microsc. Res. Tech. **43**(5), 369–378 (1998)
50. Gustafsson, M.G.L.: Nonlinear structured-illumination microscopy: Wide-field fluorescence imaging with theoretically unlimited resolution. Proc. Natl. Acad. Sci. USA **102**(37), 13081–13086 (2005)
51. Klar, T.A., Jakobs, S., Dyba, M., Egner, A., Hell, S.W.: Fluorescence microscopy with diffraction resolution barrier broken by stimulated emission. Proc. Natl. Acad. Sci. USA **97** (15), 8206–8210 (2000)
52. de Ridder, L.: Autologous confrontation of brain tumor derived spheroids with human dermal spheroids. Anticancer Res. **17**, 4119–4120 (1997)
53. Groebe, K., Mueller-Klieser, W.: On the relation between size of necrosis and diameter of tumor spheroids. Int. J. Radiat. Oncol. Biol. Phys. **34**, 395–401 (1996)
54. Hamamoto, R., Yamada, K., Kamihira, M., Iijima, S.: Differentiation and proliferation of primary rat hepatocytes cultured as spheroids. J. Biochem. (Tokyo) **124**, 972–979 (1998)
55. Hamilton, G.: Multicellular spheroids as an in vitro tumor model. Cancer Lett. **131**, 29–34 (1998)
56. Hargrave, P., Nicholson, P.W., Delpy, D.T., Firbank, M: Optical properties of multicellular tumour spheroids. Phys. Med. Biol. **41**, 1067–1072 (1996)
57. Kunz-Schughart, L.A., Kreutz, M., Knuechel, R.: Multicellular spheroids: a three-dimensional in vitro culture system to study tumour biology. Int. J. Exp. Pathol. **79**, 1–23 (1998)
58. Mueller-Klieser, W. Biophys. J. **46**, 343–348 (1984)
59. Freyer, P., Schor, P.L., Jarrett, K.A., Neeman, M., Sillerud, L.O. Cancer Res. **51**, 3831–3837 (1991)
60. Mueller-Klieser, W.: Three-dimensional cell cultures: from molecular mechanisms to clinical applications. Am. J. Physiol. **273**, 1109–1123 (1997)
61. Hyde, S.C.W., Barry, N.P., Jones, R., Dainty, J.C., French, P.M.W.: Sub-100 micron depth-resolved holographic imaging through scattering media in the near-infrared. Opt. Lett. **20**(22), 2330–2332 (1995)

62. Nolte, D.D., Jeong, K., French, P.M.W., Turek, J.J.: Holographic optical coherence imaging. In: Drexler, J.F.A.W. (ed.) Optical Coherence Tomography: Technology and Applications. Springer, Berlin (2008)
63. Vabre, L., Dubois, A., Boccara, A.C.: Thermal-light full-field optical coherence tomography. Opt. Lett. **27**(7), 530–532 (2002)
64. Jeong, K., Peng, L., Turek, J.J., Melloch, M.R., Nolte, D.D.: Fourier-domain holographic optical coherence imaging of tumor spheroids and mouse eye. Appl. Opt. **44**(10), 1798–1806 (2005)
65. Karamata, B., Leutenegger, M., Laubscher, M., Bourquin, S., Lasser, T., Lambelet, P.: Multiple scattering in optical coherence tomography. II. Experimental and theoretical investigation of cross talk in wide-field optical coherence tomography. J. Opt. Soc. Am. A Opt. Image Sci. Vis. **22**(7), 1380–1388 (2005)
66. Potton, R.J.: Reciprocity in optics. Rep. Prog. Phys. **67**(5), 717–754 (2004)
67. McDowell, E.J., Cui, M., Vellekoop, I.M., Senekerimyan, V., Yaqoob, Z., Yang, C.H.: Turbidity suppression from the ballistic to the diffusive regime in biological tissues using optical phase conjugation. J. Biomed. Opt. **15**(2) (2010)
68. Cui, M., McDowell, E.J., Yang, C.H.: An in vivo study of turbidity suppression by optical phase conjugation (TSOPC) on rabbit ear. Opt. Express **18**(1), 25–30 (2010)
69. Yu, P., Peng, L., Mustata, M., Turek, J.J., Melloch, M.R., Nolte, D.D.: Time-dependent speckle in holographic optical coherence imaging and the state of health of tumor tissue. Opt. Lett. **29**(1), 68–70 (2004)
70. Jeong, K., Turek, J.J., Nolte, D.D.: Imaging motility contrast in digital holography of tissue response to cytoskeletal anti-cancer drugs. Opt. Express **15**, 14057 (2007)
71. Boas, D.A., Bizheva, K.K., Siegel, A.M.: Using dynamic low-coherence interferometry to image Brownian motion within highly scattering media. Opt. Lett. **23**(5), 319–321 (1998)
72. Bizheva, K.K., Siegel, A.M., Boas, D.A.: Path-length-resolved dynamic light scattering in highly scattering random media: the transition to diffusing wave spectroscopy. Phys. Rev. E **58**(6), 7664–7667 (1998)
73. Wax, A., Yang, C.H., Dasari, R.R., Feld, M.S.: Path-length-resolved dynamic light scattering: modeling the transition from single to diffusive scattering. Appl. Opt. **40**(24), 4222–4227 (2001)
74. Petoukhova, A.L., Steenbergen, W., van Leeuwen, T.G., de Mul, F.F.M.: Effects of absorption on coherence domain path length resolved dynamic light scattering in the diffuse regime. Appl. Phys. Lett. **81**(4), 595–597 (2002)
75. Varghese, B., Rajan, V., Van Leeuwen, T.G., Steenbergen, W.: Path-length-resolved measurements of multiple scattered photons in static and dynamic turbid media using phase-modulated low-coherence interferometry. J. Biomed. Opt. **12**(2) (2007)

Chapter 13
Appendix: Mathematical Formulas

13.1 Gaussian Integrals

$$\int_0^\infty e^{-ax^2}\,dx = \frac{1}{2}\sqrt{\frac{\pi}{a}}$$

$$\int_{-\infty}^{\infty} e^{-x^2/2\sigma^2}\,dx = \sqrt{2\pi}\sigma$$

$$\int_{-\infty}^{\infty} e^{-x^2/\sigma^2}\,dx = \sqrt{\pi}\sigma$$

$$\int_0^\infty x^m e^{-ax^2} dx = \frac{\Gamma\left(\frac{m+1}{2}\right)}{2a^{\frac{m+1}{2}}} = \begin{cases} \frac{1}{2}\sqrt{\frac{\pi}{a}} & m = 0 \\ \frac{1}{2a} & m = 1 \\ \frac{\sqrt{\pi}}{4a^{3/2}} & m = 2 \\ \frac{1}{2a^2} & m = 3 \end{cases}$$

D.D. Nolte, *Optical Interferometry for Biology and Medicine*, Bioanalysis 1,
DOI 10.1007/978-1-4614-0890-1_13, © Springer Science+Business Media, LLC 2012

13.2 Gaussian Beams

A Gaussian beam with intensity

$$I(x,y) = S_0 \exp(-(x^2+y^2)/w_0^2)$$

has an integrated power

$$\begin{aligned}
P &= \iint I(x,y)\,\mathrm{d}x\,\mathrm{d}y \\
&= S_0 \int_0^{\infty} \mathrm{e}^{-r^2/w_0^2} 2\pi r\,\mathrm{d}r \\
&= 2\pi S_0 \int_0^{\infty} \mathrm{e}^{-r^2/2w_0^2} r\,\mathrm{d}r \\
&= 2\pi S_0 \frac{w_0^2}{2} \\
&= S_0 \pi w_0^2
\end{aligned}$$

The dimensionless field amplitude is

$$E = \sqrt{I} = \sqrt{S_0} \exp(-(x^2+y^2)/2w_0^2)$$

When Fourier transformed using a lens of focal distance f, the field at the Fourier plane is

$$E(x',y') = \sqrt{S_0}\frac{2\pi w_0^2}{f\lambda} \exp\left(-\frac{2\pi^2 w_0^2}{(f\lambda)^2}(x'^2+y'^2)\right)$$

The intensity at the Fourier plane is

$$I(x',y') = S_0 \left(\frac{2\pi}{f\lambda}\right)^2 w_0^4 \exp\left(-\left(\frac{2\pi}{f\lambda}\right)^2 w_0^2 (x'^2+y'^2)\right)$$

with an integrated power

$$P = \iint I(x',y')\,\mathrm{d}x'\,\mathrm{d}y' = S_0 \pi w_0^2$$

13.3 Fourier Transforms

Definition:

$$F(k) = \mathrm{FT}(f(x)) = \int_{-\infty}^{\infty} f(x)\mathrm{e}^{-\mathrm{i}kx}\,\mathrm{d}x$$

$$f(x) = \mathrm{FT}^{-1}(F(k)) = \frac{1}{2\pi}\int_{-\infty}^{\infty} F(k)\,\mathrm{e}^{\mathrm{i}kx}\,\mathrm{d}k = \int_{-\infty}^{\infty} F(\nu)\mathrm{e}^{\mathrm{i}2\pi\nu x}\,\mathrm{d}\nu$$

Fourier transforms of Gaussian functions:

$$\int_{-\infty}^{\infty} \mathrm{e}^{-ax^2}\mathrm{e}^{-\mathrm{i}kx}\,\mathrm{d}x = \sqrt{\frac{\pi}{a}}\mathrm{e}^{-k^2/4a}$$

$$\int_{-\infty}^{\infty} \mathrm{e}^{-(x-x_0)^2/2\sigma^2}\mathrm{e}^{-\mathrm{i}kx}\,\mathrm{d}x = \mathrm{e}^{-\mathrm{i}kx_0}\sqrt{2\pi}\sigma\mathrm{e}^{-k^2\sigma^2/2}$$

$$\frac{1}{2}\int_{-\infty}^{\infty}\left[\mathrm{e}^{-(x-x_0)^2/2\sigma^2} + \mathrm{e}^{-(x+x_0)^2/2\sigma^2}\right]\mathrm{e}^{-\mathrm{i}kx}\,\mathrm{d}x = \sqrt{2\pi}\sigma\mathrm{e}^{-k^2\sigma^2/2}\cos(kx_0)$$

$$\int_{-\infty}^{\infty}[\mathrm{e}^{-(x-x_1)^2/2\sigma^2} + \mathrm{e}^{-(x-x_2)^2/2\sigma^2}]\mathrm{e}^{-\mathrm{i}kx}\mathrm{d}x = [\mathrm{e}^{-\mathrm{i}kx_1} + \mathrm{e}^{-\mathrm{i}kx_2}]\sqrt{2\pi}\sigma\mathrm{e}^{-k^2\sigma^2/2}$$

$$= 2\sqrt{2\pi}\sigma\mathrm{e}^{-k^2\sigma^2/2}\mathrm{e}^{-\mathrm{i}k\bar{x}}\cos(k\Delta x)$$

13.3.1 Autocorrelation Relationships

$$A(\tau) = \int_{-\infty}^{\infty} f(t)f(t+\tau)\,\mathrm{d}t$$

Wiener–Khinchin theorem:

$$A(\tau) = \mathrm{FT}^{-1}(F(\omega)F^*(\omega)) = \mathrm{FT}^{-1}(S(\omega))$$

$$S(\omega) = \frac{\langle |f|^2 \rangle}{\pi} \frac{\omega_0}{\omega_0^2 + \omega^2} \quad \frac{1}{\pi} \int_{-\infty}^{\infty} \frac{\omega_0}{\omega_0^2 + \omega^2} \, \mathrm{d}\omega = 1$$

$$\mathrm{FT}(\mathrm{e}^{-t/\tau}) = \left(\frac{1}{\tau}\right) \left[\frac{1}{(1/\tau)^2 + \omega^2}\right] = \frac{\tau}{1 + (\omega\tau)^2}$$

Parseval's theorem:

$$\int_{-\infty}^{\infty} f(t) g^*(t) \, \mathrm{d}t = \int_{-\infty}^{\infty} F(\nu) G^*(\nu) \, \mathrm{d}\nu$$

Rayleigh's theorem:

$$\int_{-\infty}^{\infty} f(t) f^*(t) \, \mathrm{d}t = \int_{-\infty}^{\infty} F(\nu) F^*(\nu) \, \mathrm{d}\nu$$

13.4 Gaussian Pulses

The dimensionless field of a Gaussian pulse is

$$E(t) = \sqrt{S_0} \exp(-\mathrm{i}\omega_0 t) \exp(-t^2/2\tau^2)$$

with a total fluence given by

$$J = \int_{-\infty}^{\infty} |E(t)|^2 \, \mathrm{d}t = S_0 \int_{-\infty}^{\infty} \exp(-t^2/\tau^2) \, \mathrm{d}t$$
$$= S_0 \sqrt{\pi} \tau$$

In the spectral domain, the spectral density is

$$s(\omega) = \int_{-\infty}^{\infty} E(t)\,\mathrm{d}t = \int_{-\infty}^{\infty} \sqrt{S_0}\exp(-\mathrm{i}\omega_0 t)\exp(-t^2/2\tau^2)\,\mathrm{d}t$$
$$= \sqrt{2\pi S_0}\tau\exp(-\omega^2\tau^2/2)$$

and the total fluence is

$$J = \frac{1}{2\pi}\int_{-\infty}^{\infty} |s(\omega)|^2\,\mathrm{d}\omega$$
$$= \frac{2\pi\tau^2}{2\pi}S_0\int_{-\infty}^{\infty}\exp(-\omega^2\tau^2)\,\mathrm{d}\omega = \tau^2\sqrt{\pi}\frac{1}{\tau}$$
$$= S_0\sqrt{\pi}\tau$$

13.5 Error Function

$$\mathrm{erf}(x) = \frac{2}{\sqrt{\pi}}\int_0^x \mathrm{e}^{-t^2}\,\mathrm{d}t$$

$$\int \mathrm{erf}(x)\,\mathrm{d}x = x\,\mathrm{erf}(x) + \frac{\mathrm{e}^{-x^2}}{\sqrt{\pi}}$$

$$\int_0^p \mathrm{e}^{-x^2}\mathrm{erf}(p-x)\,\mathrm{d}x = \frac{1}{2}\sqrt{\pi}\left[\mathrm{erf}\left(\frac{1}{2}\sqrt{2}p\right)\right]^2$$

13.6 Gaussian Diffusion

The mean displacement squared for Gaussian diffusion is

$$\langle x^2\rangle = \frac{1}{\delta}\sqrt{\frac{1}{2\pi}}\int_{-\infty}^{\infty} x^2\mathrm{e}^{-x^2/2\delta^2}\,\mathrm{d}x = \delta^2$$

The squared displacement after N symmetric steps is

$$\Delta x^2 = N\delta^2$$

13.7 Probability Distribution Generation

$$\frac{\mathrm{d}y}{\mathrm{d}x} = \frac{1}{P(y)}$$

$$P(y)\,\mathrm{d}y = \mathrm{d}x$$

$$F(y) = \int P(y)\,\mathrm{d}y = x$$

$$y = F^{-1}(x)$$

Example

$$P(y) = 2(1-y)$$

$$F(y) = 2y - y^2 = x$$

$$y = \frac{2 \pm \sqrt{4-4x}}{2} = 1 - \sqrt{1-x}$$

13.8 Trigonometric Identities

$$\sin(A \pm B) = \sin A \cos B \pm \cos A \sin B$$

$$\cos(A \pm B) = \cos A \cos B \mp \sin A \sin B$$

$$\tan(A \pm B) = \frac{\tan A \pm \tan B}{1 \pm \tan A \tan B}$$

$$\sin 2A = 2 \sin A \cos A = \frac{2 \tan A}{1 + \tan^2 A}$$

$$\cos 2A = \cos^2 A - \sin^2 A = \frac{1 - \tan^2 A}{1 + \tan^2 A}$$

$$\begin{aligned}
e^{i\theta} x X^* + e^{-i\theta} x^* X &= e^{i\theta}(a + ib)(A - iB) + e^{-i\theta}(a - ib)(A + iB) \\
&= e^{i\theta}[aA + bB] + e^{-i\theta}[aA + bB] + ie^{i\theta}[bA - aB] \\
&\quad - ie^{-i\theta}[bA - aB] \\
&= 2[aA + bB] \cos\ \theta + 2[aB - bA] \sin\theta \\
&= 2[\mathrm{Re}(x)\mathrm{Re}(X) + \mathrm{Im}(x)\mathrm{Im}(X)] \cos\theta \\
&\quad + 2[\mathrm{Re}(x)\mathrm{Im}(X) - \mathrm{Im}(x)\mathrm{Im}(X)] \sin\theta
\end{aligned}$$

Index

D.D. Nolte, *Optical Interferometry for Biology and Medicine*, Bioanalysis 1, DOI 10.1007/978-1-4614-0890-1, © Springer Science+Business Media, LLC 2012

C

N

O

P

Q

T